Machine Learning and Deep Learning for Healthcare Data Processing and Analyzing

Machine Learning and Deep Learning for Healthcare Data Processing and Analyzing

Guest Editors

Rajesh K. Tripathy
Haipeng Liu

MDPI

Basel • Beijing • Wuhan • Barcelona • Belgrade • Novi Sad • Cluj • Manchester

Guest Editors

Rajesh K. Tripathy
Electrical and Electronics Engineering
BITS Pilani
Hyderabad
India

Haipeng Liu
Faculty of Health and Life Sciences
Coventry University
Coventry
UK

Editorial Office
MDPI AG
Grosspeteranlage 5
4052 Basel, Switzerland

This is a reprint of the Special Issue, published open access by the journal *Diagnostics* (ISSN 2075-4418), freely accessible at: https://www.mdpi.com/journal/diagnostics/special_issues/6S80QH4M89.

For citation purposes, cite each article independently as indicated on the article page online and as indicated below:

Lastname, A.A.; Lastname, B.B. Article Title. *Journal Name* **Year**, *Volume Number*, Page Range.

ISBN 978-3-7258-6041-8 (Hbk)
ISBN 978-3-7258-6042-5 (PDF)
https://doi.org/10.3390/books978-3-7258-6042-5

Contents

About the Editors

Rajesh K. Tripathy

Rajesh Kumar Tripathy received a Ph.D. in Machine Learning for Signal Processing from IIT Guwahati in 2017. He has also received B.Tech and M.Tech degrees in Electronics & Telecommunication and Biomedical Engineering from BPUT, Odisha, and NIT Rourkela, India. He is currently working as an associate professor at BITS Pilani Hyderabad, India, since 2024. Prior to this, he served for more than ten years as an assistant professor in reputed institutions. He has published 89 papers in reputed international journals. He has also published 13 conference papers, 3 edited books, and 9 chapters. He has supervised 3 Ph.D. students in machine learning and biomedical signal processing. He has also supervised 5 M.Tech projects and 25 B.Tech projects. He is extensively working in research areas such as biomedical signal processing, machine learning, deep learning for healthcare, time–frequency analysis, medical image processing, Internet of Things (IoT), and embedded systems for healthcare applications. He received the Outstanding Potential for Excellence in Research Award (OPERA) from BITS Pilani in 2018. He has received INR 2,280,000 in funding from BITS Pilani through an OPERA grant to conduct high-quality research on signal processing and machine learning for healthcare data analysis. He has completed one sponsored project as a co-principal investigator from the CARS project, DRDO, India. His research papers are cited more than 4369 times on Google Scholar (accessed on 05/11/2025). He is also working as the associate editor for reputed journals like *IEEE Access* and *Frontiers in Physiology*.

Haipeng Liu

Haipeng Liu received bachelor's and master's degrees in engineering from Zhejiang University, China, in 2012 and 2015, respectively, and a Doctor of Philosophy degree in medical sciences from The Chinese University of Hong Kong in 2018. From 2019 to 2020, he was a Research Fellow with the Medical Technology Research Centre, Anglia Ruskin University. From 2020 to 2024, he was a Research Fellow with Coventry University, U.K. He is currently an Assistant Professor with Coventry University, U.K. He is a Lecturer of two undergraduate course modules with Coventry University, and has been invited as a guest lecturer at the European Institute of Innovation and Technology (EIT) Summer School, University Hospitals Coventry and Warwickshire NHS Trust, Nottingham Trent University (U.K.), and Rushford Business School (Switzerland). He was a PI and a mentor of the undergraduate student internship project of Anglia Ruskin University, and a mentor of the National Medical Research Association (NMRA, U.K.). He is the author of over 100 journal articles, 30 book chapters, and 11 conference papers, with over 1700 citations and an H-index of 22 on Scopus. His research interests include computational simulation of cardio-cerebrovascular diseases, AI-enhanced diagnostics, and wearable sensor development. He was a recipient of the Travel Award of the British Heart Foundation, the Research and Innovation Support Fund of Anglia Ruskin University, Official Development Assistance (ODA) funding, and four Ph.D. grants from Coventry University. Dr Liu is an Associate Fellow of the Higher Education Academy (AFHEA, U.K.). He is a member of the World Stroke Organization (WSO), Chinese Stroke Association (CSA, China), British Society for Cardiovascular Research (BSCR, U.K.), and Cardiovascular Analytics Group (CVAG, HKSAR of China). He is an editorial member of six academic journals. He is a peer reviewer of more than 100 academic papers.

Preface

This Special Issue focuses on the development and application of machine learning and deep learning methods for the analysis and interpretation of healthcare data. The scope of the Special Issue encompasses a wide range of medical information sources, including physiological signals, medical images, electronic health records, wearable sensor data, and multimodal patient monitoring streams. The aim is to highlight approaches that can support clinical decision-making, improve diagnostic workflows, and enable more precise, timely, and personalized care.

The motivation behind this collection arises from the increasing complexity and volume of healthcare data, along with the growing need for analytical tools that are accurate, interpretable, and suitable for use in both clinical and resource-limited settings. Machine learning and deep learning have demonstrated strong potential in areas such as early disease detection, progression prediction, treatment outcome estimation, and image-based decision support. However, challenges remain in ensuring model reliability, handling heterogeneous data, and translating algorithmic advancements into practical clinical use.

The articles in this Special Issue present recent research addressing these challenges through the detection of Parkinson's disease, prediction of Alzheimer's disease progression, outcome prediction in microsurgical clipping of cerebral aneurysms, identification of lumbar disc herniation, and the development of medical image-driven diagnostic support systems. Together, these contributions illustrate how data-driven intelligence is reshaping healthcare delivery and supporting evidence-based clinical practice.

This Special Issue is intended for researchers, clinicians, biomedical engineers, computer scientists, and healthcare technology developers who are engaged in advancing intelligent medical data analysis and its real-world application.

Rajesh K. Tripathy and Haipeng Liu
Guest Editors

diagnostics

MDPI

Editorial

Machine Learning and Deep Learning for Healthcare Data Processing and Analyzing: Towards Data-Driven Decision-Making and Precise Medicine

Haipeng Liu [1,*] and Rajesh Kumar Tripathy [2,*]

1 Centre for Intelligent Healthcare, Coventry University, Coventry CV1 5RW, UK
2 Department of Electrical and Electronics Engineering, BITS-Pilani, Hyderabad Campus, Hyderabad 500078, India
* Correspondence: haipeng.liu@coventry.ac.uk (H.L.); tripathyrk@hyderabad.bits-pilani.ac.in (R.K.T.)

Artificial intelligence (AI) is reshaping the landscape of healthcare data. Alongside electronic medical records (EHRs), AI algorithms are accelerating the collection, management, processing, and analysis of healthcare data [1]. Its power was recently showcased through the realization of AI-based methods for the early detection and fine-grained severity evaluation of COVID-19. Machine learning and deep learning models enable the automatic processing of multimodal healthcare data, including EHRs, physiological signals, and medical images. In this Special Issue, "Machine Learning and Deep Learning for Healthcare Data Processing and Analyzing", we collect eleven research and opinion articles focusing on the application of AI models to different types of healthcare data. The published studies in this Special Issue provide up-to-date examples of AI uses in a healthcare context (Table 1).

Machine learning can comprehensively analyze different features and estimate the risk, enhancing diagnoses and prognoses. Based on a longitudinal dataset of 150 people, Alshamlan et al. compared some common machine learning models, including support vector machine (SVM), random forest (RF), and logistic regression (LR) approaches, in predicting Alzheimer's disease. Minimum redundancy maximum relevance (mRMR) and mutual information (MI) were employed for feature selection, and LR combined with mRMR achieved the highest accuracy of 99.08% in predicting Alzheimer's disease. Their results highlighted the role of AI in disease prediction and clinical decision-making.

Toader et al. used machine learning models to predict clinical outcomes in microsurgical clipping treatments of cerebral aneurysms, based on a dataset of 344 patients' preoperative characteristics. Validating prediction outcomes on the Glasgow Outcome Scale (GOS), their extreme gradient boosting (XGB) model outperformed the others and achieved an area under the receiver operating characteristic curve (AUC ROC) of 0.72 ± 0.03 for specific GOS outcome prediction, and an AUC ROC of 0.78 ± 0.02 for the binary classification of outcomes. These results demonstrate the potential of machine learning as a tool for predicting the surgical outcomes of ruptured cerebral aneurysm treatments. Moreover, the study underscores the need for high-quality, large-scale datasets and external validation in order to enhance the reliability and generalizability of machine learning models.

Deep learning enables the in-depth analysis of clinical images, making automatic classification and fine-grained feature analysis possible. In their contribution, Hadj-Alouane et al. propose an AI framework for the diagnosis and severity classification of Parkinson's Disease using video data captured in uncontrolled environments. Deep learning models, including a convolutional neural network (CNN), residual network (ResNet), and vision

 https://doi.org/10.3390/diagnostics15081051

transformer (ViT), were deployed in gait analysis based on skeleton energy images (SEIs) of gait sequences. Although only 167 subjects were included, 90% accuracy was achieved. Their proposed framework provides a new potential pathway for the cost-effective early detection of Parkinson's Disease in normal healthcare settings.

Mudavadkar et al. used ensemble models that mixed different deep learning architectures to improve the performance of AI in detecting gastric cancer. The models were trained and tested on a database of 600 pictures of stomach cancer pathology, with the pictures cropped to different sizes. With their proposed ensemble model achieving over 90% accuracy at different resolutions, the authors concluded that ensemble models may be able to accurately detect some pathological features from smaller picture patches, empowering pathologists to diagnose gastric cancer at an early stage.

The quantification of interested features is important in diagnostics. With this in mind, Nair et al. applied AI-based quantitative lung texture analysis to high-resolution computed tomography (HRCT) images of 45 patients in order to detect subtle lung parenchymal involvement in bronchiectasis. Among the patterns detected by AI, hyperlucency was suggestive of air-trapping and alveolar destruction, while ground-glass opacity suggested active alveolitis and early interstitial inflammation. The results of this pilot study showed that AI is able to identify some minor features from HRCT that are often neglected in human-led visual assessments, offering new insights into the pathophysiology of bronchiectasis and other lung diseases.

Guo et al. developed two-dimensional (2D) and three-dimensional (3D) no new U-Net (nnU-Net) networks to segment contrast-enhanced computed tomography (CT) images and quantify the morphological features required to detect Type A aortic dissection. Of the two networks, their 3D nnU-Net architectures displayed a better performance on the CT datasets of 24 patients. While further large-scale validation is needed, the nnU-Net architectures showed the potential of the automatic segmentation and quantification of aortic structures for rapid diagnosis, surgical planning, and the subsequent biomechanical simulation of the aorta.

Bendella et al. proposed AI-based MRI brain volumetrics to distinguish between patients with idiopathic normal pressure hydrocephalus (iNPH), Alzheimer's disease, and age- and sex-matched healthy controls by evaluating cortical, subcortical, and ventricular volumes. The study was conducted retrospectively on 123 age- and sex-matched subjects, with 41 in each group. The authors found that iNPH patients exhibited ventricular enlargement and changes in gray and white matter compared to the other two groups, with the most significant differences observed in total ventricular volume (+67%) and the lateral (+68%), third (+38%), and fourth (+31%) ventricles compared to the controls. Their AI-based MRI volumetry approach provided the quantitative evidence used to investigate the pathology of iNPH and improve patient management.

Deep learning can be used for processing physiological signals and multimodal data. Chin et al. introduced a neural network model for continuously estimating patient respiratory rate from photoplethysmogram signals with a reduced window size and lower processing requirements. The algorithm was validated on two datasets, and the best-performing model achieved a mean absolute error of 2 breaths/min at a window size of 7 s. This study demonstrates the superior performance of an AI model with a smaller window size, showing the potential for the quick, AI-based evaluation of the respiratory system.

Table 1. Overview of 12 studies on artificial intelligence in healthcare.

Theme	Authors and DOI	Study Title	Key Findings
Machine learning for clinical data analysis	Alshamlan et al., 10.3390/diagnostics14192237	Improving Alzheimer's Disease Prediction with Different Machine Learning Approaches and Feature Selection Techniques	Machine learning may offer more accurate disease prognosis for clinical decision-making.
Machine learning for clinical data analysis	Toader et al., 10.3390/diagnostics14192156	Machine Learning-Based Prediction of Clinical Outcomes in Microsurgical Clipping Treatments of Cerebral Aneurysms	High-quality large-scale datasets and external validation are essential to enhance model reliability and generalizability.
Deep learning for medical image analysis	Hadj-Alouane et al., 10.3390/diagnostics14232685	Severity Classification of Parkinson's Disease via Synthesis of Energy Skeleton Images from Videos Produced in Uncontrolled Environments	Deep learning may enable the cost-effective early detection of Parkinson's Disease in various healthcare settings.
Deep learning for medical image analysis	Mudavadkar et al., 10.3390/diagnostics14161746	Gastric Cancer Detection with Ensemble Learning on Digital Pathology: Use Case of Gastric Cancer on GasHisSDB Dataset	Ensemble deep learning may detect some pathological features from smaller picture patches, enabling the early diagnosis of gastric cancer.
Quantification of image features	Nair et al., 10.3390/diagnostics14242883	Artificial Intelligence Unveils the Unseen: Mapping Novel Lung Patterns in Bronchiectasis via Texture Analysis	AI-based quantified lung texture analysis provides valuable insights into the diagnosis of bronchiectasis and other lung diseases.
Quantification of image features	Guo et al., 10.3390/diagnostics14131332	Automatic Segmentation of Type A Aortic Dissection on Computed Tomography Images Using Deep Learning Approach	nnU-Net architectures enable the automatic segmentation and quantification of aorta for rapid diagnosis, surgical planning, and biomechanical simulation.
Quantification of image features	Bendella et al., 10.3390/diagnostics14131422	Brain and Ventricle Volume Alterations in Idiopathic Normal Pressure Hydrocephalus Determined by Artificial Intelligence-Based MRI Volumetry	Integrating AI volumetry with traditional radiologic measures can reveal new pathological features involving the supratentorial white matter, aiding in the identification of iNPH and patient management.
Deep learning for physiological signal analysis	Chin et al. 10.3390/diagnostics14030284	A Novel Respiratory Rate Estimation Algorithm from Photoplethysmogram Using Deep Learning Model	Deep learning can estimate respiratory rate from photoplethysmography signals with short window sizes for continuous monitoring.
Data preprocessing in machine learning	Manir and Deshpande, 10.3390/diagnostics14100984	Critical Risk Assessment, Diagnosis, and Survival Analysis of Breast Cancer	Preprocessing is important in enabling AI-enhanced individualized approaches to the management and treatment of breast cancer.
AI tools for clinical use	Lohaj et al., 10.3390/diagnostics14090917	Conceptually Funded Usability Evaluation of an Application for Leveraging Descriptive Data Analysis Models for Cardiovascular Research	Software usability should be evaluated in different dimensions, and can be improved through measures like a user manual and clear error messages for efficient feedback
AI tools for clinical use	Badahman et al., 10.3390/diagnostics14171870	Validating the Accuracy of a Patient-Facing Clinical Decision Support System in Predicting Lumbar Disc Herniation: Diagnostic Accuracy Study	AI-enhanced CDSSs provide a reasonable level of efficacy and may largely reduce the time and cost of screening of patients with lumbar disk herniation.
Summarization of machine learning models	Pinton, 10.3390/diagnostics14131324	Machine Learning for Predicting Biologic Agent Efficacy in Ulcerative Colitis: An Analysis for Generalizability and Combination with Computational Models	Machine learning models based on multiple pathways, multiple ethnicities, and real-world and clinical trial data are needed for data-driven decision-making and precision medicine. Data quality and quantity, overfitting, generalization, and interpretability are all unmet challenges.

Data preprocessing is important for optimizing the performance of AI models. Manir and Deshpande used different machine learning models to investigate the relationship between the risk factors of breast cancer incidence and survivability. Performing resampling and principal component analysis on the training dataset to enhance the performance of the classifiers, three breast cancer datasets were examined using a variety of preprocessing approaches and classification models. The best accuracies achieved were above 85% on all the datasets. The authors observed that resampling can worsen the accuracy scores of the test data, even when the training data accuracy is increased. Their results emphasize the significance of individualized approaches in the management and treatment of breast cancer, showcasing the versatility of machine learning in incorporating phenotypic variations and recognizing the heterogeneity of a disease when estimating clinical risk.

Despite these emerging AI models, there is a great need for specified AI tools for clinical use. Lohaj et al. developed a Shiny dashboard application named DESSFOCA (Decision Support System For Cardiologists) using the R language, structured around three core functionalities: discovering association rules, applying clustering methods, and identifying association rules within predefined clusters. The application was tested based on the feedback of end users. Based on the evaluation results, the authors proposed recommendations to further improve usability, such as adding a user manual and improving error messages to provide efficient feedback.

Badahman et al. compared the performance of a piece of AI-enhanced web-based software incorporated into a clinical decision support system (CDSS) with that of magnetic resonance imaging (MRI) in predicting lumbar disk herniation in a cross-sectional study of 100 patients. The software showed significant diagnostic accuracy, with an AUC ROC of 0.84, a sensitivity of 88%, and a specificity of 80%. The AI-enhanced CDSS may achieve comparable performance with MRI, which will largely reduce how time-consuming and costly the screening of patients with lumbar disk herniation is.

Although machine learning can offer efficient, automatic, and personalized diagnosis, thereby providing support for clinical decision-making, there are unmet challenges in terms of data, algorithms, and their application. Pinton compared the performance of two models in predicting the efficacy of biologic agents in ulcerative colitis. The author suggested that machine learning models based on multiple pathways, multiple ethnicities, and real-world and clinical trial data are required for data-driven decision-making and precision medicine. The author also pointed out that data quality and quantity, overfitting, generalization, and interpretability are major limitations in many machine learning models.

In summary, this Special Issue covers the major applications of AI in healthcare data analysis, including data-driven risk evaluation, medical image analysis, the quantification of image features, and physiological signal analysis (Figure 1). Besides its applications in data processing and analysis, AI-enhanced CDSSs improve clinical practice regarding decision-making, care delivery, and treatment selection, offering higher diagnostic accuracy and lower risks of medical error [2,3]. In addition to the technological challenges, there are ethical concerns surrounding AI use, specifically regarding data security issues, underrepresented minorities, and data-access disparities [4].

It should be noted that there are still unmet challenges in terms of the availability high-quality data, which limits large-scale validation [5,6]; the lack of standardized and efficient data preprocessing [7]; and the low explainability of AI approaches, which is a main obstacle towards their real-world clinical application [8]. In future research, the collection of high-quality data, the development of advanced data preprocessing frameworks (e.g., edge computing for big data [9]), and explainable AI techniques (e.g., 'spread of relevance' at the layer level of neural networks [10]) will accelerate the integration of AI models into medical systems [11], further reshaping the landscape of modern healthcare towards data-driven decision-making and precise medicine.

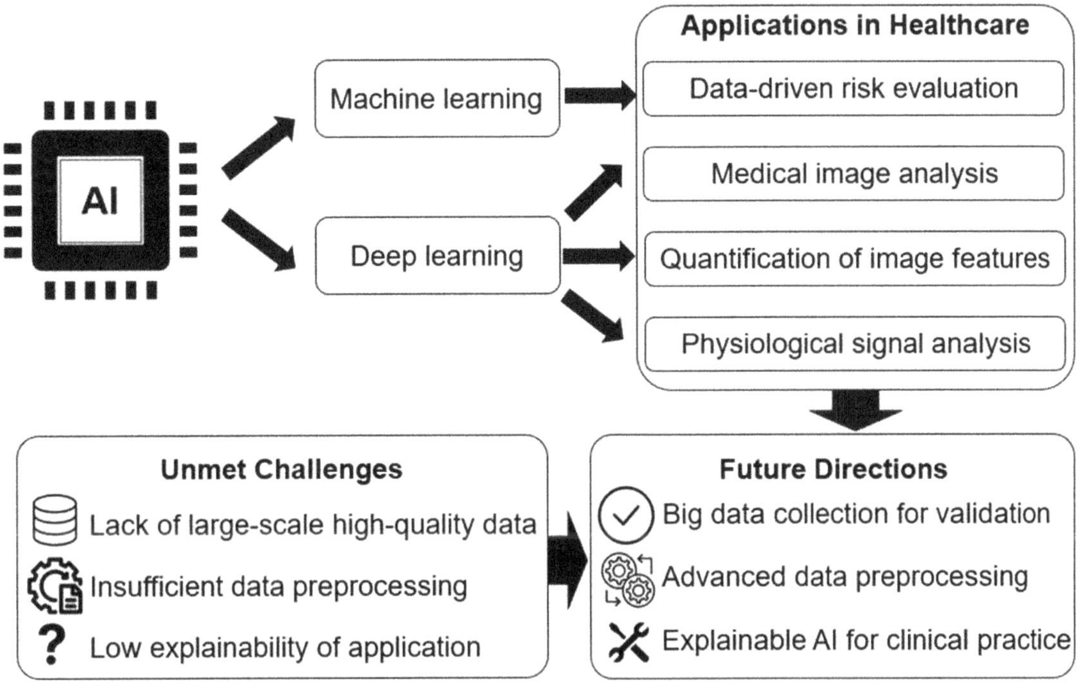

Figure 1. Applications of machine learning and deep learning in modern healthcare, unmet challenges, and future directions.

We appreciate the contributors to this Special Issue.

Author Contributions: Conceptualization, H.L.; methodology, H.L.; writing—original draft preparation, H.L.; writing—review and editing, R.K.T.; visualization, H.L. and R.K.T.; supervision, H.L. and R.K.T. All authors have read and agreed to the published version of the manuscript.

Conflicts of Interest: The authors declare no conflict of interest.

References

1. Tse, G.; Lee, Q.; Chou, O.H.I.; Chung, C.T.; Lee, S.; Chan, J.S.K.; Li, G.; Kaur, N.; Roever, L.; Liu, H.; et al. Healthcare Big Data in Hong Kong: Development and Implementation of Artificial Intelligence-Enhanced Predictive Models for Risk Stratification. *Curr. Probl. Cardiol.* **2024**, *49*, 102168. [CrossRef] [PubMed]
2. Ouanes, K.; Farhah, N. Effectiveness of Artificial Intelligence (AI) in Clinical Decision Support Systems and Care Delivery. *J. Med. Syst.* **2024**, *48*, 74. [CrossRef] [PubMed]
3. Dai, L.; Wu, Z.; Pan, X.; Zheng, D.; Kang, M.; Zhou, M.; Chen, G.; Liu, H.; Tian, X. Design and implementation of an automatic nursing assessment system based on CDSS technology. *Int. J. Med. Inform.* **2024**, *183*, 105323. [CrossRef] [PubMed]
4. Goldberg, C.B.; Adams, L.; Blumenthal, D.; Brennan, P.F.; Brown, N.; Butte, A.J.; Cheatham, M.; deBronkart, D.; Dixon, J.; Drazen, J.; et al. To do no harm—And the most good—With AI in health care. *Nat. Med.* **2024**, *30*, 623–627. [CrossRef] [PubMed]
5. Silcox, C.; Zimlichmann, E.; Huber, K.; Rowen, N.; Saunders, R.; McClellan, M.; Kahn, C.N.; Salzberg, C.A.; Bates, D.W. The potential for artificial intelligence to transform healthcare: Perspectives from international health leaders. *npj Digit. Med.* **2024**, *7*, 88. [CrossRef] [PubMed]
6. Enshaei, N.; Naderkhani, F. The Role of Data Quality for Reliable AI Performance in Medical Applications. *IEEE Reliab. Mag.* **2024**, *1*, 24–28. [CrossRef]

7. Ortiz, B.L.; Gupta, V.; Kumar, R.; Jalin, A.; Cao, X.; Ziegenbein, C.; Singhal, A.; Tewari, M.; Choi, S.W. Data Preprocessing Techniques for AI and Machine Learning Readiness: Scoping Review of Wearable Sensor Data in Cancer Care. *JMIR Mhealth Uhealth* **2024**, *12*, e59587. [CrossRef] [PubMed]
8. Misra, R.; Keane, P.A.; Hogg, H.D.J. How should we train clinicians for artificial intelligence in healthcare? *Future Healthc. J.* **2024**, *11*, 100162. [CrossRef] [PubMed]
9. Yu, K.; Chakraborty, C.; Xu, D.; Zhang, T.; Zhu, H.; Alfarraj, O.; Tolba, A. Hybrid quantum classical optimization for low-carbon sustainable edge architecture in RIS-assisted AIoT healthcare systems. *IEEE Internet Things J.* **2024**, *11*, 38987–38998. [CrossRef]
10. Albahri, A.S.; Duhaim, A.M.; Fadhel, M.A.; Alnoor, A.; Baqer, N.S.; Alzubaidi, L.; Albahri, O.S.; Alamoodi, A.H.; Bai, J.; Salhi, A.; et al. A systematic review of trustworthy and explainable artificial intelligence in healthcare: Assessment of quality, bias risk, and data fusion. *Inf. Fusion* **2023**, *96*, 156–191. [CrossRef]
11. Iqbal, T.; Masud, M.; Amin, B.; Feely, C.; Faherty, M.; Jones, T.; Tierney, M.; Shahzad, A.; Vazquez, P. Towards integration of artificial intelligence into medical devices as a real-time recommender system for personalised healthcare: State-of-the-art and future prospects. *Health Sci. Rev.* **2024**, *10*, 100150. [CrossRef]

diagnostics

MDPI

Article

Artificial Intelligence Unveils the Unseen: Mapping Novel Lung Patterns in Bronchiectasis via Texture Analysis

Athira Nair [1], Rakesh Mohan [2], Mandya Venkateshmurthy Greeshma [3], Deepak Benny [4], Vikram Patil [4], SubbaRao V. Madhunapantula [3], Biligere Siddaiah Jayaraj [1], Sindaghatta Krishnarao Chaya [1], Suhail Azam Khan [1], Komarla Sundararaja Lokesh [1], Muhlisa Muhammaed Ali Laila [1], Vadde Vijayalakshmi [1], Sivasubramaniam Karunakaran [1], Shreya Sathish [5] and Padukudru Anand Mahesh [1,*]

1 Department of Respiratory Medicine, JSS Medical College, JSS Academy of Higher Education & Research (JSS AHER), Mysore 570004, Karnataka, India; athiranair94@gmail.com (A.N.); drjayarajbs@yahoo.com (B.S.J.); chaya.sindghatta@gmail.com (S.K.C.); khan2201@gmail.com (S.A.K.); lokeshpulmo@gmail.com (K.S.L.); muhlisamr@gmail.com (M.M.A.L.); vijjichowdary6@gmail.com (V.V.); sivasubramaniam1810@gmail.com (S.K.)

2 Department of Community Medicine, JSS Medical College, JSS Academy of Higher Education & Research (JSS AHER), Mysore 570004, Karnataka, India; rakeshm468@gmail.com

3 Center of Excellence in Molecular Biology and Regenerative Medicine (CEMR) Laboratory (DST-FIST Supported Center and ICMR Collaborating Center of Excellence—ICMR-CCoE), Department of Biochemistry (DST-FIST Supported Department), JSS Medical College, JSS Academy of Higher Education & Research (JSS AHER), Mysuru 570015, Karnataka, India; greeshmamv@jssuni.edu.in (M.V.G.); madhunapantulas@yahoo.com (S.V.M.)

4 Department of Radiology, JSS Medical College, JSS Academy of Higher Education & Research (JSS AHER), Mysore 570004, Karnataka, India; drdeepakbenny@gmail.com (D.B.); vikrampatil@jssuni.edu.in (V.P.)

5 Father Muller Medical College, Mangaluru 575002, Karnataka, India; shreyarathna26@gmail.com

* Correspondence: mahesh1971in@yahoo.com

Abstract: Background and Objectives: Thin-section CT (TSCT) is currently the most sensitive imaging modality for detecting bronchiectasis. However, conventional TSCT or HRCT may overlook subtle lung involvement such as alveolar and interstitial changes. Artificial Intelligence (AI)-based analysis offers the potential to identify novel information on lung parenchymal involvement that is not easily detectable with traditional imaging techniques. This study aimed to assess lung involvement in patients with bronchiectasis using the Bronchiectasis Radiologically Indexed CT Score (BRICS) and AI-based quantitative lung texture analysis software (IMBIO, Version 2.2.0). **Methods:** A cross-sectional study was conducted on 45 subjects diagnosed with bronchiectasis. The BRICS severity score was used to classify the severity of bronchiectasis into four categories: Mild, Moderate, Severe, and tractional bronchiectasis. Lung texture mapping using the IMBIO AI software tool was performed to identify abnormal lung textures, specifically focusing on detecting alveolar and interstitial involvement. **Results:** Based on the Bronchiectasis Radiologically Indexed CT Score (BRICS), the severity of bronchiectasis was classified as Mild in 4 (8.9%) participants, Moderate in 14 (31.1%), Severe in 11 (24.4%), and tractional in 16 (35.6%). AI-based lung texture analysis using IMBIO identified significant alveolar and interstitial abnormalities, offering insights beyond conventional HRCT findings. This study revealed trends in lung hyperlucency, ground-glass opacity, reticular changes, and honeycombing across severity levels, with advanced disease stages showing more pronounced structural and vascular alterations. Elevated pulmonary vascular volume (PVV) was noted in cases with higher BRICSs, suggesting increased vascular remodeling in severe and tractional types. **Conclusions:** AI-based lung texture analysis provides valuable insights into lung parenchymal involvement in bronchiectasis that may not be detectable through conventional HRCT. Identifying significant alveolar and interstitial abnormalities underscores the potential impact of AI on improving the understanding of disease pathology and disease progression, and guiding future therapeutic strategies.

Keywords: bronchiectasis; IMBIO; lung texture analysis; alveolar

1. Introduction

Bronchiectasis is a chronic pulmonary disorder characterized by the permanent dilation and destruction of the bronchial walls, leading to impaired mucociliary clearance and recurrent infections [1]. Despite advancements in respiratory medicine, bronchiectasis remains underdiagnosed and poses significant challenges to patient management, severely affecting quality of life. Once considered rare, recent epidemiological data indicate a rising prevalence, especially among older adults and individuals with comorbid respiratory conditions such as chronic obstructive pulmonary disease [COPD] and cystic fibrosis. In Europe and North America, the prevalence of bronchiectasis has been reported to range from 67 to 566.1 per 100,000 inhabitants [2–5], and, in China, among those aged 40 years and older, it reaches as high as 1200 per 100,000 [6].

The European Multicenter Bronchiectasis Audit and Research Collaboration [EMBARC] India registry highlights that males are more frequently affected than females, with a median patient age of 56 years [IQR 41–66]. Compared to Western registries, pulmonary tuberculosis [PTB] remains a leading cause of bronchiectasis in India, while allergic bronchopulmonary aspergillosis [ABPA] contributes significantly [8.9%] to disease etiology. Pseudomonas aeruginosa is commonly isolated in sputum cultures, affecting 13.7% of cases [7].

The pathogenesis of bronchiectasis is multifactorial, involving a complex interplay between infections, immune responses, and airway structural abnormalities. Etiologies include post-infectious complications, genetic disorders such as cystic fibrosis, immunodeficiency, and environmental exposures [8]. Despite the advent of high-resolution computed tomography [HRCT], which has revolutionized the detection and characterization of bronchiectasis, a disconnect persists between HRCT findings and clinical parameters, such as symptom severity, exacerbation frequency, and prognosis. HRCT often serves as the diagnostic gold standard, yet its visual interpretation remains subjective, with limited predictive value regarding treatment outcomes and disease progression [9].

Computer-aided diagnostic [CAD] systems aim to overcome these limitations by objectively quantifying lung abnormalities [10]. Previous research has demonstrated that certain HRCT features, such as Cystic bronchiectasis, correlate with worse lung function and HRCT scores [11–13]. However, studies have also shown a weak or absent correlation between HRCT scores and clinical parameters, such as lung function and quality of life assessments [12–15]. This discordance highlights the inadequacy of conventional HRCT in capturing the full complexity of bronchiectasis pathophysiology. As Artificial Intelligence [AI] continues to evolve, its application in medical imaging offers significant potential to augment HRCT interpretation. AI algorithms can enhance diagnostic precision, reveal latent disease patterns, objectively and quantitatively assess various spectrum of lung pathologies that can be identified in HRCT, and provide a more nuanced understanding of disease progression. By integrating AI into clinical practice, healthcare providers may improve diagnostic accuracy, optimize treatment decisions, and ultimately enhance patient outcomes [16]. This study aims to compare the capacity of AI to detect subtle and latent abnormalities in bronchiectasis, quantify radiologic abnormalities compared to traditional HRCT interpretations, and compare AI scoring with traditional bronchiectasis clinical scoring [BRICS].

2. Materials and Methods

2.1. Patients Selection

This was an observational retrospective study conducted at JSS Medical College and Hospital, Mysuru, Karnataka. This study was approved by the Institutional Ethics Committee [approval number: JSS/MC/PG/IEC-105/2022-23].

A total of 45 patients were enrolled in the Department of Respiratory Medicine, JSS Hospital, Mysuru, between June 2022 and June 2023, based on predefined inclusion criteria. Inclusion criteria were those with HRCT-confirmed bronchiectasis J47 [as per ICD 10]. As per exclusion criteria, those with active tuberculosis and active malignancy were excluded from the study. Comprehensive data regarding the patients' general characteristics, comorbidities, smoking history, respiratory symptoms, and HRCT results were collected. The three experienced radiologists collated the Bronchiectasis Severity Score.

2.2. CT Scoring

The Bronchiectasis Radiologically Indexed CT Score [BRICS] is a severity assessment score for bronchiectasis that was created from a cohort of patients with idiopathic and post-infectious bronchiectasis by thoracic radiologists and chest physicians in Edinburgh, Scotland. As per the BRICS severity score assessment, subjects were classified into Mild [1], Moderate [2,3], and Severe [4,5] based on bronchial dilation and the number of bronchopulmonary segments affected by emphysema. The BRICS was chosen as it was simple, accurate, and predictive of illness severity as compared to other existing scores [17].

2.3. CT Scan Technique and Image Analysis

HRCT scans for the studies were acquired using the Philips Ingenuity 128-slice MDCT scanner. Only axial volumetric multidetector computed tomography [CT] scans taken in the supine position with a slice thickness of 3 mm, then reconstructed to 1 mm, were included for analysis.

QA [quantitative analysis] was performed in the set of images with a standard reconstruction algorithm by using Imbio LTA [based on the algorithm of CALIPER—Computer-aided Lung Informatics for Pathology Evaluation and Rating], a computational platform for the near-real-time characterization and quantification of lung parenchymal patterns on CT scans [18,19]. Lung parenchymal patterns include normal, hyperlucent, ground glass [GG], reticular, and honeycombing. Imbio LTA provided the Total Lung Volume, relative volumes, and absolute volumes of lung patterns, and their regional distribution within three different lung zones [upper, middle, lower] in each lung. The platform provided a fast output of the DICOM series, as well as a PDF file in which each lung pattern was colored differently, allowing the evaluation of disease extent, composition, and location at a glance. Each QA was performed automatically for up to 20 min [20]. Among the patterns detected by AI, hyperlucency is suggestive of air-trapping and alveolar destruction, while ground-glass opacity suggests active alveolitis and early interstitial inflammation. Both reticular and honeycombing are findings associated with fibrosis, while the former suggests interlobular interstitial thickening and the latter interstitial fibrosis with lung destruction and dilation of peripheral airspaces. The pulmonary vascular volume [PVV] refers to the total volume of all visible arteries and veins in the lungs, encompassing both blood and vessel walls. PVV is an important metric for assessing the severity of various lung disorders, such as pulmonary hypertension [PH] and other pulmonary vascular conditions [21,22].

2.4. Definitions

According to the WHO classification for BMI among Asians, people were classified as underweight, normal, overweight, and obese E66 [as per ICD 10]. The comorbidities such as type 2 diabetes mellitus E11 [as per ICD 10] and hypertension I10 [as per ICD 10] have been defined by WHO, while the definitions for congestive cardiac failure [CCF] and ischemic heart disease [IHD] have been given by the American College of Cardiology [ACC] and the American Heart Association [AHA]. We have also considered 5 types of bronchiectasis—Cystic, Saccular, varicose, Tubular or cylindrical, and traction bronchiectasis as per Reid classification [23]. The emphysema-dominant phenotype is an imaging-derived phenotype that is characterized by hyperlucency $\geq$ 20% [24]. Fibrosis's predominant phenotype is often associated with fibrotic changes exceeding $\geq$20% in lung parenchyma [25]. Combined fibrosis and emphysema were subjects with a combination of both emphysema [≥20%] and fibrosis [≥20%] on chest CT scans [26].

2.5. Data Analysis

Data were entered into Microsoft Excel 2019 and analyzed using SPSS Version 26.0 [IBM Corp, released 2019, IBM SPSS Statistics for Windows, Version 26.0. Armonk, NY, USA: IBM Corp.]. The percentage was calculated for descriptive variables. The chi-square test was applied to analyze qualitative data. Mean and standard deviation were used to describe quantitative data. Pearson's correlation was applied to analyze the correlation between BRICS and parenchymal changes detected by Artificial Intelligence. ANOVA was used to test the difference between parenchymal changes detected by Artificial Intelligence and the severity and type of Bronchiectasis. A p-value of <0.05 was considered to be statistically significant.

3. Results

Forty-five patients were included in this study. The majority were aged ≥60 years [48.9%] and the gender distribution was relatively balanced. More than one-third of the subjects were obese [19, 42.2%] with most participants [36, 80%] having comorbidities. Tractional bronchiectasis was the most common [35.6%], followed by nearly one-third [31.1%] with Moderate bronchiectasis, while 24.4% had Severe bronchiectasis and 8.9% had Mild bronchiectasis (Table 1).

Table 1. Socio-demographic profile of the study participants [n = 45].

Socio-Demographic Profile		Number [%]
Age [in years]	≤40	4 [8.9%]
	40–60	19 [42.2%]
	≥60	22 [48.9%]
Gender	Male	23 [51.1%]
	Female	22 [48.9%]
BMI [kg/m^2] [As per WHO]	Underweight	6 [13.3%]
	Normal	11 [24.4%]
	Overweight	9 [20.0%]
	Obese	19 [42.2%]
Comorbidities	Hypertension [ICD I10]	11 [24.4%]
	Type 2 DM [ICD E11.9]	14 [31.1%]
	Hypertension and Type 2 DM	5 [11.1%]
	Congestive Cardiac Failure [ICD I50]	1 [2.2%]
	Hypothyroid [ICD E03.9]	2 [4.4%]
	Ischemic Heart Disease [ICD I25]	3 [6.6%]
	No Comorbidities	9 [20%]
Severity of Bronchiectasis based on Bronchiectasis Radiologically Indexed CT Score [BRICS]	Mild	4 [8.9%]
	Moderate	14 [31.1%]
	Severe	11 [24.4%]
	Tractional	16 [35.6%]
Smoking Status	Smoker	14 [31.1%]
	Non-smoker	31 [68.9%]
Symptoms	Cough	29 [64.4%]
	Fever	16 [35.5%]
	Shortness of Breath	22 [48.8%]
	Hemoptysis	2 [4.44%]

Lung volume increases progressively from Mild [2.95 L] to Severe [3.85 L] bronchiectasis. The differences across severity levels are statistically significant [p = 0.038], suggesting a clear relationship between lung volume and disease severity and could suggest air trapping due to airway obstruction or emphysema. Hyperlucency, which reflects areas of lung tissue that appear abnormally transparent on imaging, increases significantly with severity, from 8.50% in Mild cases to 24.64% in Severe cases. The p-value of 0.025 indicates a statistically significant increase in lung hyperlucency as severity increases. Mild cases show the highest ground-glass opacity [22.50%], and the values for Moderate [14.93%] and Severe [16.27%] are lower. The *p*-value [0.477] indicates no statistically significant differences across severity levels for ground-glass opacity. Reticular opacity, which suggests fibrosis or scarring, increases with disease severity. Mild cases show 4.75%, while Severe cases reach 17.45%. Honeycombing, associated with advanced fibrosis, increases with severity, from 2.75% in Mild cases to 12.18% in Severe cases. PVV increases significantly [p = 0.04] as severity progresses, from 95 cm^3 in Mild cases to 122.82 cm^3 in Severe cases. Lung characteristics such as hyperlucency, reticular opacity, and PVV significantly increase with disease severity, indicating that these factors correlate with the progression of lung volume and increase progressively from Mild [2.95 L] to Severe [3.85 L] bronchiectasis. The differences across severity levels are statistically significant [p = 0.038], suggesting a clear relationship between lung volume and disease severity and could suggest air trapping due to airway obstruction or emphysema. Hyperlucency, which reflects areas of lung tissue that appear abnormally transparent on imaging, increases significantly with severity, from 8.50% in Mild cases to 24.64% in Severe cases. The *p*-value of 0.025 indicates a statistically significant increase in lung hyperlucency as severity increases. Mild cases show the highest ground-glass opacity [22.50%], and the values for Moderate [14.93%] and Severe [16.27%] are lower. The *p*-value [0.477] indicates no statistically significant differences across severity levels for ground-glass opacity. Reticular opacity, which suggests fibrosis or scarring, increases with disease severity. Mild cases show 4.75%, while Severe cases reach 17.45%. Honeycombing, associated with advanced fibrosis, increases with severity, from 2.75% in Mild cases to 12.18% in Severe cases. PVV increases significantly [p = 0.04] as severity progresses, from 95 cm^3 in Mild cases to 122.82 cm^3 in Severe cases. Lung characteristics such as hyperlucency, reticular opacity, and PVV significantly increase with disease severity, indicating that these factors correlate with the progression of lung pathology. Lung volume and honeycombing also increase, but the differences in honeycombing are not statistically significant. Ground-glass opacity does not show a clear trend or significant change with severity (Table 2).

Table 2. Relationship between various parenchymal pathologies detected by Artificial Intelligence and severity of Bronchiectasis.

Severity		Mean ± Std. Deviation	Std. Error	95% Confidence Interval for Mean	*p*-Value
Total Lung Volume [L]	Mild	2.950 ± 1.0214	0.5107	1.325–4.575	0.038
	Moderate	3.200 ± 0.8806	0.2353	2.692–3.708	
	Severe	3.845 ± 2.0863	0.6290	2.444–5.247	
	Tractional	2.969 ± 1.0562	0.2641	2.406–3.532	
Total Lung Hyperlucency [%]	Mild	8.50 ± 15.695	7.848	−16.47–33.47	0.025
	Moderate	9.79 ± 11.696	3.126	3.03–16.54	
	Severe	24.64 ± 31.646	9.542	3.38–45.90	
	Tractional	16.19 ± 13.949	3.487	8.75–23.62	
Total Lung Ground-glass Opacity [%]	Mild	22.50 ± 19.261	9.631	−8.15–53.15	0.477
	Moderate	14.93 ± 18.805	5.026	4.07–25.79	
	Severe	16.27 ± 14.158	7.887	10.692–24.85	
	Tractional	7.63 ± 9.966	2.491	2.31–12.94	

Table 2. *Cont.*

Severity		Mean ± Std. Deviation	Std. Error	95% Confidence Interval for Mean	*p*-Value
Total Lung Reticular Opacity [%]	Mild	4.75 ± 2.754	1.377	0.37–9.13	0.05
	Moderate	7.21 ± 6.302	1.684	3.58–10.85	
	Severe	17.45 ± 15.885	5.393	5.585–25.47	
	Tractional	17.38 ± 15.850	4.712	4.04–23.42	
Total Lung Honeycombing [%]	Mild	2.75 ± 5.500	2.750	−6.00–11.50	0.328
	Moderate	9.86 ± 4.646	2.578	5.784–10.43	
	Severe	12.18 ± 9.040	3.630	8.23–17.27	
	Tractional	12.88 ± 17.366	4.342	3.62–22.13	
Total Lung PVV [cm^3]	Mild	95.00 ± 44.803	22.402	23.71–166.29	0.04
	Moderate	117.50 ± 49.640	13.267	88.84–146.16	
	Severe	122.82 ± 53.961	16.270	86.57–159.07	
	Tractional	110.00 ± 37.932	9.483	89.79–130.21	

Total Lung Volume [L] is highest in Tubular bronchiectasis [3.875 L] and lowest in Saccular bronchiectasis [2.963 L], suggesting greater air trapping or hyperinflation in Tubular types. Total Lung Hyperlucency follows a similar trend, being highest in the Tubular [23.5%] and lowest in the Saccular [6.62%], with elevated hyperlucency also observed in Cystic types. Ground-glass opacity [GGO] is most prominent in Saccular bronchiectasis [26.87%] and least in Tubular [1.5%], potentially reflecting active inflammation or early fibrosis in Saccular types. Reticular opacity is highest in Cystic bronchiectasis [13.27%] and lowest in Tubular [3.5%], while honeycombing—a marker of advanced fibrosis—is most pronounced in Tubular [13.25%] and least in Saccular [2.87%]. Pulmonary Vessel Volume [PVV], reflecting blood vessel volume, is elevated in Cystic bronchiectasis, suggesting increased vascular involvement. Overall, Cystic and Tubular bronchiectasis are associated with higher lung volumes, hyperlucency, and PVV, which is indicative of more advanced lung changes, while Saccular and tractional types show relatively lower values. These findings reveal emerging patterns of lung alterations specific to bronchiectasis types, highlighting a trend toward more severe structural changes in Cystic and Tubular bronchiectasis (Table 3).

Table 3. Relationship between various parenchymal pathologies detected by Artificial Intelligence and type of bronchiectasis.

Type of Bronchiectasis			Mean ± Std. Deviation	Std. Error	95% Confidence Interval for Mean	*p*-Value
Total Lung Volume [L]	Saccular	8	2.963 ± 1.0501	0.3713	2.085–3.840	0.641
	Cystic	11	3.555 ± 1.8976	0.5721	2.280–4.829	
	Tubular	4	3.875 ± 0.8098	0.4049	2.586–5.164	
	Varicoid	6	3.433 ± 1.5423	0.6296	1.815–5.052	
	Tractional	16	2.969 ± 1.0562	0.2641	2.406–3.532	
Total Lung Hyperlucency [%]	Saccular	8	6.625 ± 9.6501	3.4118	−1.443–14.693	0.446
	Cystic	11	21.455 ± 32.2222	9.7154	−0.193–43.102	
	Tubular	4	23.500 ± 17.2916	8.6458	−4.015–51.015	
	Varicoid	6	9.833 ± 11.8561	4.8402	−2.609–22.276	
	Tractional	16	16.188 ± 13.9486	3.4871	8.755–23.620	

Table 3. *Cont.*

Type of Bronchiectasis			Mean ± Std. Deviation	Std. Error	95% Confidence Interval for Mean	*p*-Value
Total Lung Ground-glass Opacity [%]	Saccular	8	26.875 ± 31.1698	11.0202	0.816–52.934	0.106
	Cystic	11	13.727 ± 15.4861	4.6692	3.324–24.131	
	Tubular	4	1.500 ± 3.0000	1.5000	−3.274–6.274	
	Varicoid	6	12.167 ± 18.2254	7.4405	−6.960–31.293	
	Tractional	16	7.625 ± 9.9658	2.4914	2.315–12.935	
Total Lung Reticular Opacity [%]	Saccular	8	8.625 ± 8.8307	3.1221	1.242–16.008	0.671
	Cystic	11	13.273 ± 17.2921	5.2138	1.656–24.890	
	Tubular	4	3.500 ± 2.6458	1.3229	−0.710–7.710	
	Varicoid	6	6.500 ± 5.5767	2.2767	0.648–12.352	
	Tractional	16	13.375 ± 18.8498	4.7125	3.331–23.419	
Total Lung Honeycombing [%]	Saccular	8	2.875 ± 8.1317	2.8750	−3.923–9.673	0.398
	Cystic	11	6.091 ± 6.4568	1.9468	1.753–10.429	
	Tubular	4	13.250 ± 15.1079	7.5540	−10.790–37.290	
	Varicoid	6	6.167 ± 14.6208	5.9689	−9.177–21.510	
	Tractional	16	12.875 ± 17.3662	4.3415	3.621–22.129	
Total Lung PVV [cm^3]	Saccular	8	104.250 ± 35.1110	12.4136	74.896–133.604	0.561
	Cystic	11	134.545 ± 59.5472	17.9542	94.541–174.550	
	Tubular	4	101.500 ± 58.2666	29.1333	8.785–194.215	
	Varicoid	6	109.333 ± 42.9216	17.5227	64.290–154.377	
	Tractional	16	110.000 ± 37.9315	9.4829	89.788–130.212	

AI appears to detect abnormalities at a higher rate than traditional radiology in all categories. The most significant difference is in the detection of lung hyperlucency, where AI detects 88.9% compared to radiology's 24.4%, while for ground-glass opacity, the detection rate of AI is 82.2% compared to conventional radiology [26.7%] (Table 4).

Table 4. Pathological parameters detected by radiology and AI.

	Radiology		AI	
	Detected	Not Detected	Detected	Not Detected
Total Lung Hyperlucency	11 [24.4%]	34 [75.6%]	40 [88.9%]	5 [11.1%]
Total Ground Glass	12 [26.7%]	33 [73.3%]	37 [82.2%]	8 [17.8%]
Total Reticular Opacity	3 [6.7%]	43 [93.3%]	25 [55.6%]	20 [44.4%]
Total Honeycombing	6 [13.3%]	39 [86.7%]	27 [60%]	18 [40%]

The AI consistently detected more cases of parenchymal pathologies compared to the radiologist, especially in more Severe and tractional cases (Figure 1).

The correlation between AI-detected pathologies and BRICSs is weak (Figure 2).

In our study, we identified 16 [35.5%] patients with hyperlucency exceeding 20% [smokers—6 patients, non-smokers—10 patients], and this can be classified as an emphysema-predominant phenotype. Additionally, 7 [15%] patients presented with combined reticular shadows and honeycombing above 20%, and can be classified as a fibrosis-predominant phenotype, and 2 [4.4%] of the subjects satisfied the criteria for combined emphysema and fibrosis [24–28]. About 33% of patients had significant ground-glass opacities and follow-up studies are needed to understand the natural history of these patients (Figure 3).

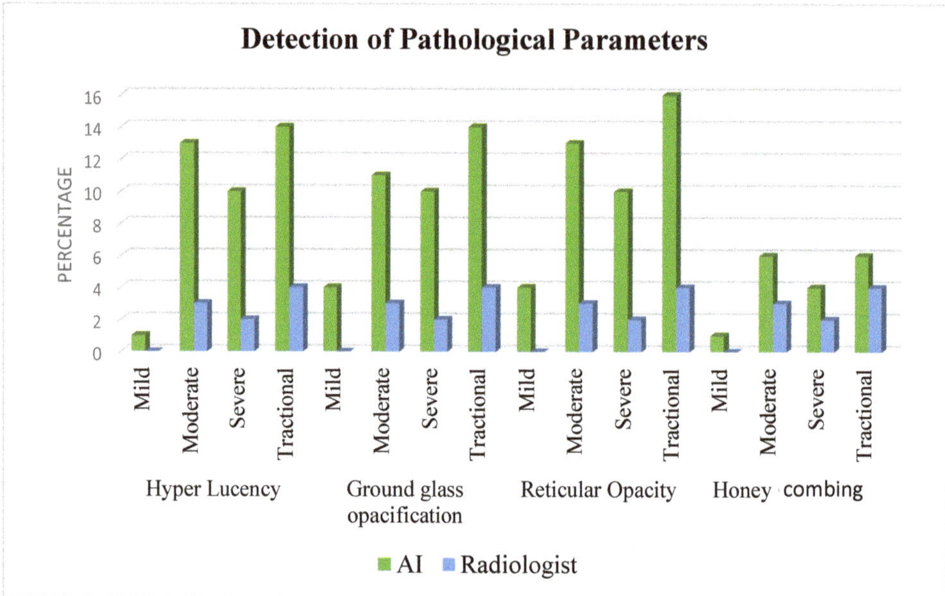

Figure 1. Detection of parenchymal pathologies by AI and radiologist.

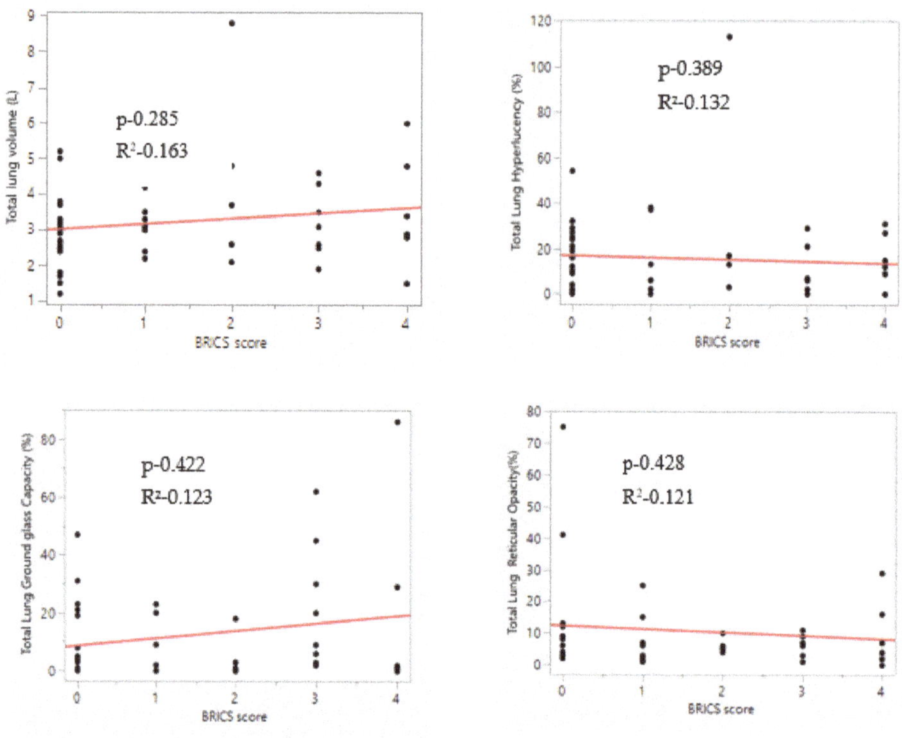

Figure 2. *Cont.*

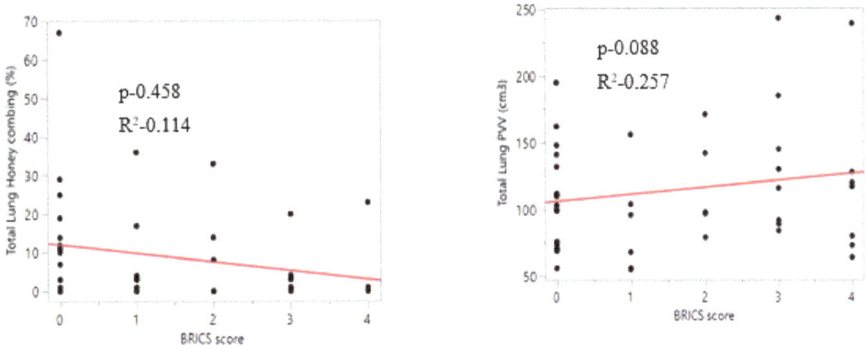

Figure 2. Correlation between various parenchymal pathologies detected by Artificial Intelligence and BRICS.

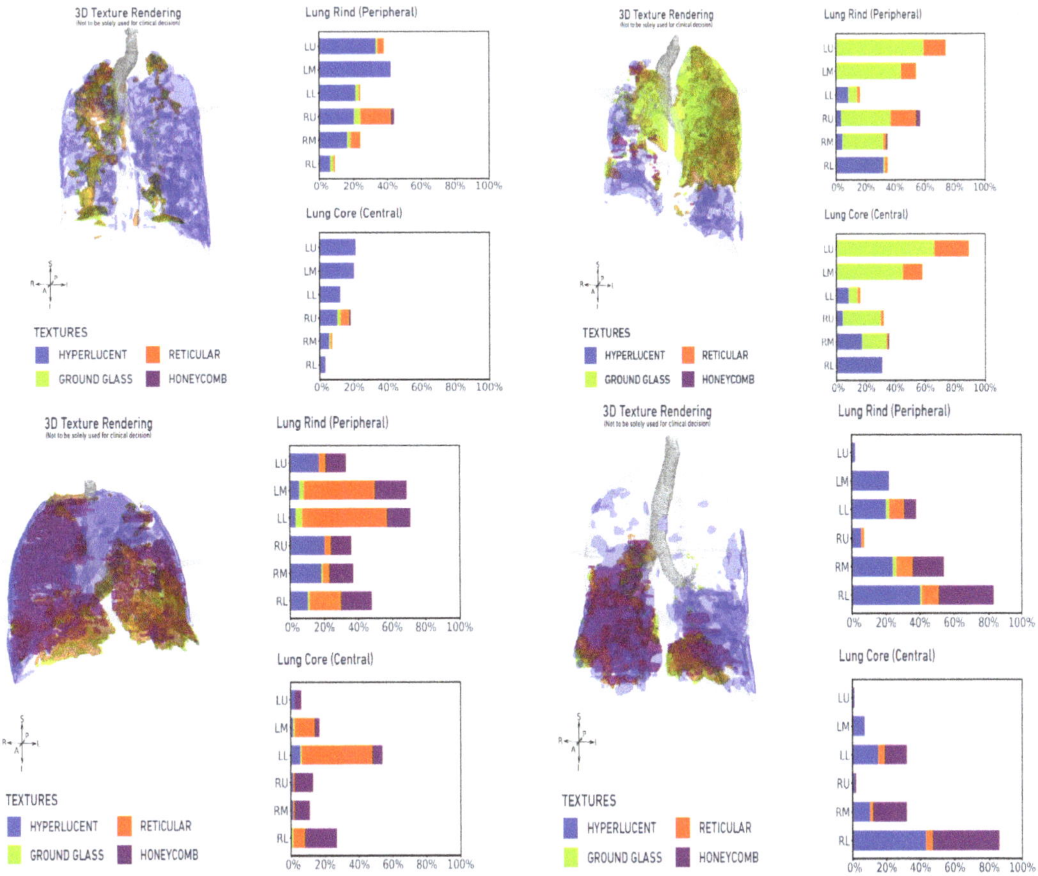

Figure 3. Lung texture analysis report: The Dicom representative image of each category of pattern. These were chosen based on the predominance of lung parenchymal patterns including hyperlucent,

ground glass [GG], reticular, and honeycombing. Imbio LTA provided their regional distribution within three different lung zones [upper, middle, lower] in each lung. The platform provided an image in which each lung pattern was colored differently, allowing the evaluation of disease extent, composition, and location at a glance. Specific images were chosen because of the predominance of one of the patterns in each of them; at the same time, they had other patterns, also showing that all pathologies could coexist in a single case of bronchiectasis, yet have a predominance of a single pattern that may have a bearing on their clinical symptoms.

4. Discussion

This study provides an important analysis of the use of Artificial Intelligence [AI] in interpreting HRCT scans of bronchiectasis. While numerous studies have examined the relationships between functional status, spirometry, quality of life, and HRCT findings in bronchiectasis, to the best of our knowledge, none have explored the potential of AI to enhance the diagnostic and prognostic capabilities of bronchiectasis. Our findings demonstrate that AI can significantly augment the detection of radiological phenotypes in bronchiectasis patients and provide quantitative estimation compared to traditional radiology alone, making this study novel and impactful. In line with findings from the EMBARC India Registry, a significant portion of bronchiectasis cases in our cohort were post-infectious, primarily following tuberculosis. AI-driven lung texture analysis identified specific radiological patterns—hyperlucency, ground-glass opacity [GGO], reticular shadows, and honeycombing—at higher rates than those observed by radiologists. Additionally, this AI-based analysis provided quantitative estimates of the proportion of lungs affected by each of these patterns. These radiological patterns offer valuable insights into the underlying pathophysiology of bronchiectasis and its associated pulmonary abnormalities.

Hyperlucency in the lungs in bronchiectasis can arise from a variety of structural and pathological changes, including emphysema, and vascular anomalies [29]. AI detected hyperlucency in 88.9% of cases, with the highest incidence in Tubular bronchiectasis [23.5%]. This rate was significantly higher than the 24.4% detection rate by radiologists, underscoring AI's potential to capture subtle emphysematous alterations. Supporting studies by Loubeyre et al. suggest that these emphysematous changes may result from bronchiolar inflammation, which disproportionately affects peripheral airways and may increase susceptibility to emphysema in advanced bronchiectasis [30]. Increased neutrophil elastase in the sputum or BAL is associated with increased hyperlucency [31–33]. The severity of the emphysema appears to be the most important determinant of the chronic deterioration of airflow [34–36].

Ground-glass opacity [GGO], often indicative of decreased air content in the lung parenchyma without complete alveolar obstruction, is associated with reversible conditions such as minimal alveolar septal thickening or partial fluid accumulation. In our study, AI detected GGO in 82.2% of cases, with the highest prevalence in Saccular bronchiectasis [26.85%]. The presence of GGO may reflect an ongoing inflammatory response or infectious process, highlighting AI's capacity to detect early, potentially reversible changes in bronchiectasis pathology [37].

Reticular opacities—characterized by a network of linear densities often linked with interstitial lung diseases—were identified in 55.6% of cases by AI, in contrast to only 6.7% detected by three radiologists. This substantial discrepancy demonstrates AI's heightened sensitivity in identifying interstitial changes, which could have significant implications for diagnosing fibrotic conditions in bronchiectasis. While reticular shadows appeared across all bronchiectasis types, slightly higher values were noted in cystic and tractional forms [38].

Honeycombing, a hallmark of end-stage fibrotic lung disease, presents as clustered cystic airspaces. AI detected honeycombing in 60% of cases compared to 13.3% by radiologists. This notable difference suggests that AI could play a critical role in identifying advanced fibrotic changes that may be challenging to detect visually, even by experienced radiologists [38]. The AI-detected patterns align with the histopathological features of bronchiectasis, such as alveolar collapse and dilation of alveolar ducts, further validating AI's utility in identifying advanced disease stages [39].

While HRCT can assess the type and severity of bronchiectasis, AI algorithms offer the capacity to analyze vast datasets, uncovering patterns and associations that may elude human observers. By integrating data from multiple sources, including imaging and clinical records, AI provides a more comprehensive view of bronchiectasis, aiding in the identification of disease phenotypes. In our study, we identified the following three main radiological phenotypes: bronchiectasis with emphysema and bronchiectasis with interstitial involvement [evidenced by ground-glass opacity, reticular patterns, and honeycombing], and the third with both emphysema and fibrosis. Recognizing bronchiectasis as a heterogeneous disease and not just airway disease helps explain the broad spectrum of clinical symptoms and variability in severity, which is potentially linked to these distinct radiological phenotypes, and may inform future clinical studies to address these pathologies in bronchiectasis [40].

Advancements in AI-based algorithms for monitoring and assessing bronchiectasis align with broader progress across medical fields. For instance, a study by Díaz et al. utilizing AI in chest CT imaging from the COPD Gene cohort examined bronchiectasis in COPD, particularly focusing on the airway-to-artery ratio [AAR] [41]. This study found that a higher AAR [greater than 1] correlated with increased exacerbation frequency in COPD patients, suggesting AI's potential in early bronchiectasis detection and exacerbation risk prediction. However, our study uniquely employs the IMBIO software's lung texture analysis tool to detect subtle patterns that may be overlooked by radiologists, highlighting the novelty of our approach [42].

Our study's strengths include a bronchiectasis-specific cohort, enabling an in-depth analysis of the links between radiographic abnormalities and clinically meaningful outcomes. Nonetheless, there are limitations to consider. Radiographic scores represent a single time point and do not capture the dynamic nature of bronchiectasis, as radiographic abnormalities can vary over time. Additionally, the widespread application of these scoring methods is challenging due to the need for specialized expensive software. AI technologies are still in an early stage of development and are yet to be fully integrated into routine clinical practice [41,43].

The integration of AI into clinical workflows also requires thoughtful planning to ensure seamless alignment with existing radiological practices. Effective implementation will necessitate training for radiologists and healthcare professionals to use AI tools in a way that complements, rather than disrupts, the diagnostic process [44]. Addressing these challenges will require a collaborative effort among researchers, radiologists, and clinicians to facilitate the adoption of AI in a comprehensive evaluation of bronchiectasis. With further research, AI holds the potential to transform the early detection and prognosis of bronchiectasis, ultimately enhancing patient outcomes [45]. Additionally, further clinical research is needed to identify the ideal therapeutic modalities for the two main radiologic phenotypes identified in this study, the emphysema-predominant bronchiectasis [role of neutrophil elastase inhibitors], and the interstitial fibrosis-predominant bronchiectasis [role of anti-fibrotics]. The natural course of ground-glass opacity is quite unclear and further longitudinal studies would be needed to conclude the phenotype into which it can be eventually categorized.

In summary, our findings underscore AI's potential to enhance the diagnostic precision of HRCT in bronchiectasis by delivering more consistent, detailed quantitative measurements and objective evaluations of disease-specific radiological characteristics. AI's superior performance in detecting hyperlucency, GGO, reticular shadows, and honeycombing suggests that it could serve as a valuable adjunct to traditional radiology, potentially guiding more tailored therapeutic strategies. Future research should further explore AI's role in assessing disease progression and response to treatment, which could ultimately enhance patient outcomes and quality of life.

Author Contributions: Conceptualization, A.N., V.P. and P.A.M.; Methodology, A.N., M.V.G., D.B. and P.A.M.; Software, A.N., R.M., M.V.G., D.B. and P.A.M.; Validation, A.N., R.M. and P.A.M.; Formal analysis, R.M. and P.A.M.; Investigation, P.A.M.; Resources, A.N. and P.A.M.; Data curation, A.N., R.M. and D.B.; Writing—original draft, A.N., R.M. and P.A.M.; Writing—review & editing, M.V.G., S.V.M., B.S.J., S.K.C., S.A.K., K.S.L., M.M.A.L., V.V., S.K., S.S. and P.A.M.; Visualization, R.M., M.V.G., V.P., S.V.M., B.S.J., S.A.K., K.S.L., M.M.A.L., V.V., S.K., S.S. and P.A.M.; Supervision, V.P., S.V.M., B.S.J., S.K.C., K.S.L. and M.M.A.L. All authors have read and agreed to the published version of the manuscript.

Funding: This research received no external funding.

Institutional Review Board Statement: The study was conducted in accordance with the Declaration of Helsinki, and approved by the Institutional Review Board (or Ethics Committee) of JSS Medical College and Hospital, Mysuru, Karnataka (approval number: JSS/MC/PG/IEC-105/2022-23, approval date: 22 August 2022).

Informed Consent Statement: Informed consent was obtained from all subjects involved in the study.

Data Availability Statement: The original contributions presented in this study are included in the article. Further inquiries can be directed to the corresponding author.

Acknowledgments: The authors would like to thank all the staff in the Department of Respiratory Medicine and Department of Radiology, JSS Hospital, Mysuru.

Conflicts of Interest: The authors declare no conflicts of interest. The funders had no role in the design of the study; in the collection, analyses, or interpretation of data; in the writing of the manuscript; or in the decision to publish the results.

References

1. Contarini, M.; Finch, S.; Chalmers, J.D. Bronchiectasis: A case-based approach to investigation and management. *Eur. Respir. Rev.* **2018**, *27*, 180016. [CrossRef] [PubMed]
2. Ringshausen, F.C.; de Roux, A.; Diel, R.; Hohmann, D.; Welte, T.; Rademacher, J. Bronchiectasis in Germany:A population-based estimation of disease prevalence. *Eur. Respir. J.* **2015**, *46*, 1805–1807. [CrossRef] [PubMed]
3. Monteagudo, M.; Rodríguez-Blanco, T.; Barrecheguren, M.; Simonet, P.; Miravitlles, M. Prevalence and incidence of bronchiectasis in Catalonia, Spain: A population-based study. *Respir. Med.* **2016**, *121*, 26–31. [CrossRef]
4. Quint, J.K.; Millett, E.R.; Joshi, M.; Navaratnam, V.; Thomas, S.L.; Hurst, J.R.; Smeeth, L.; Brown, J.S. Changes in the incidence, prevalence and mortality of bronchiectasis in the UK from 2004 to 2013:A population-based cohort study. *Eur. Respir. J.* **2016**, *47*, 186–193. [CrossRef] [PubMed]
5. Weycker, D.; Hansen, G.L.; Seifer, F.D. Prevalence and incidence of noncystic fibrosis bronchiectasis among US adults in 2013. *Chron. Respir. Dis.* **2017**, *14*, 377–384. [CrossRef]
6. Lin, J.L.; Xu, J.F.; Qu, J.M. Bronchiectasis in China. *Ann. Am. Thorac. Soc.* **2016**, *13*, 609–616. [CrossRef] [PubMed]
7. Hariprasad, K.; Krishnan, S.; Mehta, R.M. Bronchiectasis in India: Results from the EMBARC and Respiratory Research Network of India Registry. *Natl. Med. J. India* **2020**, *33*, 99–101. [CrossRef] [PubMed]
8. Whitwell, F. A Study of the Pathology and Pathogenesis of Bronchiectasis. *Thorax* **1952**, *7*, 213–239. [CrossRef]
9. Shteinberg, M.; Flume, P.A.; Chalmers, J.D. Is bronchiectasis really a disease? *Eur. Respir. Rev.* **2020**, *29*, 190051. [CrossRef]
10. Tolouee, A.; Abrishami-Moghaddam, H.; Garnavi, R.; Forouzanfar, M.; Giti, M. Texture Analysis in lung HRCT images. In Proceedings of the Digital Image Computing: Techniques and Applications [DICTA2008], Canberra, Australia, 1–3 December 2008; pp. 305–311.
11. Alzeer, A.H. HRCT score in bronchiectasis: Correlation with pulmonary function tests and pulmonary artery pressure. *Ann. Thorac. Med.* **2008**, *3*, 82–86. [CrossRef]
12. Kang, E.Y.; Miller, R.R.; Müller, N.L. Bronchiectasis: Comparison of preoperative thin-section CT and pathologic findings in resected specimens. *Radiology* **1995**, *195*, 649–654. [CrossRef]
13. Edwards, E.; Narang, I.; Li, A.; Hansell, D.; Rosenthal, M.; Bush, A. HRCT lung abnormalities are not a surrogate for exercise limitation in bronchiectasis. *Eur. Respir. J.* **2004**, *24*, 538–544. [CrossRef]
14. Eshed, I.; Minski, I.; Katz, R.; Jones, P.W.; Priel, I.E. Bronchiectasis: Correlation of high-resolution CT findings with health-related quality of life. *Clin. Radiol.* **2007**, *62*, 152–159. [CrossRef] [PubMed]
15. García, M.M.; Tordera, M.P.; Sánchez, P.R.; Cataluña, S. Internal consistency and validity of the Spanish version of the St George's respiratory questionnaire for use in patients with clinically stable bronchiectasis. *Arch. Bronconeumol.* **2005**, *41*, 110–117. [CrossRef] [PubMed]
16. Tiddens, H.A.; Meerburg, J.J.; Van Der Eerden, M.M.; Ciet, P. The radiological diagnosis of bronchiectasis: What's in a name? *Eur. Respir. Rev.* **2020**, *29*, 190120. [CrossRef] [PubMed]

17. Bedi, P.; Chalmers, J.D.; Goeminne, P.C.; Mai, C.; Saravanamuthu, P.; Velu, P.P.; Cartlidge, M.K.; Loebinger, M.R.; Jacob, J.; Kamal, F.; et al. The BRICS [Bronchiectasis Radiologically Indexed CT Score]. *Chest J.* **2018**, *153*, 1177–1186. [CrossRef]
18. Bartholmai, B.J.; Raghunath, S.; Karwoski, R.A.; Moua, T.; Rajagopalan, S.; Maldonado, F.; Decker, P.A.; Robb, R.A. Quantitative computed tomography imaging of interstitial lung diseases. *J. Thorac. Imaging* **2013**, *28*, 298–307. [CrossRef] [PubMed] [PubMed Central]
19. Maldonado, F.; Moua, T.; Rajagopalan, S.; Karwoski, R.A.; Raghunath, S.; Decker, P.A.; Hartman, T.E.; Bartholmai, B.J.; Robb, R.A.; Ryu, J.H. Automated quantification of radiological patterns predicts survival in idiopathic pulmonary fibrosis. *Eur. Respir. J.* **2014**, *43*, 204–212. [CrossRef] [PubMed]
20. Occhipinti, M.; Bosello, S.; Sisti, L.G.; Cicchetti, G.; De Waure, C.; Pirronti, T.; Ferraccioli, G.; Gremese, E.; Larici, A.R. Quantitative and semi-quantitative computed tomography analysis of interstitial lung disease associated with systemic sclerosis: A longitudinal evaluation of pulmonary parenchyma and vessels. *PLoS ONE* **2019**, *14*, e0213444. [CrossRef] [PubMed]
21. John, J.; Clark, A.R.; Kumar, H.; Vandal, A.C.; Burrowes, K.S.; Wilsher, M.L.; Milne, D.G.; Bartholmai, B.; Levin, D.L.; Karwoski, R.; et al. Pulmonary vessel volume in idiopathic pulmonary fibrosis compared with healthy controls aged > 50 years. *Sci. Rep.* **2023**, *13*, 4422. [CrossRef]
22. Alkhanfar, D.; Shahin, Y.; Alandejani, F.; Dwivedi, K.; Alabed, S.; Johns, C.; Lawrie, A.; Thompson, A.R.; Rothman, A.M.; Tschirren, J.; et al. Severe pulmonary hypertension associated with lung disease is characterised by a loss of small pulmonary vessels on quantitative computed tomography. *ERJ Open Res.* **2022**, *8*, 00503–02021. [CrossRef]
23. King, P.T. The pathophysiology of bronchiectasis. *Int. J. Chron. Obs. Pulmon Dis.* **2009**, *4*, 411–419. [CrossRef] [PubMed] [PubMed Central]
24. Subramanian, D.R.; Gupta, S.; Burggraf, D.; Vom Silberberg, S.J.; Heimbeck, I.; Heiss-Neumann, M.S.; Haeussinger, K.; Newby, C.; Hargadon, B.; Raj, V.; et al. Emphysema- and airway-dominant COPD phenotypes defined by standardised quantitative computed tomography. *Eur. Respir. J.* **2016**, *48*, 92–103. [CrossRef] [PubMed]
25. Amariei, D.E.; Dodia, N.; Deepak, J.; Hines, S.E.; Galvin, J.R.; Atamas, S.P.; Todd, N.W. Combined Pulmonary Fibrosis and Emphysema: Pulmonary Function Testing and a Pathophysiology Perspective. *Medicina* **2019**, *55*, 580. [CrossRef]
26. Cottin, V.; Nunes, H.; Brillet, P.-Y.; Delaval, P.; Devouassoux, G.; Tillie-Leblond, I.; Israel-Biet, D.; Court-Fortune, I.; Valeyre, D.; Cordier, J.-F. Combined pulmonary fibrosis and emphysema: A distinct underrecognised entity. *Eur. Respir. J.* **2005**, *26*, 586–593. [CrossRef] [PubMed]
27. Fernandes, L.; Fernandes, Y.; Mesquita, A. Quantitative computed tomography imaging in chronic obstructive pulmonary disease. *Lung India* **2016**, *33*, 646. [CrossRef]
28. Zhang, H.; Li, X.; Zhang, X.; Yuan, Y.; Zhao, C.; Zhang, J. Quantitative CT analysis of idiopathic pulmonary fibrosis and correlation with lung function study. *BMC Pulm. Med.* **2024**, *24*, 437. [CrossRef]
29. Cherian, S.V.; Girvin, F.; Naidich, D.P.; Machnicki, S.; Brown, K.K.; Ryu, J.H.; Gupta, N.; Mehta, V.; Estrada-Y-Martin, R.M.; Narasimhan, M.; et al. Lung Hyperlucency: A Clinical-Radiologic Algorithmic Approach to Diagnosis. *Chest* **2020**, *157*, 119–141. [CrossRef] [PubMed]
30. Loubeyre, P.; Paret, M.; Revel, D.; Wiesendanger, T.; Brune, J. Thin-Section CT Detection of Emphysema Associated With Bronchiectasis and Correlation With Pulmonary Function Tests. *Chest* **1996**, *109*, 360–365. [CrossRef] [PubMed]
31. Bhalla, M.; Turcios, N.; Aponte, V.; Jenkins, M.; Leitman, B.S.; McCauley, D.I.; Naidich, D.P. Cystic fibrosis: Scoring system with thin-section CT. *Radiology* **1991**, *179*, 783–788. [CrossRef] [PubMed]
32. Klein, J.S.; Gamsu, G.; Webb, W.R.; Golden, J.A.; Müller, N.L. High-resolution CT diagnosis of emphysema in symptomatic patients with normal chest radiographs and isolated low diffusing capacity. *Radiology* **1992**, *182*, 817–821. [CrossRef] [PubMed]
33. Reid, L.M. Reduction in bronchial subdivision in bronchiectasis. *Thorax* **1950**, *5*, 233–247. [CrossRef]
34. Willems, L.N.; Kramps, J.A.; Stijnen, T.; Sterk, P.J.; Weening, J.J.; Dijkman, J.H. Relation between small airways disease and parenchymal destruction in surgical lung specimens. *Thorax* **1990**, *45*, 89–94. [CrossRef] [PubMed]
35. Wiggs, B.R.; Bosken, C.; Paré, P.D.; James, A.; Hogg, J.C. A model of airway narrowing in asthma and in chronic obstructive pulmonary disease. *Am. Rev. Respir. Dis.* **1992**, *145*, 1251–1258. [CrossRef] [PubMed]
36. Jeffery, P.K. Morphology of the airway wall in asthma and in chronic obstructive pulmonary disease. *Am. Rev. Respir. Dis.* **1991**, *143*, 1152–1158. [CrossRef] [PubMed]
37. Engeler, C.E.; Tashjian, J.H.; Trenkner, S.W.; Walsh, J.W. Ground-Glass Opacity of the Lung Parenchyma: A Guide to Analysis with High-Resolution CT. *AJR* **1993**, *160*, 249–251. [CrossRef]
38. Elicker, B.; De Castro Pereira, C.A.; Webb, R.; Leslie, K.O. High-resolution computed tomography patterns of diffuse interstitial lung disease with clinical and pathological correlation. *J. Bras. Pneumol.* **2008**, *34*, 715–744. [CrossRef] [PubMed]
39. Johkoh, T.; Sakai, F.; Noma, S.; Akira, M.; Fujimoto, K.; Watadani, T.; Sugiyama, Y. Honeycombing on CT; its definition, pathologic correlation, and future direction of its diagnosis. *Eur. J. Radiol.* **2014**, *83*, 27–31. [CrossRef]
40. Chotirmall, S.H.; Chalmers, J.D. The Precision Medicine Era of Bronchiectasis. *Am. J. Respir. Crit. Care Med.* **2024**, *210*, 24–34. [CrossRef]
41. Díaz, A.A.; Nardelli, P.; Wang, W.; Estépar, R.S.J.; Yen, A.; Kligerman, S.; Maselli, D.J.; Dolliver, W.R.; Tsao, A.; Orejas, J.L.; et al. Artificial Intelligence–based CT Assessment of Bronchiectasis: The COPDGene Study. *Radiology* **2023**, *307*, e221109. [CrossRef] [PubMed]

42. Amisha; Malik, P.; Pathania, M.; Rathaur, V.K. Overview of artificial intelligence in medicine. *J. Fam. Med. Prim. Care* **2019**, *8*, 2328. [CrossRef]
43. Park, J.; Kim, S.; Lee, Y.J.; Park, J.S.; Cho, Y.; Yoon, H.I.; Lee, K.; Lee, C.; Lee, J.H. Factors associated with radiologic progression of non-cystic fibrosis bronchiectasis during long-term follow-up. *Respirology* **2016**, *21*, 1049–1054. [CrossRef] [PubMed]
44. Gramegna, A.; Amati, F.; Terranova, L.; Sotgiu, G.; Tarsia, P.; Miglietta, D.; Calderazzo, M.A.; Aliberti, S.; Blasi, F. Neutrophil elastase in bronchiectasis. *Respir. Res.* **2017**, *18*, 211. [CrossRef] [PubMed]
45. Willis, B.C.; Borok, Z. TGF-β-induced EMT: Mechanisms and implications for fibrotic lung disease. *Am. J. Physiol. -Lung Cell. Mol. Physiol.* **2007**, *293*, L525–L534. [CrossRef]

diagnostics

MDPI

Article

Severity Classification of Parkinson's Disease via Synthesis of Energy Skeleton Images from Videos Produced in Uncontrolled Environments

Nejib Ben Hadj-Alouane [1,*], Arav Dhoot [2], Monia Turki-Hadj Alouane [3,*] and Vinod Pangracious [1]

[1] Electrical and Computer Engineering Department, American University in Dubai, Dubai P. O. Box 28282, United Arab Emirates; vpangracious@aud.edu
[2] Columbia College, Columbia University, New York, NY 10027, USA; arav.dhoot@columbia.edu
[3] College of Computer Science, King Khalid University, Abha 62521, Saudi Arabia
* Correspondence: nalouane@aud.edu (N.B.H.-A.); malouane@kku.edu.sa (M.T.-H.A.)

Abstract: Background/Objectives: Parkinson's Disease is a prevalent neurodegenerative disorder affecting millions worldwide, primarily marked by motor and non-motor symptoms due to the degeneration of dopamine-producing neurons. Despite the absence of a cure, current treatments focus on symptom management, often relying on pharmacotherapy and surgical interventions. Early diagnosis remains a critical challenge, particularly in underserved areas, as existing diagnostic protocols lack standardization and accessibility. This paper proposes a novel framework for the diagnosis and severity classification of PD using video data captured in uncontrolled environments. **Methods:** Leveraging deep learning techniques, our approach synthesizes Skeleton Energy Images (SEIs) from gait sequences and employs three advanced models—a Convolutional Neural Network (CNN), a Residual Network (ResNet), and a Vision Transformer (ViT)—to analyze these images. Our methodology allows for the accurate detection of PD and differentiation of its severity without requiring specialized equipment or professional oversight. The dataset used consists of labeled videos capturing the early stages of the disease, facilitating the potential for timely intervention. **Results:** The four models performed very accurately during the training phase. In fact, an accuracy higher than 99% was achieved by the ViT and ResNet models. Moreover, a lesser accuracy of 90% was achieved by the CNN five-layer model. During the test phase, only the best-performing models from the training experiments were tested. The ResNet-18 model has achieved a 100% accuracy. However, the ViT and the CNN five-layer models have achieved, respectively, 99.96% and 96.40% test accuracy. **Conclusions:** The results demonstrate high accuracy, highlighting the framework's capabilities, and in particular the effectiveness of the workflow used for generating the SEI images. Given the nature of the dataset used, the proposed framework stands to function as a cost-effective and accessible tool for early PD detection in various healthcare settings. This study contributes to the advancement of mobile health technologies, aiming to enhance early diagnosis and monitoring of Parkinson's disease.

Keywords: Parkinson's diagnosis; classification; CNN; ViT; ResNets; Skeleton Energy Image (SEI)

1. Introduction

Parkinson's Disease (PD) is an incurable neurodegenerative condition affecting an estimated 10 million individuals worldwide [1]. The onset of PD is characterized by a decline in muscular function, primarily due to the degeneration of dopamine-producing neurons in the brain. While the exact cause remains uncertain, a combination of genetic, environmental, and lifestyle factors is believed to contribute to its onset. Aging is a significant risk factor, with most cases occurring in individuals over the age of 60.

The primary symptoms of PD include tremors, bradykinesia (slowness of movement), rigidity, and postural instability. Beyond these motor symptoms, PD can also lead to non-motor symptoms such as cognitive impairment, depression, and sleep disturbances, complicating the overall clinical picture [2].

Current treatments for PD focus on managing symptoms, as there is no cure for the condition. Common strategies include pharmacotherapy, which often involves medications designed to replenish or mimic dopamine levels in the brain to alleviate motor symptoms. Additionally, deep brain stimulation has emerged as a surgical intervention for individuals with advanced PD. However, physical therapy is considered the gold standard for non-invasive management, particularly when initiated early in the disease course. Engaging in physical therapy can help reduce tremor severity and improve the overall quality of life for patients.

Diagnosing PD presents significant challenges, as there is currently no universal protocol. Physicians rely on a combination of laboratory tests, blood analyses, and physiological assessments to render a diagnosis. Unfortunately, many patients do not receive a PD diagnosis until the disease has progressed to advanced stages. This diagnostic challenge is particularly acute for residents in rural and remote areas, who often have limited access to specialized healthcare resources. Furthermore, non-Parkinsonian gaits may sometimes be misclassified as Parkinsonian, leading to inappropriate therapies and medications.

With the recent advancement in computational power and increased access to technology, research is being conducted to develop computer-based solutions assisting in the diagnosis of PD [3]. These solutions aim to develop tools accessible in diverse environments, to facilitate its early and accurate detection. For example, mobile systems hold promise for assisting healthcare professionals with the diagnosis and management of PD. These technologies could offer tools for preliminary assessments, aiding in the early identification of symptoms and facilitating more timely interventions.

Advanced algorithms and artificial intelligence can analyze vast and diverse datasets, aiding in the identification of hidden patterns associated with PD. In fact, research for the effective application of machine learning (ML) and deep learning (DL) to develop systems to help with the diagnosis of PD is ongoing. Over the past decade, multiple ML and DL systems have been developed to detect PD based on multiple media datasets [3].

In particular, early systems have been developed for the diagnosis of PD based on the analysis of handwriting [4], speech [5], and other types of data. More modern systems have leveraged video data and focused on using feature engineering techniques combined with ML algorithms [6]. Most of these, however, require video datasets developed in controlled environments or under the supervision of professionals. Obviously, this limits their availability and reduces their impact for the above stated cause linked to early diagnosis and availability in environments with limited resources.

The methodology developed in this paper for the diagnosis and severity detection of PD is based on video data produced mostly in uncontrolled environments. These videos involve the subjects simply walking in different situations and turning in different directions. In fact the dataset we use in this paper is the same used in [7] and has been collected from various open sources. The resulting gaits are automatically analyzed to generate a PD diagnosis. This diagnosis technique, based on a person's gait, is known to be reliable and is commonly used by neurologists and physicians [8].

Currently, most video-based systems necessitate a significant level of human intervention for both recording the videos and generating the diagnoses [9]. For instance, the systems described in [10,11] require the placement of physical sensors on the subjects, while [12] relies on digital markers applied to the subject's body. Consequently, these systems are not more accessible than traditional methods that do not utilize video data. Furthermore, even studies employing video data combined with various feature engineering techniques and ML, without needing specialized artifacts, mandate that videos be captured in controlled environments [13]. This requirement enhances the effectiveness of their feature engineering algorithms [14].

In contrast, the framework developed in this paper employs modern deep learning techniques to facilitate accurate detection and severity classification of Parkinson's Disease.

For configuring our deep learning models, we utilize a generic video dataset recorded in uncontrolled environments, without any specific directions or professional assistance [7]. This dataset is labeled with four severity levels according to the classification standard set by the Unified Parkinson's Disease Rating Scale (UPDRS) [15]. It includes videos of subjects in the early "mild" and "moderate" stages of the disease, allowing our framework to effectively capture and distinguish these early stages of PD. Thus, the methodology developed in this paper is designed to function in various environments using standard cameras, without requiring professional intervention. This enables doctors to use it as a diagnostic aid in their offices or while remotely connecting with patients in their homes, even in remote areas. Furthermore, the proposed framework, subject to the availability of an adequate database, can be applied to detect other types of gait abnormalities related to neurological diseases. Many neurological diseases, such as Alzheimer's disease, cerebellar ataxia, Huntington's disease, etc., impact the human gait, yielding various gait disorders [16].

More specifically, the framework includes a workflow for pre-processing the video dataset that uses DL-based techniques to synthesize Skeleton Energy Images (SEIs) [17]. These SEIs involve sequences of images capturing the gaits of the respective subjects. Three deep learning models—a Convolutional Neural Network (CNN), a Residual Network (ResNet), and a Vision Transformer (ViT)—are trained and applied to the classification task. All models demonstrate effective performance with the workflow, achieving high accuracy in diagnosing PD and classifying its severity.

The primary objective of this study is to leverage deep learning techniques in a computationally efficient manner to develop a system capable of detecting Parkinsonism from simple video data. Considering the main use case of wide availability and early detection, our system is cost-effective and less complex compared to existing methods. By optimizing computational efficiency, this research aims to contribute to the advancement of mobile health technologies, providing a cost-effective and widely accessible solution for the early detection and monitoring of Parkinson's Disease.

This paper is organized as follows: The next section presents an extensive literature review highlighting significant works in the field. Section 3 introduces and describes the dataset used in our study. Section 4 details the workflows of our methodology, particularly discussing the pre-processing applied to the dataset and the synthesis of SEIs using human pose estimation techniques. Section 5 evaluates and discusses the effectiveness of the deep learning models employed for the classification task. Section 6 presents our comprehensive experimental study, detailing our results and evaluations with thorough discussions. Finally, Section 7 concludes this paper.

2. Literature Review

Numerous approaches have been explored in the literature for the automated detection and diagnosis of Parkinson's Disease. Many studies have employed various features, including handwriting [4], eye movements [18], and biomarker data [19]. However, methods utilizing image and video analysis are particularly effective, as they provide a direct visual representation of the motor symptoms associated with PD, closely resembling a doctor's diagnosis.

A review in [20] highlighted the connection between human gait and PD. Early-stage PD is characterized by movement irregularities, reduced arm swing, and limited motion in the lower limb joints (ankle, knee, and hip). As the disease progresses to mild and moderate stages, patients exhibit more pronounced impairments, such as stooped posture during walking and fragmented turning. In advanced stages, symptoms like freezing gait and severe postural issues become evident. The severity of PD is assessed using UPDRS, which rates gait on a scale from 0 to 4, where 0 indicates a normal gait and 4 signifies a complete lack of movement. Consequently, vision-based gait analysis techniques are essential for the automated diagnosis of Parkinson's Disease. Recent advancements in computing hardware and the development of more efficient deep learning and machine learning (ML) models have accelerated investigations into automated PD diagnosis using video data [21].

Many studies combine feature extraction from videos and images with ML techniques to enhance PD detection. In [14], a computer vision technique is proposed to differentiate between normal and Parkinsonian gaits. They employed an a Region-Based Convolutional Neural Network (R-CNN) to extract human silhouettes from RGB video frames, generating Gait Energy Images (GEIs) for further processing. This method, evaluated on a dataset of 228 normal gait videos and 73 PD gait videos collected in a controlled environment, achieved an accuracy of 97.33%. Similarly, in [13], a 2D pose estimation is utilized to extract gait features and identify four PD stages using various ML algorithms, including Support Vector Machine (SVM), achieving an accuracy ranging from 96% to 99%.

A similar approach is proposed in [21], which involves extracting 3D key points for significant body joints using OpenPose software. This method allows for the extraction of features characterizing the gait of individuals with Parkinson's disease, healthy older adults, and individuals with multiple sclerosis (MS). The study employed Residual Neural Networks (RNNs) and 1D Convolutional Neural Networks (CNNs), achieving an accuracy of 78.1%. The dataset comprised locally collected videos from 33 participants, including 10 with MS, 9 with PD, and 14 healthy individuals.

Four additional leading studies in the field of PD diagnosis utilize video data and feature extraction approaches. The first, presented in [7], analyzes gait patterns from video recordings using an ML model, namely, Random Forest (RF), on key-point features obtained via AlphaPose, a human pose estimation model. This study highlights the advantages of human pose estimation for computationally efficient feature extraction. Notably, it employs a self-collected dataset consisting of videos sourced from uncontrolled environments in open-access domains, which is also used in this study.

The second approach, as proposed in [22], focuses on gait features to identify Parkinson's stages in individuals with dementia, assessed using the UPDRS gait scale. This method begins with pose estimation, followed by extracting body joint coordinates using MediaPipe for gait analysis. Various machine learning algorithms, including Extra Trees, achieved an accuracy of 90.1%. The dataset included videos from 14 elderly individuals exhibiting diverse gait patterns.

The third method leverages motion features from 3D skeleton estimation, particularly using pelvic position coordinates to detect freezing gait in patients. The fourth study [23] focuses on detecting Freezing of Gait (FoG), a specific PD symptom, in uncontrolled clinical environments. This study combines a rule-based model utilizing pelvic coordinates to detect walking stops with a machine learning model based on a 1D-CNN linked to a Long Short-Term Memory (LSTM) network. This hybrid approach enables accurate FoG detection from videos recorded during routine clinical practice, even with multiple individuals present. The system demonstrated high reliability, with an intraclass correlation coefficient ranging from 0.75 to 0.94 compared to expert annotations. This innovation offers potential for monitoring FoG in home environments using standard video cameras, such as those on smartphones. However, challenges remain in cluttered scenes and variations in lighting or camera angles that may impact detection accuracy.

With advancements in computing capabilities, including powerful GPUs, video serves as a promising medium for PD detection [24]. However, the literature on the intersection of deep learning approaches and video data for PD detection and severity classification remains sparse [25]. A recent comprehensive study in this area [24] highlighted the necessity and importance of such tools, discussing challenges like the need for quality video datasets. It noted that deep learning solutions for PD diagnosis based on video data often exhibit significantly lower accuracy compared to traditional machine learning approaches. Furthermore, many proposed video-based systems require access to specialized software and hardware, posing accessibility barriers for widespread adoption in normal clinical settings or at home.

In [26], a complex, entirely DL-based approach is proposed for the automated diagnosis of gait impairments in PD. Silhouettes and 2D skeletons are extracted from gait videos, from which pose graphs are constructed. Long-Term Gait Energy Images (LT-GEIs) are computed from silhouettes, defined as a composite image of GEIs extracted from three gait cycles. The two streams of pose graphs and LT-GEIs are fused using a Spatio-Temporal Graph Convolutional Network (STGCN), achieving an overall classification accuracy of 71.25% on a controlled dataset comprising 54 PD patients and 26 healthy subjects.

Additionally, [27] introduced a novel dataset called GAIT-IT, consisting of videos captured in controlled settings, featuring 21 patients displaying five different gait types, not limited to PD, with only two severity levels. The authors noted the lack of existing video data and surveyed four existing small datasets, primarily composed of synthesized video data. They discussed two key methods for generating synthetic images from videos. The first, originating in [28], involves synthesizing silhouette images from video frames and aggregating gait sequences into Gait Energy Images (GEIs). The second, more recent approach discussed in [17], utilizes OpenPose to synthesize skeleton frames from images, aggregating gait sequences into Skeleton Energy Images (SEIs). Furthermore, this study compared the SEI approach with the GEI approach using a CNN on one of the small available datasets, concluding that SEIs yield better results.

Lastly, the Gait Abnormality Video Dataset (GAVD), a relatively large dataset containing 1,874 normal, abnormal, and pathological gait sequences, not all related to PD, is discussed in [29]. The GAVD includes clinically annotated RGB data sourced from publicly available online content, comprising more than 400 subjects captured in various settings, such as urban uncontrolled outdoor environments and hospital clinics. This dataset has been employed with Temporal Segment Networks and SlowFast networks, achieving over 90% accuracy in abnormality detection.

In summary, our literature review highlights the scarcity of raw video datasets for the automated diagnosis of Parkinson's Disease. The few available datasets have been developed in controlled settings and do not necessarily specialize just in Parkinson's, and often encompass multiple neurological diseases. Moreover, the publicly available datasets primarily consist of synthetic artifacts (silhouettes, skeletons or key points) rather than actual video recordings. Most datasets used in the studied papers are collected locally. Thus, it is practically difficult and even impossible to compare the state-of-the-art findings as they are not related to our dataset or Parkinson's Disease.

While deep learning approaches are gaining traction for the diagnosis of Parkinson's Disease, due to the above hurdles, machine learning methods based on feature engineering techniques remain the most prevalent. Indeed, few papers in the current state of the art have proposed approaches based on deep learning models. However, gait analysis combined with deep learning techniques can be interesting in understanding and interpreting gait abnormalities, allowing an efficient prediction and severity classification of Parkinson's Disease and also the distinguishing between neurological diseases. It is very important to notice that the choice of the best deep neural network architecture may depend on the size of the dataset and the aims of the research. The dataset size and the video recording context are important factors in selecting the appropriate deep neural network architecture.

3. Dataset Description

The dataset utilized for this study is the same as the one used in [7]. It comprises 292 videos divided into four classes: **normal, mild, moderate, and severe** . In fact, these classes correspond to stages 0 to 3 of the UPDRS [15] standard five stages of PD. It is generally viewed that in Stage 4 the patient is unable to move, and this is obviously why it is not represented in the dataset.

The videos in our dataset were obtained from open-access domains and included recordings of 167 subjects, comprising 93 healthy individuals and 74 with Parkinson's Disease. Notably, the videos of patients with PD were primarily captured during therapeutic sessions. To ensure privacy, an anonymized version of the dataset was released to us. This anonymization was performed in two ways.

The first subset of videos featured only the subjects, who were visible in all frames, as illustrated in the left image of Figure 1, where the corresponding subject's face is blurred. The second subset involved severe cases of PD, where medical assistance was required. In this subset, any medical personnel present, in addition to the subject, were fully masked, as shown in the right image of Figure 1.

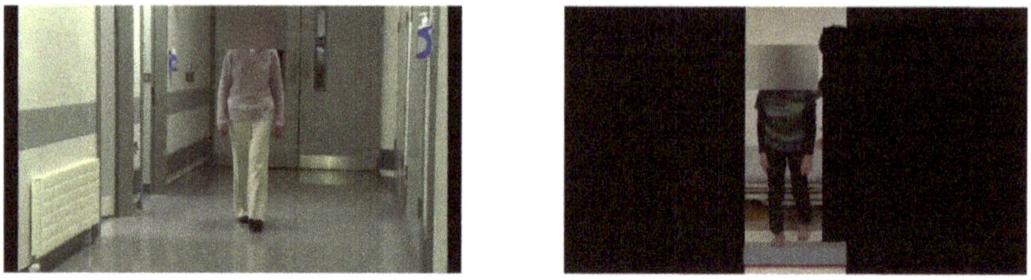

(a) Subject with blurred face.

(b) Subject with blurred face and totally masked assistant.

Figure 1. Different blurring situations.

The dataset encompasses recordings captured both indoors and outdoors, under various environmental and lighting conditions. Additionally, different camera angles relative to the subject's position were utilized. This diversity presents both a challenge and a robust foundation for training the image classification algorithms we employed, as it offers sufficient variability in the visual inputs. Considering these factors, it is clear that the dataset effectively supports the development and evaluation of algorithms and applications aimed at the rapid, non-invasive diagnosis and severity assessment of Parkinson's Disease.

One can easily envision a scenario where a camera is placed in the office of a medical practitioner, such as a neurologist, equipped with a tool that provides insights into the patient's condition as they walk in, even before the session begins.

Examining the distribution of the four classes in the dataset, as shown in Figure 2, reveals a notable imbalance. The "normal" class, representing healthy subjects, contains significantly more videos than the other classes. Moreover, within the categories of subjects labeled as having varying degrees of PD, the "mild" class has only half as many videos as each of the other two classes. At this stage, the data remain in their raw form; as we will discuss later, several techniques will be employed to introduce balance among the classes.

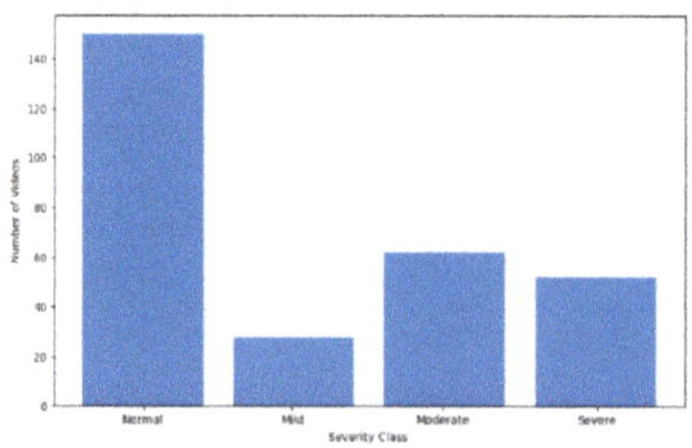

Figure 2. Distribution of videos by severity of Parkinson's Disease (by class label).

4. Methodology and Workflow

Our methodology involves generating Skeleton Energy Images (SEIs) from skeleton images (SIs). These SIs are obtained through Human Pose Estimation (HPE) techniques, applied to our dataset following appropriate pre-processing. The goal is to create a training and testing dataset composed of SEIs [26], which is subsequently used to parameterize the various models proposed in this paper for the identification and diagnosis of Parkinson's disease.

As a matter of fact, we generate two SEI datasets. The first is designated as the normal SEI dataset, and the second is designated as the key-frames SEI dataset. For simplicity purposes, we chose, in what follows, to describe in detail the process for generating the first, normal-SEI dataset. The last subsection details the modifications in processing, which are minor, to also obtain the second key-frames SEI dataset.

This section details the two workflows needed for producing the normal SEI dataset from the original dataset, consisting of the repository of videos described in the previous section. The first subsection describes the pre-possessing workflow, yielding subsets of sequential frames, form each video. The next subsection deals with using HPE, yielding the skeleton frames. The last subsection describes how several skeleton images are combined into an SEI frame, thereby producing the target SEI dataset.

4.1. Dataset Pre-Processing

The main objective of the pre-processing is to enable the computation of the Skeleton Energy Images. Moreover, the videos in the original dataset were recorded in uncontrolled environments but do not carry any noise, so a denoising process is not needed for this dataset.

To generate a Skeleton Energy Image (SEI), the required skeleton images (SIs) must be derived from successive frames within a single video. Additionally, these frames should either capture the subject facing a single direction or transitioning between directions. This requirement helps in minimizing any noise during the training phase, which could negatively impact the generalization capabilities of our target models. In order to efficiently produce such SIs, the videos in our original dataset described in the previous section need to be processed through two workflows.

The steps of the first workflow, illustrated in the diagram shown in Figure 3 and designated as the pre-processing workflow, are described in this section. Note that the steps framed in blue are performed manually; those framed in green are automated (through a script); and the steps framed in red involve discarding certain information, such as sets of frames.

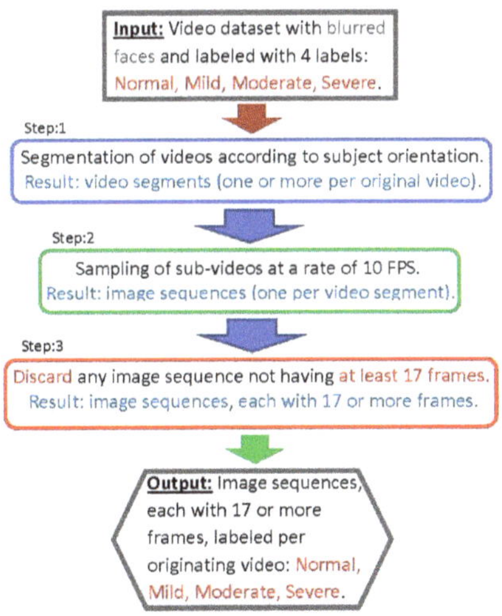

Figure 3. Pre-processing workflow producing sequences of images from videos.

The first step involves manually segmenting each video into one or more clips. Each clip captures the subject in one of four possible orientations: (1) facing forwards, (2) facing backwards, (3) facing sideways, or (4) transitioning between orientations. In the second step, each video segment is sampled at a rate of 10 frames per second, resulting in correspondingly labeled frame/image subsets from the segments.

The third step involves discarding segments that yield fewer than 17 frames after sampling. Frames from each segment are kept separately and labeled according to their originating sequence and/or video. It is important to note that most videos result in frame sequences that capture the subject in the various orientations listed above. Proper segmentation of the videos is crucial for generating a relevant and de-noised SEI dataset [30]. This dataset is used in the subsequent sections of this study.

Figure 4 displays representative frames depicting subjects in the four described orientations.

(**a**) Subject facing backwards.

(**b**) Subject facing forwards.

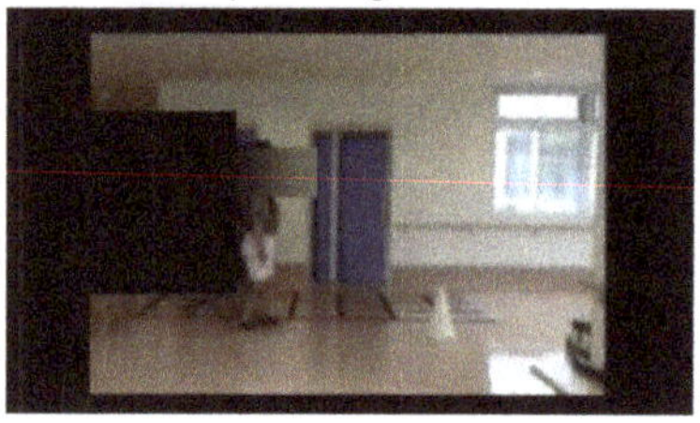

(**c**) Subject oriented sideways.

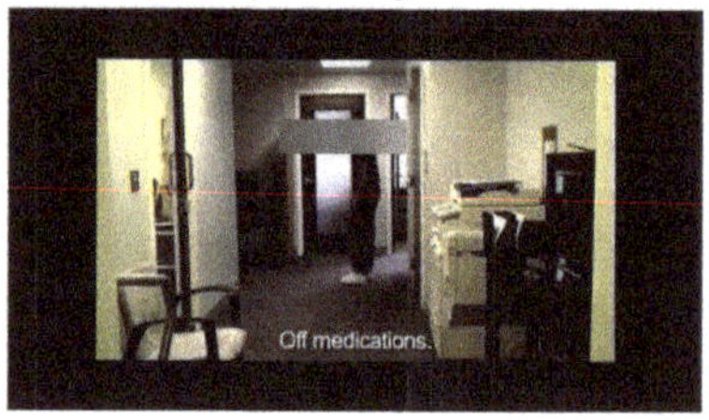

(**d**) Subject turning between orientations.

Figure 4. Representative frames from four possible video segmentation situations.

4.2. Human Pose Estimation

There is a well-established consensus in the literature on ML-based diagnosis that understanding and analyzing the movements and trajectories of key human joints (such as the arms, legs, and shoulders) is crucial. Our proposed method for the automatic diagnosis of Parkinson's Disease relies on extracting these key points, which represent the joints, from frames obtained after Step 3 of our pre-processing workflow. This extraction is performed automatically using Human Pose Estimation.

This subsection details our second workflow, illustrated in Figure 5, and designated as the SI workflow. Its objective is to apply humain pose estimation on the sequences of images resulting from the pre-processing workflow of the previous subsection. This results in sequences of SIs corresponding the the image sequences. Note that the steps of the workflow have frame colors indicating their types of function as indicated in the previous subsection.

The main tool used in this SI workflow is the pre-trained and highly effective YOLOv8m HPE model, which comprises 26.4 million parameters [31]. In step 1 of our SI workflow, each input image is processed by the YOLOv8m model to produce a frame with the corresponding key points. If this operation is successful, the following 12 key points are extracted [31]: right shoulder, left shoulder, right elbow, left elbow, right wrist, left wrist, right hip, left hip, right knee, left knee, right foot, and left foot.

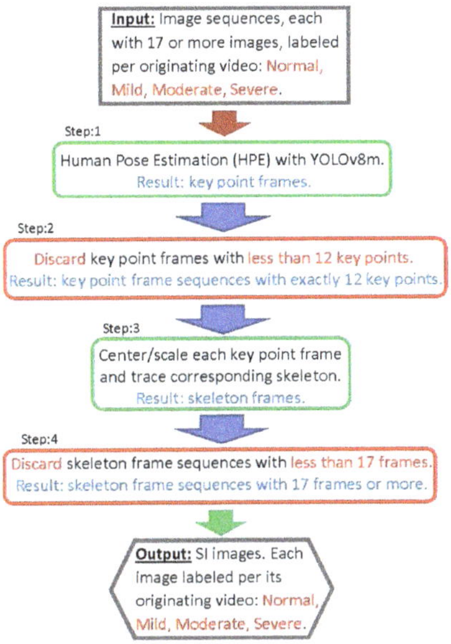

Figure 5. SI workflow producing sequences of skeleton images.

Figure 6 illustrates the joints corresponding to a human skeleton, returned as key points by the HPE model. To understand how the HPE model works, we designate these key points with the following notation,

$$\{(x_1, y_1, p_1), (x_2, y_2, p_2), \ldots, (x_{12}, y_{12}, p_{12})\},$$

where x_i and y_i denote the 2D coordinates of the $i$the key point, and p_i represents the confidence score of the corresponding key point's 2D coordinates. A key point coordinate is considered valid and is retained if it corresponding probability score is $p_i > 0.5$. In fact, in step 2 of our SI workflow, each image that results in fewer than 12 key points is discarded.

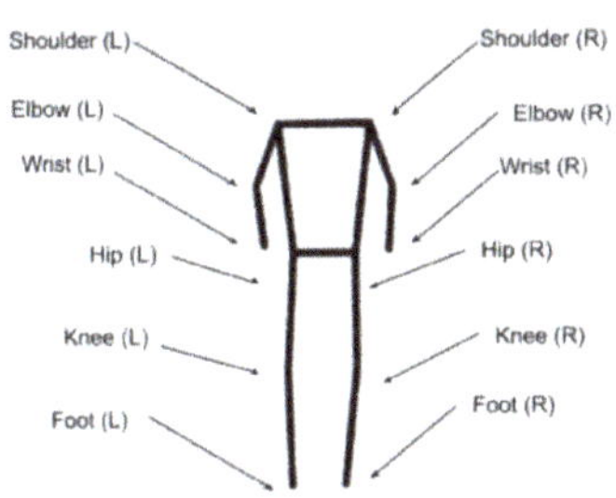

Figure 6. List of joints/key points returned by HPE.

We should also recall that each key point frame produced in step 2 corresponds to an image that is part of a sequence of images,and this image information regarding its order within the sequence and its class label, corresponding to its originating video. Obviously, the resulting key point frames keep this information and forward it to corresponding results in the SI workflow.

Step 3 in our SI workflow, consists of two sub-steps. The first sub-step involves centering and scaling the 12 key points in each frame to a height of 360 pixels. This process is crucial for maintaining spatial consistency and preventing distortions in the resulting dataset that could arise from variations in the relative positions of subjects within the original videos.

The second sub-step involves tracing a skeleton from the 12 key points, thereby converting each centered and scaled key point frame into an SI, i.e., skeleton image. This skeleton image/frame represents a headless human skeleton of the corresponding subject, as heads and faces were blurred in the original video dataset to protect the anonymity of the subjects.

Figure 7 illustrates a sample image originating from one of the videos alongside its corresponding SI.

Figure 7. Image from video with corresponding SI.

In step 4 of the SI workflow, we discard any sequence of SIs not having at least 17 SIs. Obviously, some sequences resulting from a corresponding input sequence of 17 or more images are discarded, but some of these images resulted in key point frames that were discarded in step 2 as they had less than the 12 required key points.

4.3. Energy Images

The concept of the Skeleton Energy Image (SEI), as utilized in this study, is based on the energy image concept outlined by [27]. An energy image is essentially the average of multiple consecutive frames, designed to extract key features from the movements and joint trajectories of the subject. Additionally, it accentuates and isolates these movements and trajectories, which facilitates the effective detection of early forms of PD [32].

While [27] used a variable number of frames to produce a single energy image, our study generates a Skeleton Energy Image (SEI) from each of the consecutive 17 SIs , which empirically corresponds to approximately 1.5 gait cycles. This approach aligns with analysis criteria developed in prior works that base effective PD detection on gait [20]. The technique employed is described by the following equation, where $E_{(i,j)}$ represents the pixel value at position (i,j) in the SEI. It is computed by summing the corresponding pixels of each of the 17 skeleton images and dividing this sum by 17:

$$E_{(i,j)} = \frac{1}{17}\sum_{n=1}^{17} S^{n}_{(i,j)}$$

Our approach further differs from [27] by employing a rolling kernel to generate multiple SEIs from a sequence of skeleton images. This process is illustrated in Figure 8. When the sequence contains more than 17 skeleton images, the first 17 images are used as the kernel to produce the initial SEI. The kernel is then moved to start with the next image, and so on, until the last kernel captures the final 17 skeleton images in the sequence. In the case the sequence contains exactly 17 frames, obviously, only one SEI is produced from it. Moreover, recall that sequences with fewer than 17 frames were entirely discarded during the pre-processing stage.

In addition to augmenting the SEI dataset, this method increases the amount of information made available to our image classification models and enhances their generalization capability. This is because the same joint trajectory is captured by multiple SEIs, each starting from different coordinates.

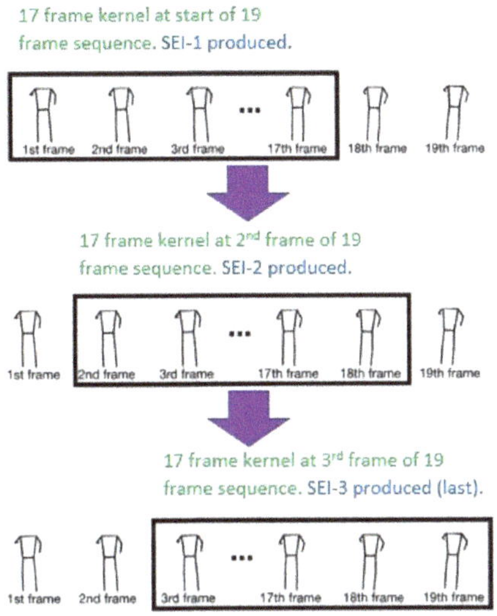

Figure 8. Rolling kernel used for creating SEI.

4.4. Processing Differences for the Generation of the Key-SEI Dataset

It was noted earlier in Section 3 that, in fact, for the purposes of our experiments, described later, two SEI datasets were synthesized from our original dataset. The difference in the processing is not significant, and, for purposes of simplification, we deal with it in this subsection.

In fact, step 3 of the first pre-processing workflow (the sampling step) is executed in two different ways, leading to obtaining each of the two SEI datasets. This is described below.

A frame is classified as a key frame if it has a difference in value that is 1.5 times higher than the average of the previous 10 frames [33]. This method for selecting candidate frames to be examined was utilized within the context of emotion detection in [34]. The main objective was to avoid training the model on similar images.

The first frame in a video segment is always designated as a key frame; then, all other key frames following it are identified. For any key frame, the 16 frames following it are taken as part of its sequence. If fewer than 16 frames follow a key frame in the corresponding video segment, then no frame sequence starting with this key frame is formed. Therefore, a key-frame sequence has exactly 17 images, whereas the normal frame sequence, corresponding to the video segment, consists of all the frames. Moreover, a video segment may generate zero or more frame sequences.

For the rest of the steps, while key-frame sequences are treated separately from the normal frame sequences, the steps of the workflows are applied to both in the same way. Note, for example, that an SI key-frame sequence may result from one key-frame sequence, and in that case, one key-frame SEI is generated from it. However, if during the HPE step (step 2) of the SI-workflow, a key-point sequence is rejected because less that 12 key points were generated from the corresponding frame, then the entire key-frame sequence generates no SEI.

5. Selection of Deep Learning Models

In this section, we deal with the selection of a candidate set of deep learning models for the purpose of using them in the task of diagnosis of PD. This is in fact a task which involves the classification of the various videos in our original dataset, based on the synthesized

SEI dataset produced from them. We have discussed in the previous two sections how the original dataset is processed and how these synthetic SEI images are produced from each corresponding video.

In our analysis of the synthesized SEI dataset, we concluded that effective classification requires a focus on the global aspects of SEI images rather than on specific detailed features. This is illustrated by the SEI images in Figure 9, which represent three classes from the dataset: normal, moderate, and severe. Notably, posture is a key differentiator; the normal class displays a more upright stance compared to the others. Additionally, the severe class is characterized by slow-moving feet. Overall, the normal SEI exhibits correct posture with sharp, fast limb movements, while the moderate and severe classes display slower movements and more stooped postures.

(**a**) Normal class SEI. (**b**) Moderate class SEI. (**c**) Severe class SEI.

Figure 9. SEIs from front-facing subjects.

These observations prompted us to explore multiple models for the classification task rather than relying on a single architecture. In this study, we utilize and compare results from three prominent deep learning architectures: Convolutional Neural Networks (CNNs), Residual Networks (ResNets), and Vision Transformers (ViTs). We anticipate that the last two architectures will be particularly well suited to our classification objectives.

This section is organized as follows. The first three subsections reviewing the above-mentioned three deep learning architectures and give indications about the suitability of each architecture for our classification task. In the last section we summarize our evaluation and shed light on particular parameters that we shall be using for each model/architecture.

5.1. Convolutional Neural Network (CNN) Architectures

CNNs are a class of deep learning algorithm that operates on the principle of hierarchical feature learning [35]. Initially, the convolutional layers capture low-level features like edges and textures from the raw pixel data. As the data progress through the deeper layers, increasingly complex features and patterns are recognized, culminating in high-level representations of the input data.

CNNs are popularly used, in their 2D format, for tasks involving the classification of images. And since our SEIs can be considered as images showing a skeleton in a particular trajectory, it is only natural that we consider using CNNs for our classification task.

When designing the CNNs for this study, however, we pondered the fact that the SEIs, in fact, have relatively simple features, and, therefore, would go better with shallow CNNs. Moreover, along with the important convolutional, pooling, and fully connected layers that we included, we found it important to incorporate batch normalization and dropout layers following each of the above layers of the network. This allows the model to proceed directly and speedily to the relevant details due to its relatively shallow depth.

Batch normalization is used both in feed-forward and convolutionnal neural networks with two main advantages [36]. First, it helps to mitigate the problem of the internal co-variate shift of the network, where the distribution of the inputs to each layer can change during training, and thus has a stabilizing effect on the learning process. Second, it is believed to have a regularization effect, since it is computed over the batches and not over the entire dataset

Dropout is a also a regularization technique designed to prevent overfitting [37]. During training, dropout randomly sets a fraction of the activations to zero in a given layer, effectively preventing the network from relying too heavily on any particular neurons. This enables the network to develop redundant representations, thereby improving its ability to generalize to new, unseen data.

5.2. Residual Networks (ResNets)

ResNets, introduced in [38], address the degradation problem prevalent in deep neural networks, where increasing depth can lead to a higher training error. The key innovation of ResNets is the incorporation of residual learning through shortcut connections. These connections enable the network to learn residual functions relative to the inputs of each layer, rather than attempting to learn unreferenced functions directly.

Residual functions are defined as the difference between the desired output and the input to the layer, allowing the model to focus on learning the changes needed to reach the target output. Mathematically, if $F(x)$represents the desired mapping to be learned, the residual learning framework transforms this into learning the residual, i.e., $F(x) - x$. This allows the network to adjust its outputs based on how far the learned function deviates from the identity mapping and simplifies the optimization process.

This departure from CNNs also facilitates the flow of gradients during backpropagation, alleviating the vanishing gradient problem and allowing for the effective training of networks with hundreds or even thousands of layers. The residual connections provide a direct path for gradients to flow back through the network, enabling deeper architectures to maintain performance without suffering from the typical pitfalls associated with increased depth.

The architecture of a ResNet is characterized by its use of residual blocks, which combine convolutional layers with batch normalization and ReLU activation functions. Within these blocks, the shortcut connections bypass one or more layers, allowing the original input to be added to the output of internal convolutional layers. This additive operation empowers the network to learn identity mappings, ensuring that performance does not degrade as depth increases.

By explicitly allowing layers to copy their inputs directly to their outputs, ResNets streamline the optimization of very deep networks. This structure encourages the layers to focus on learning adjustments to the identity function rather than recreating outputs from scratch. Consequently, even if some layers learn suboptimal transformations, the overall network can still preserve performance, as the identity mapping is easily accessible through the shortcut connections.

In summary, the concept of residual functions is central to the success of ResNets, enabling deeper networks to train effectively by focusing on learning modifications to established patterns rather than starting from zero. This not only enhances the efficiency of training but also improves the generalization capabilities of deep learning models in various applications.

5.3. Vision Transformers (ViTs)

Vision Transformer models (ViTs), introduced in [39], represent a significant paradigm shift in image recognition tasks by leveraging the transformer architecture originally designed for Natural Language Processing (NLP) [40]. In this innovative approach, ViTs partition an image into fixed-size patches, treating these patches similarly to sequences of tokens in NLP, akin to words in a sentence. This methodology utilizes the self-attention mechanism inherent in transformers, which effectively models complex, long-range dependencies and relationships between different parts of the image. This alleviates the loss of detail that can occur early in Convolutional Neural Networks (CNNs), where the focus is predominantly on local features and their synthesis.

The ViT architecture functions by linearly embedding the fixed-size patches into vector representations, augmented by positional embeddings that retain spatial information. This allows the model to understand the arrangement of patches within the original image. These embeddings are then processed through multiple layers of Transformer encoders, each comprising multi-head self-attention mechanisms and feed-forward neural networks. Unlike CNN layers, which prioritize local interactions, the self-attention mechanism in ViTs enables the model to dynamically attend to various parts of the image, capturing global context effectively.

Recent studies have shown that ViTs outperform traditional CNNs [39,41] across various benchmarks, particularly when ample data and computational resources are available. The capacity of ViTs to leverage large datasets enhances their learning, making them particularly powerful for complex image recognition tasks. As a result, they are increasingly adopted in state-of-the-art image classification applications, demonstrating their potential to redefine the landscape of computer vision.

5.4. Evaluation of the Architectures

To understand the performance variations related to the number of learnable parameters for each type of model/architecture and their adaptability our task, we conducted preliminary experiments. These simple experiments allowed us to determine the appropriate depths to be used with each model type.

For the CNN architecture, we selected the 5-layer and 6-layer configurations; for the ResNets, we selected 18-layer and 34-layer variants; and, for the ViTs, we selected three sizes, the base size, the small size, and the tiny size.

Notably, the selected model depths/sizes are shallower than those typically used in conventional studies. For example, while ResNet-50 is widely used for image classification tasks, we only employed ResNet-18 and ResNet-34. Similarly, although ViT-Base is commonly used as the baseline model, we used it as the largest ViT model in our study. The SEI images that we attempt to classify, as shown in Figure 9, do not have particularly complex or numerous features. Thus, the use of shallower models enhances computational efficiency and makes the models more suitable for deployment on smartphones and other devices without dedicated GPUs. This consideration makes the models trained in this study suited for this intended use case, particularly in developing nations, thereby ensuring broader accessibility and usability. Moreover, our expectations were realized in this preliminary study as the ResNets and ViT models we used outperformed the CNNs. It also became clear that that increasing the depth of the CNN model does not improve its results. This shall be addressed in more details in the next section.

6. Experimental Design, Results, and Discussion

In this section, we discuss the experiments that we utilize to test the models extensively. Specifically, we describe our rationale in designing the experiments and the specific parameters that bound our experiments, and, finally, this section ends by discussing the results of each of the experiments in great detail. We have an in-depth conversation about what the results entail and offer insights about which model might perform depending on the application.

6.1. Design of the Experiments

Based on our preliminary study, discussed in the previous section, we have selected the architectures and corresponding models used in the main comprehensive study that we conducted and discuss them in this section. The models involved are named as shown in Table 1.

In fact, we systematically conducted four experiments and evaluated the accuracy of each trained version of the selected models for the task of PD diagnosis and classification.

These experiments used the two SEI datasets composed of Skeleton Energy Images and synthesized through the process described in the previous sections from our original dataset composed of videos. We recall that both SEI datasets, as is the case for the original dataset (see Figure 9), are imbalanced with respect to the four classes. So, a third, class-balanced dataset is produced from the normal SEI dataset using the random oversampling technique [42]. With oversampling, randomly selected images in the underrepresented classes were artificially duplicated within the respective class subset to achieve class balancing. Therefore, a total of three datasets, given in Table 2, are used for our comprehensive study.

Table 1. Selected architectures/models.

Architecture	Model Name	Model Parameter
CNN	CNN-5	5 Conv. Layers
	CNN-6	6 Conv. Layers
ResNet	ResNet-18	Depth 18
	ResNet-34	Depth 34
ViT	Vit-Tiny	Tiny Patch
	Vit-Small	Small Patch
	Vit-Base	Base Patch

Table 2. Number of SEIs in datasets/subsets employed in experiments.

	Total	Training	Validation	Testing
Normal SEI Dataset	17,038	13,628	1705	1705
Key-Frame SEI Dataset	513	410	51	52
Balanced Normal SEI Dataset	23,580	18,864	2358	2358

We should note that while the original dataset is composed of 292 videos, they were used, with the workflow presented in Section 4.3, to synthesize 17,038 SEI images. Each SEI is computed using a rolling kernel with size 17 frames over one video subsequence (see Figure 8). The sizes of the normal SEI and balanced normal SEI datasets shown in Table 2 are relatively large to enable good learning of the models and to provide them with high generalization ability. However, the sizes of the key-frame SEI dataset is relatively small, which may limit the models' generalization ability. Moreover, as we have mentioned in the literature survey section, raw video datasets in this domain are scarce and not available publicly [27,29]. In fact, the dataset we use in this paper [7] is among the largest.

All SEI datasets are split into respective training, validation, and testing sub-datasets using an 8:1:1 ratio via random sampling. The contents of the respective sub-datasets remained consistent throughout the four experiments. Moreover, we recall that the label associated with each SEI image in the training, test and validation SEI sub-datasets falls into one of four classes: normal, mild, moderate, and severe. In fact, this is the same label associated with the corresponding video from the original dataset from which the SEI image is synthesized/generated.

The four experiments were divided into two distinct categories. The first category comprised two experiments aimed at fine-tuning the training techniques to optimize model performance. The second category involved testing the trained models under diverse conditions to assess their robustness and generalization capabilities. The last category of experiments, is crucial for understanding how well the models performed under adverse scenarios; this underlined their ability in maintaining their corresponding accuracies across all situations.

The following describes the four experiments which we conducted for the purposes of our comprehensive study.

- **Experiment 1:** The training sub-set corresponding to the normal SEI dataset is used for the training all the models given in Table 1. The class distribution of the SEI images were proportional to those given in Figure 9. The model was evaluated on the corresponding test subset (for the normal SEI dataset).
- **Experiment 2:** The training on all the models given in Figure 9 is performed using the training subset corresponding to the key-frames SEI dataset. In this experiment, we aim to analyze and compare the results obtained from the generated models with the results of the models generated in the previous experiment. We note that while the key-frames SEI dataset has fewer frames than the normal dataset, its images should carry greater facial cues.
- **Experiment 3:** This experiment concerns the normal SEI dataset. Since it is unbalanced, with respect to the four classes, this experiment aims to determine whether or not the generated models, in Experiment 1, memorized the uneven class distribution of the corresponding training dataset. For this purpose, the corresponding test subset is supplied class by class, and not completely at once, to these models. The results are compared to those produced in Experiment 1.
- **Experiment 4:** This experiment concerns the balanced normal SEI dataset. It aims to understand how the different models would perform in a scenario where the subsets used are balanced with respect to the four classes. The best-performing models, from each architecture used in Experiment 1, are trained to generate corresponding models. These models are tested on the test subset of the balanced normal SEI test subset. The results are compared with those obtained in Experiment 1.

6.2. Experimental Parameters

In this study, the training protocol we used for all the models does not impose an upper limit on the number of epochs. Instead, a minimum threshold of 25 epochs is established to ensure adequate initial training. To optimize the training process and prevent overfitting, we employ an early stopping mechanism [43]. This mechanism monitors the validation performance, at each epoch, and suspends any further training if there is no improvement in the validation loss for the last 20 consecutive epochs. The model parameters corresponding to the epoch with the best validation performance (within the last 20 epochs) are retained. The early stopping approach ensures that the model is sufficiently trained, without using unnecessary computational burden. To summarize, the minimum epoch limit guarantees that the model has ample opportunity to learn from the data, while the early stopping mechanism helps in identifying, early on, the point of convergence, thereby avoiding overfitting.

To ensure consistency, all the deep learning hyper-parameters for training are kept consistent throughout all the experiments executed with the models used. These hyper-parameters are listed in Table 3, given below.

We note that all our experimental work, consisting of training and testing, is performed using NVIDIA A100 cloud GPUs.

Table 3. Hyper-parameters for deep learning models used for experiments.

Hyper-Parameter	Value
Weight Initialization	Random
Optimizer	AdamW
Loss Function	Cross Entropy Loss
Learning Rate	5×10^{-5}
Batch Size	64

6.3. Results and Discussion

In this subsection, we report on the results obtained from performing our comprehensive computational study. We also discuss the performance of the various architectures and models we used. The obtained results are summarized in Table 4, and from a quick glance at it, we can easily perceive that the methodology developed in this paper outperforms existing system surveyed the the literature.

Table 4. Accuracy scores for corresponding generated model instances.

Models	Experiment 1	Experiment 2	Experiment 3		Experiment 4
			Average	Weighted Average	
ViT-Tiny	99.71%	32.69%	99.74%	99.71%	
ViT-Small	99.76%	32.69%	99.81%	99.76%	99.96%
ViT-Base	99.70%	59.63%	99.43%	99.70%	
ResNet-18	**99.88%**	**67.30%**	**99.89%**	**99.88%**	**100.00%**
ResNet-34	**99.88%**	63.46%	**99.89%**	**99.88%**	
CNN 5-layer	91.61%	30.77%	88.41%	90.73%	96.40%
CNN 6-layer	87.22%	7.69%	67.86%	82.99%	

In fact, we have conducted four types of computational experiments, and each experiment involves all or most of the models but with a different parts of the datasets discussed in the previous sections. The following three subsections comment on the results presented in Table 4 for each corresponding experiment. The last subsection gives global cross-comparison comments, insights, and some lessons learned.

It is worth noting at this point that globally our results showed that there are three models that performed best in our experiments, namely, the five-layer CNN, ResNet-18, and ViT-Small. So, in what follows, we present detailed results with these models for the relevant experiments.

6.3.1. Experiment 1 Results

In Experiment 1, all models performed significantly better than most existing deep learning and machine learning models/systems used for tasks similar to the task of our paper. In particular, the ViTs and ResNets achieved diagnostic and classification accuracy scores that were close to perfect, attaining state-of-the-art scores. This observation holds true for most of the experiments conducted in our comprehensive study. On the other hand, the CNN models exhibited slightly lower accuracy performance. One hypothesis to explain these findings is the inherent architectural differences between the models. ViTs and ResNets incorporate advanced features such as self-attention mechanisms and skip connections, which might allow for more effective learning of global and less obvious features and patterns present in our SEI datasets. Furthermore, even though shallower than the models utilized conventionally, the ViT and ResNet architectures were significantly shallower than the CNN architectures and still had more learnable parameters.

It is interesting to note that an increase in a model's size does not correlate with an increase in performance, as might intuitively be expected. For instance, in the case of the ResNets, the performance remains similar between ResNet-18 and ResNet-34, even though ResNet-34 has almost twice as many learnable parameters. Similarly, while there is a slight increase in performance between ViT-Tiny and ViT-Small, the ViT-Base actually has a lower, yet still appreciable, accuracy score.

There are two possible reasons for the above observation: First, the scores of the shallower models are already high due to effective data processing and the distilled information content being communicated (SEI images carry exactly the gait and, economically, not more). Consequently, deeper models cannot significantly outperform the shallower models as their capacity to capture more features is not needed. Second, the deeper models might not be learning new information but rather memorizing labels. However, this is unlikely since the observed drops in performance were negligible.

Figures 10–12, given above, respectively show for the models five-layer CNN, Resnet-18 and ViT-Small, the training and validation curves, the confusion matrices, and the classification reports. For these three models, the learning curves show interesting training and validation accuracies, at the end of the training period. Furthermore, the models are expected to have high generalization ability, as no overfitting is observed. The confusion matrices and classification reports confirm the good performance of the models when applied to the test dataset. Nevertheless, the five-layer CNN has low accuracy in detecting the mild class. In terms of the confusion matrices, Figure 11, the best performing models are the ResNet-18 and ViT-Small as they show near-perfect per-class percentages. The CNN five-layer model suffers a bit in that area, especially for the least represented mild class. These same observations are confirmed by the precision and recall figures given in the classification reports Figure 12.

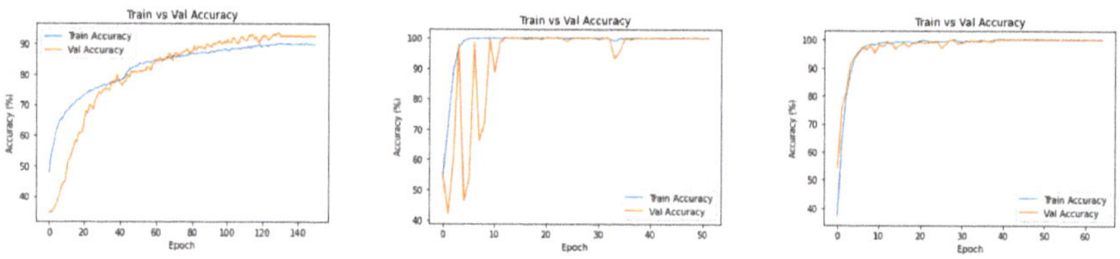

(**a**) Five-layer CNN. (**b**) ResNet-18. (**c**) ViT-Small.

Figure 10. Experiment 1 training/validation accuracies for selected models.

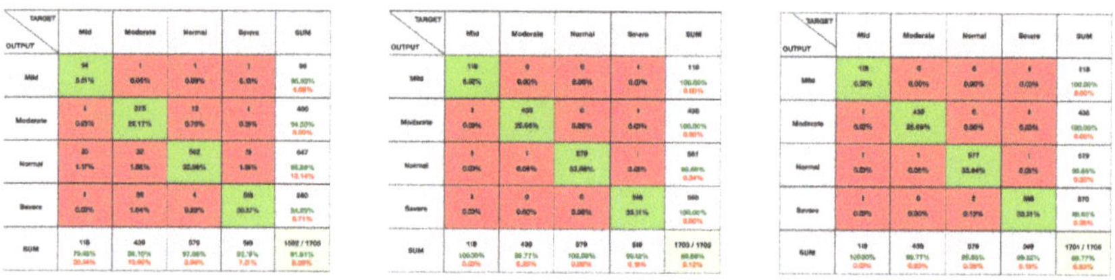

(**a**) Five-layer CNN. (**b**) ResNet-18. (**c**) ViT-Small.

Figure 11. Experiment 1 confusion matrices for selected models.

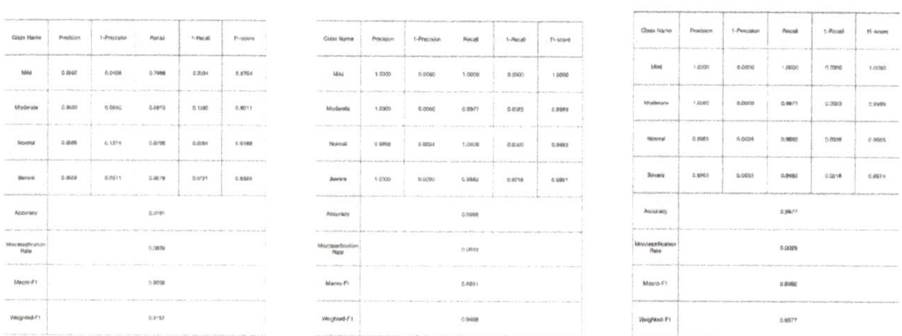

(**a**) Five-layer CNN. (**b**) ResNet-18. (**c**) ViT-Small.

Figure 12. Experiment 1 classification reports for selected models.

6.3.2. Experiment 2 Results

There is a considerable drop in the accuracy scores across the board in this second experiment. This decline can be attributed to the paucity of data when utilizing the key-frames SEI dataset. In fact, there were fewer frames available for extraction and training than in the case of the normal SEI dataset. This results in models unable to generalize effectively across the dataset. Unlike emotion detection, where a single key frame might provide sufficient information, this study necessitates multiple sequences of frames, even containing repeated ones, underscoring and highlighting the type of gait for the corresponding subject.

Once again, ResNets and ViTs outperformed the CNN architectures. However, several interesting patterns emerged that warrant further exploration. Firstly, while the performances of ViT-Tiny and ViT-Small are similar, ViT-Base performs significantly better than the two smaller models. One explanation for this observation is that ViT-Base, with its greater capacity and more complex architecture, might be better equipped to capture and utilize the temporal dependencies present in the time series data, leading to superior performance.

Secondly, we observe that the five-layer CNN performs much better than the six-layer CNN, which performs worse than random guessing. A possible explanation for this could be that the six-layer CNN is overfitting to the limited data available, leading to poor generalization. Alternatively, the additional layer might be disrupting the model's ability to learn effective representations, suggesting that the optimal architecture for this task might be shallower.

Finally, both ResNet models achieved the highest levels of performance, with accuracy scores that are within a few percentage points of each other. However, the shallower ResNet-18 performs slightly better.

6.3.3. Experiment 3 Results

In the third experiment, the patterns observed in the first experiment persisted, as expected. Notably, the ResNets and ViTs maintained the same weighted accuracies across both experiments. This consistency is a positive indication that these models did not memorize the labels of the images. In contrast, both CNN models experienced a slight drop in performance in the third experiment compared to the first. This decline could be attributed to the models, with their reduced number of parameters, attempting to memorize the class labels. However, overfitting did not appear to be a significant issue as the decrease in accuracy score was minor.

6.3.4. Experiment 4 Results

In the fourth experiment, only the best-performing models from the previous three experiments were tested. As expected, all the learned models exhibited near-perfect accuracy, with ResNet-18 achieving a perfect accuracy. However, it is anticipated that ResNet-18 may exhibit some inaccuracies, as the size of the test data increases.

Figure 13 shows the training and validation curves for the top three performing models. At the end of the training period, all models present significant training and validation accuracies. Moreover, Figures 14 and 15 present the confusion matrices and the classification report for the three models. All the evaluation criteria values, precision, recall, and F1-score, are significant for all classes for the Vit-Small and ResNet-18 models. For the five-layer CNN, the evaluation criteria values are lower than for the other models. The accuracy for the least represented mild class improved a bit with the oversampling we used in this experiment.

In particular, the performance of the three models is interesting compared the state of the art. Indeed, in [7], a machine learning technique using a gait-based feature engineering approach was applied to the dataset we used in our paper. A maximum accuracy of 97% was obtained with a Random Forest classifier with 250 trees. Moreover, in [26], an overall classification accuracy of 71.25% was achieved using a complex approach that is entirely based on deep learning applied to a locally collected controlled dataset comprising 54 Parkinson's Disease patients and 26 healthy subjects.

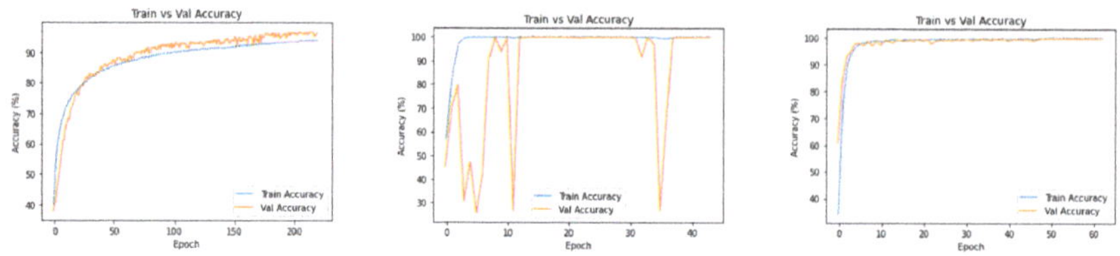

(a) Five-layer CNN. (b) ResNet-18. (c) ViT-Small.

Figure 13. Experiment 4 training/validation accuracies for selected models.

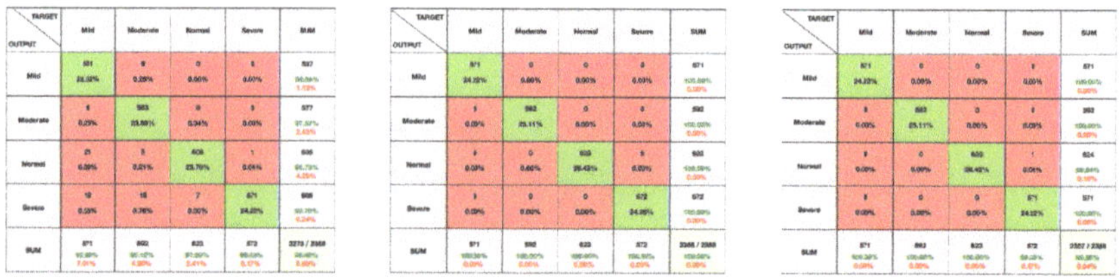

(a) Five-layer CNN. (b) ResNet-18. (c) ViT-Small.

Figure 14. Experiment 4 confusion matrices for selected models.

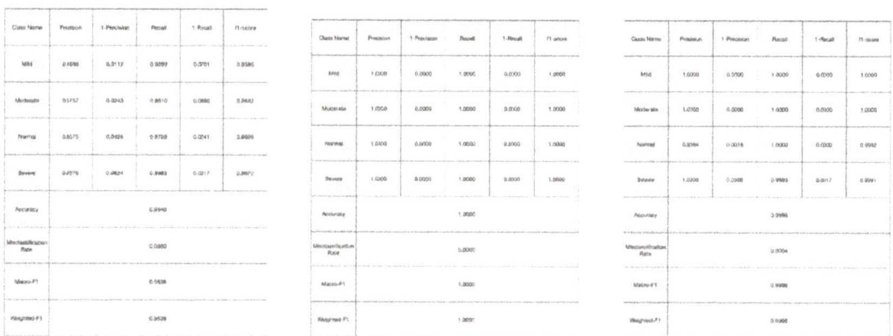

(a) Five-layer CNN. (b) ResNet-18. (c) ViT-Small.

Figure 15. Experiment 4 classification reports for selected models.

6.3.5. Overview of Results of Experiments

Across all experiments, ResNet-18 consistently demonstrates superior accuracy performance. ResNet-18, although less complex and less computationally intensive, outperforms Vision Transformers (ViTs), which demand even greater computational resources. Since the system is envisioned for use in a medical setting, where accuracy is of paramount importance, the ResNet architecture would be best suited for this application.

The residuals and the bypass connections, which are the backbone of the ResNet architecture and seem to capture even the faint patterns, could be essential in the accuracy scores attained by the architecture. In comparison, Vision Transformers (ViTs) require extensive computational resources due to their self-attention mechanisms, which capture long-range dependencies in the data. While ViTs excel in capturing global context, they often require larger datasets and more computational power to train effectively. Due to the structure of energy images, many patches input to the ViT were completely or almost completely white, with limited features to extract. Thus, the model would likely have performed better with a larger dataset.

As a summary and a global overview of the results of the four experiments we conducted, it is worth mentioning that, while the choices of models used are definitely important in delivering the results displayed, the approach has proven good robustness and promises to be the basis for an applicable PD classification framework. Firstly, our approach, does not rely on any specific feature engineering method or on an any precise definition of gait; it rather generates SEIs from sequences of almost any contiguous 17 frames from the raw videos. In many ways, this SEI synthetization technique functions as an effective dataset enhancement method that is somehow expanding the size of the dataset and enhancing the generalization capabilities and robustness of the models. This can be easily seen by overviewing the results of experiments 1, 2, 3, and 4 in the following manner.

Recall that Experiment 1 used the normal SEI dataset, while Experiment 2 used the more restricted key-frame SEI dataset, based on generating SEIs that start with a key frame. The key-frame SEI dataset has far many fewer SEI frames than the Experiment 1 dataset. However, both are based on the same original video dataset. The results of Experiment 2 are inferior to those of the near-perfect Experiment 1 results. We have further attempted to confirm the results of Experiment 1 by performing Experiment 3 on the same normal SEI dataset, and supplying the classes one by one to the model in the testing phase. This allowed us to confirm that the results of Experiment 1 did not hinge on any model perception of the class distributions. This definitely proved that our SEI sythetization approach is working, and that the greater number of SEIs generated with our approach, not originating just from key frames, enhanced the generalization capabilities of our models. This has been further confirmed by Experiment 4, which performed more data augmentation using a standard random oversampling technique to achieve some class balancing. In fact, Experiment 4 enhanced class accuracies, especially for the under-represented classes.

In retrospect, the SEIs generated based on skeletons present a powerful motion abstraction technique, giving the model a way of perceiving the motion of the person's body independent of the original video. Our method for generating the SEIs, by not limiting them to start from a key frame, captures even more motion patters and enhances the generalization capabilities, as proven with the test datasets results for the four experiments.

7. Conclusions and Perspectives

In this study, we successfully devised a deep learning-based approach to detect and classify the severity of Parkinson's Disease utilizing an open-source, relatively large dataset of videos recorded in diverse and uncontrolled environments and labeled according to the well-accepted UPDRS standard. We applied an automatic workflow based on state-of-the-art pose estimation techniques to the dataset of videos to generate Skeleton Energy Images to effectively capture the gaits of the subjects. Then, we devised a comprehensive and systematic set of experiments to train and test our carefully selected set of advanced deep learning models. The results demonstrated highly accurate classification of the videos of patients, achieving state-of-the-art performance.

Furthermore, the good results shown by the models presented in this paper can be used as a basis for the development of software that can made available in the offices of doctors treating PD patients. By placing a few regular cameras strategically in a given doctor's office, the software can capture a patient as he/she walks in and can be used to alert and advise the doctor as to how the patient is responding to a given treatment. We see this as practical, since our approach for synthesizing SEIs can be easily applied to any subset of videos. In fact, recent transfer learning techniques/experiences [44] in similar fields and with similar models could be leveraged to perform model fine-tunings to specific perimeters and conditions.

In fact, we believe the above ideas carry good prospects for improving the lives of many patients, especially in areas with limited financial capabilities. However, a regulated study needs to be performed to prove the generalization capabilities of our selected models and to seek further enhancement directions for our methodology's workflows. Moreover, creating a large video dataset of patients with Parkinson's Disease is important to the field and could prompt a revisit to this study. Furthermore, our proposed methodology, which is based on gait analysis, can be extended to detect many other neurological diseases such as Alzheimer's disease, Huntington's disease, etc. Thus, applying it to other neurological diseases is one of our future perspectives.

In future studies, we aim to realize the above and to generalize our framework to a broader range of video-based detection tasks that necessitate continuous data, such as sports analysis.

Author Contributions: Conceptualization, N.B.H.-A. and M.T.-H.A.; Methodology, N.B.H.-A., M.T.-H.A., A.D. and V.P.; Software, N.B.H.-A. and A.D.; Formal analysis, M.T.-H.A., N.B.H.-A. and A.D.; Investigation, M.T.-H.A., N.B.H.-A. and V.P.; Writing—original draft, M.T.-H.A., N.B.H.-A. and A.D.; Writing—review & editing, M.T.-H.A., N.B.H.-A. and V.P. All authors have read and agreed to the published version of the manuscript.

Funding: This research was funded by Deanship of Research and Graduate Studies at King Khalid University (RGP1/71/45).

Institutional Review Board Statement: Not applicable.

Informed Consent Statement: Not applicable.

Data Availability Statement: The raw data supporting the conclusions of this article will be made available by the authors on request.

Acknowledgments: The authors extend their appreciation to the Deanship of Research and Graduate Studies at King Khalid University for funding this work through small-group research under grant number RGP1/71/45.

Conflicts of Interest: The authors declare no conflict of interest.

References

1. Ajroud, A.; Khalfallah, S.; Ghribi, A. Parkinson's Disease: A review of the current literature. *J. Neurol.* **2019**, *266*.
2. Brink, L.; Franks, R. Non-Motor Symptoms in Parkinson's Disease: The Underappreciated Side. *Mov. Disord.* **2022**, *37*.
3. Dixit, S.; Bohre, K.; Singh, Y.; Himeur, Y.; Mansoor, W.; Atalla, S.; Srinivasan, K. A Comprehensive Review on AI-Enabled Models for Parkinson's Disease Diagnosis. *Electronics* **2023**, *12*, 783. [CrossRef]
4. Ibtissame, A.; Ghizlane, K.; Mrabti, M. A literature review of online handwriting analysis to detect Parkinson's disease at an early stage. *Multimed. Tools Appl.* **2022**, *82*, 11923–11948.
5. Sharma, S.R.; Singh, B.; Kaur, M. Classification of Parkinson disease using binary Rao optimization algorithms. *Expert Syst.* **2021**, *38*, e12674. [CrossRef]
6. Kour, N.; Arora, S. Computer-Vision Based Diagnosis of Parkinson's Disease via Gait: A Survey. *IEEE Access* **2019**, *7*, 156620–156645. [CrossRef]
7. Connie, T.; Aderinola, T.B.; Ong, T.S.; Goh, M.K.O.; Erfianto, B.; Purnama, B. Pose-Based Gait Analysis for Diagnosis of Parkinson's Disease. *Algorithms* **2022**, *15*, 474. [CrossRef]
8. Di Biase, L.; Di Santo, A.; Caminiti, M.L.; De Liso, A.; Shah, S.A.; Ricci, L.; Di Lazzaro, V. Gait Analysis in Parkinson's Disease: An Overview of the Most Accurate Markers for Diagnosis and Symptoms Monitoring. *Sensors* **2022**, *20*, 3529. [CrossRef]
9. Celik, E.; Omurca, S.I. Improving Parkinson's Disease diagnosis with machine learning methods. In Proceedings of the 2019 Scientific Meeting on Electrical-Electronics & Biomedical Engineering and Computer Science (EBBT), Istanbul, Turkey, 24–26 April 2019.
10. Muro, A.; Zapirain, B.; Mendez-Zorrilla, A. Gait Analysis Methods: An Overview of Wearable and Non-Wearable Systems, Highlighting Clinical Applications. *Sensors* **2014**, *14*, 3362–3394. [CrossRef]
11. Nieto-Hidalgo, M.; Ferrandez, J.; Valdivieso-Sarabia, R.; Mora-Pascual, J.; García-Chamizo, J. Vision Based Extraction of Dynamic Gait Features Focused on Feet Movement Using RGB Camera. In *Ambient Intelligence for Health: First International Conference, AmIHEALTH 2015, Puerto Varas, Chile, 1–4 December 2015, Proceedings 1*; Springer International Publishing: Berlin/Heidelberg, Germany, 2015; pp. 155–166.
12. Albert, L.; Potheegadoo, J.; Herbelin, B.; Bernasconi, F.; Blanke, O. Digital-robotic markers for hallucinations in Parkinson's Disease. *bioRxiv* **2023**. [CrossRef]
13. Marquez Chavez, J.; Tang, W. A Vision-Based System for Stage Classification of Parkinsonian Gait Using Machine Learning and Synthetic Data. *Sensors* **2022**, *22*, 4463. [CrossRef] [PubMed]
14. Gong, L.; Li, J.; Yu, M.; Zhu, M.; Clifford, R. A novel computer vision based gait analysis technique for normal and Parkinson's gaits classification. In Proceedings of the 2020 IEEE International Conference on Dependable, Autonomic & Secure Computing, Calgary, AB, Canada, 22–26 June 2020.
15. Goetz, C. The Unified Parkinson's Disease Rating Scale (UPDRS): Status and recommendations. *Mov. Disord.* **2003**, *18*, 738–750.
16. Moon, Y.; Sung, J.; An, R.; Hernandez, M.E.; Sosnoff, J.J. Gait variability in people with neurological disorders: A systematic review and meta-analysis. *Hum. Mov. Sci.* **2016**, *47*, 197–208. [CrossRef] [PubMed]

17. Loureiro, J.; Correia, P. Using a Skeleton Gait Energy Image for Pathological Gait Classification. In Proceedings of the 15th IEEE International Conference on Automatic Face and Gesture Recognition (FG 2020), Buenos Aires, Argentina, 16–20 November 2020; pp. 503–507.
18. Antoniades, C.; Spering, M. Eye movements in Parkinson's Disease: From neurophysiological mechanisms to diagnostic tools. *Trends Neurosci.* **2023**, *47*. [CrossRef]
19. Das, S.; Ramteke, H. A Comprehensive Review of the Role of Biomarkers in Early Diagnosis of Parkinson's Disease. *Cureus* **2024**, *16*, e54337. [CrossRef]
20. Guo, Y.; Yang, J.; Liu, Y.; Chen, X.; Yang, G.Z. Detection and assessment of Parkinson's Disease based on gait analysis: A survey. *Front. Aging Neurosci.* **2022**, *14*, 916971. [CrossRef]
21. Kaur, R.; Motl, R.W.; Sowers, R.; Hernandez, M.E. A Vision-Based Framework for Predicting Multiple Sclerosis and Parkinson's Disease Gait Dysfunctions-A Deep Learning Approach. *IEEE J. Biomed. Health Inform.* **2023**, *27*, 190–201. [CrossRef]
22. Islam, E.; Medhat, M.; Waleed, F.; Deif, M.A.; Alkhazaleh, S.; Hafez, M. Parkinsonian Gait Severity in Older Adults with Dementia Using Natural Gait Video Analysis. In Proceedings of the 2024 International Telecommunications Conference (ITC-Egypt), Cairo, Egypt, 22–25 July 2024; pp. 493–498.
23. Kondo, Y.; Bando, K.; Suzuki, I.; Miyazaki, Y.; Nishida, D.; Hara, T.; Kadone, H.; Suzuki, K. Video-Based Detection of Freezing of Gait in Daily Clinical Practice in Patients With Parkinsonism. *IEEE Trans. Neural Syst. Rehabil. Eng. Publ. IEEE Eng. Med. Biol. Society* **2024**, *32*, 2250–2260. [CrossRef]
24. Sibley, K.G.; Girges, C.; Hoque, E.; Foltynie, T. Video-based analyses of Parkinson's disease severity: A brief review. *J. Park. Dis.* **2021**, *11*, S83–S93. [CrossRef] [PubMed]
25. Shaban, M. Deep learning for Parkinson's disease diagnosis: A short survey. *Computers* **2023**, *12*, 58. [CrossRef]
26. Zeng, Q.; Liu, P.; Yu, N.; Wu, J.; Huo, W.; Han, J. Video-Based Quantification of Gait Impairments in Parkinson's Disease Using Skeleton-Silhouette Fusion Convolution Network. *IEEE Trans. Neural Syst. Rehabil. Eng. Publ. IEEE Eng. Med. Biol. Soc.* **2023**, *31*, 2912–2922. [CrossRef] [PubMed]
27. Albuquerque, P.; Machado, J.P.; Verlekar, T.T.; Correia, P.L.; Soares, L.D. Remote Gait type classification system using markerless 2D video. *Diagnostics* **2021**, *11*, 1824. [CrossRef] [PubMed]
28. Han, J.; Bhanu, B. Individual Recognition Using Gait Energy Image. *IEEE Trans. Pattern Anal. Mach. Intell.* **2006**, *28*, 316–322. [CrossRef]
29. Ranjan, R.; Ahmedt-Aristizabal, D.; Armin, M.A.; Kim, J. Computer Vision for Clinical Gait Analysis: A Gait Abnormality Video Dataset. *arXiv* **2024**, arXiv:2407.04190.
30. Shen, W.; Zhao, K.; Jiang, Y.; Wang, Y.; Zhang, Z.; Bai, X. Object skeleton extraction in natural images by fusing scale-associated deep side outputs. In Proceedings of the IEEE Conference on Computer Vision and Pattern Recognition, Las Vegas, NV, USA, 27–30 June 2016.
31. Terven, J.; Cordova-Esparza, D.M.; Romero-González, J.A. A Comprehensive Review of YOLO Architectures in Computer Vision: From YOLOv1 to YOLOv8 and YOLO-NAS. *Mach. Learn. Knowl. Extr.* **2023**, *5*, 1680–1716. [CrossRef]
32. Rand, M.K.; Stelmach, G.E.; Bloedel, J.R. Movement accuracy constraints in Parkinson's disease patients. *Neuropsychologia* **2000**, *38*, 203–212. [CrossRef]
33. Hossain, M.S.; Muhammad, G. Emotion recognition using deep learning approach from audio–visual emotional big data. *Inf. Fusion* **2019**, *49*, 69–78. [CrossRef]
34. Avots, E.; Sapiński, T.; Bachmann, M.; Kamińska, D. Audiovisual emotion recognition in wild. *Mach. Vis. Appl.* **2019**, *30*, 975–985. [CrossRef]
35. Ghosh, A.; Sufian, A.; Sultana, F.; Chakrabarti, A.; De, D. *Fundamental Concepts of Convolutional Neural Network*; Springer: Berlin/Heidelberg, Germany, 2020; pp. 519–567.
36. Ioffe, S.; Szegedy, C. Batch Normalization: Accelerating Deep Network Training by Reducing Internal Covariate Shift. *arXiv* **2015**, arXiv:1502.03167.
37. Srivastava, N.; Hinton, G.; Krizhevsky, A.; Sutskever, I.; Salakhutdinov, R. Dropout: A Simple Way to Prevent Neural Networks from Overfitting. *J. Mach. Learn. Res.* **2014**, *15*, 1929–1958.
38. He, K.; Zhang, X.; Ren, S.; Sun, J. Deep Residual Learning for Image Recognition. 2015. Available online: https://ieeexplore.ieee.org/document/7780459 (accessed on 17 September 2024).
39. Dosovitskiy, A.; Beyer, L.; Kolesnikov, A.; Weissenborn, D.; Zhai, X.; Unterthiner, T.; Hohenauer, M.; de Vries, H.; T, G. Transformers for Image Recognition at Scale. In Proceedings of the International Conference on Learning Representations (ICLR), Vienna, Austria, 3–7 May 2021.
40. Vaswani, A.; Shardlow, M.; Parmar, N.; Uszkoreit, J.; Jones, L.; Gomez, A.N.; Kaiser, Ł.; Polosukhin, I.; Wu, Y.; Yang, N.; et al. Attention Is All You Need. In Proceedings of the Advances in Neural Information Processing Systems, Long Beach, CA, USA, 4–9 December 2017; Volume 30.
41. Han, K.; Wang, Y.; Chen, H.; Chen, X.; Guo, J.; Liu, Z.; Tang, Y.; Xiao, A.; Xu, C.; Xu, Y.; et al. A Survey on Vision Transformer. *IEEE Trans. Pattern Anal. Mach. Intell.* **2023**, *45*, 87–110. [CrossRef]
42. Yedida, R.; Menzies, T. On the Value of Oversampling for Deep Learning in Software Defect Prediction. *IEEE Trans. Softw. Eng.* **2021**, *48*, 3103–3116. [CrossRef]

43. Yao, Y.; Rosasco, L.; Caponnetto, A. On early stopping in gradient descent learning. *Constr. Approx.* **2007**, *26*, 289–315. [CrossRef]
44. Nassar, S.; Yasser, I.; Amer, H.; Mohamed, M. Comparative analysis of vision transformers and fine-tuned transfer learning models for brain tumor classification. *Imaging Radiat. Res.* **2024**, *7*, 8514. [CrossRef]

diagnostics

MDPI

Article

Improving Alzheimer's Disease Prediction with Different Machine Learning Approaches and Feature Selection Techniques

Hala Alshamlan *, Arwa Alwassel, Atheer Banafa and Layan Alsaleem

Department of Information Technology, College of Computer and Information Sciences, King Saud University, P.O. Box 51178, Riyadh 11543, Saudi Arabia; 442202188@student.ksu.edu.sa (A.A.); 442203479@student.ksu.edu.sa (A.B.); 442201879@student.ksu.edu.sa (L.A.)

* Correspondence: halshamlan@ksu.edu.sa

Abstract: Machine learning (ML) has increasingly been utilized in healthcare to facilitate disease diagnosis and prediction. This study focuses on predicting Alzheimer's disease (AD) through the development and comparison of ML models using Support Vector Machine (SVM), Random Forest (RF), and Logistic Regression (LR) algorithms. Additionally, feature selection techniques including Minimum Redundancy Maximum Relevance (mRMR) and Mutual Information (MI) were employed to enhance the model performance. The research methodology involved training and testing these models on the OASIS-2 dataset, evaluating their predictive accuracies. Notably, LR combined with mRMR achieved the highest accuracy of 99.08% in predicting AD. These findings underscore the efficacy of ML algorithms in AD prediction and highlight the utility of the feature selection methods in improving the model performance. This study contributes to the ongoing efforts to leverage ML for more accurate disease prognosis and underscores the potential of these techniques in advancing clinical decision-making.

Keywords: machine learning; OASIS; prediction; Alzheimer's disease (AD); feature selection

1. Introduction

Dementia, a cognitive impairment characterized by deficits in thinking, remembering, and reasoning abilities, as well as behavioral impairments, hinders an individual's daily functioning and activities. Its severity varies from an incipient stage that marginally affects functioning to an advanced stage where the individual becomes entirely reliant on others for the basic activities of daily living. Neurological alterations may commence a decade or more before the manifestation of clinical symptoms in Alzheimer's disease. During this early stage, a series of toxic changes occur in the brain, including the abnormal aggregation of proteins leading to the formation of amyloid plaques and tau tangles. Healthy neurons cease to function properly, lose connections with their neighboring neurons, and ultimately degenerate. Numerous other intricate modifications in the brain are believed to contribute to the pathology of Alzheimer's disease [1].

Alzheimer's disease (AD) is the most common type of dementia. It is a progressive disease beginning with mild memory loss and possibly leading to the loss of the ability to carry on a conversation and respond to the environment. However, early accurate diagnosis can affect the patient's outcomes and slow the decline in memory. More focus has been shifted towards risk reduction, leading to new discoveries and solutions. The research indicates that individuals with healthy lifestyle habits are less likely to develop dementia. Also, behaviors proven to prevent diseases such as cancer and diabetes may also lower the risk of cognitive decline [2].

As a division of artificial intelligence (AI) and computer science, machine learning (ML) involves using data and algorithms to enable AI to replicate human learning behaviors [3]. The increased adoption of machine learning in healthcare has created new possibilities for disease diagnosis and treatment [4]. Relevant data and patterns can be identified by applying machine learning to patient datasets stored in electronic health records [5].

Machine learning algorithms have been playing a vital role in solving complex, highly nonlinear classification and prediction problems [6].

There is a pressing need for more research on the causes of Alzheimer's disease (AD), as well as on the early diagnosis and treatment strategies. Our study focuses on using machine learning to aid in the early diagnosis of AD and to evaluate the effectiveness of various algorithms. Given the increasing rates of AD cases, effective early detection strategies are needed. By employing machine learning, we aim to enhance early detection.

The objective of this study is to investigate various machine learning (ML) models such as Support Vector Machine, Logistic Regression, and Random Forest, along with different feature selection methods such as Minimum Redundancy Maximum Relevancy and Mutual Information, aimed at evaluating the effectiveness of each model and eventually advancing their application in healthcare through supervised learning methodologies. By the end of this study, we develop a total of 72 different models with accuracies ranging from 91.43% to 99.08%. The structure of this paper is organized as follows: Section 2 provides the background on the techniques utilized in this study, including Support Vector Machine, Random Forest, and Logistic Regression, along with feature selection methods such as Minimum Redundancy Maximum Relevance and Mutual Information. Section 3 presents a literature review of the relevant research conducted on machine learning models for predicting dementia and Alzheimer's disease. Section 4 outlines the materials and methods employed in this study, detailing the data collection, preparation, and analysis processes. Section 5 discusses the results obtained from the various machine learning models, including a comparison of their performances. Finally, Section 6 concludes the paper with a discussion regarding the findings and their implications for future research.

2. Background

The techniques used in this study, such as SVM, Random Forest, Logistic Regression, Mutual Information, and mRMR, will be discussed in this section.

Support Vector Machine (SVM) is a supervised learning algorithm in machine learning, effective for tasks like classification and regression. It excels particularly in binary classification, where it categorizes data into two groups [7].

Logistic Regression (LR) is a supervised machine learning algorithm utilized for classification tasks, where the objective is to predict the probability of an instance belonging to a particular class. It is well-suited for binary classification problems [8]. The Random Forest (RF) algorithm consists of an ensemble of Decision Trees. Each tree in the ensemble is constructed using a bootstrap sample, drawn with replacement from the training set. This algorithm is known for its robustness and ability to handle complex datasets [9].

mRMR, which stands for 'Minimum Redundancy Maximum Relevancy', is a feature selection method that aims to select the most relevant attributes from the dataset while minimizing redundancy among them [10].

The relevance of each feature is evaluated based on its correlation and relevance with the target variable and redundancy among other features. Features with high relevancy and low redundancy are prioritized and selected [10].

mRMR is usually formulated as follows [11]:

$$\text{score}_i(f) = \frac{F(f, \text{target})}{\sum_{s \in \text{features selected until } i-1} \frac{|\text{corr}(f,s)|}{i-1}} \tag{1}$$

The correlation coefficient quantifies the strength of the linear relationship between two variables, ranging from −1 to 1. −1 indicates a perfect inverse correlation (when a value in one series increases, another value decreases), 1 indicates a perfect positive correlation (when a value in one series increases, another value rises), and 0 denotes no linear relationship [12].

The Pearson correlation coefficient, denoted as r, is calculated using the following formula [13]:

$$r = \frac{n(\sum xy) - (\sum x)(\sum y)}{\sqrt{[n\sum x^2 - (\sum x)^2][n\sum y^2 - (\sum y)^2]}} \quad (2)$$

Mutual Information (MI) estimates relationships between fixed categories in classification problems and continuous variables in regression. It is based on the entropy of the variables involved. MI is a measure of how much two random variables depend on each other. It is a non-negative value that equals zero when the variables are independent, meaning they do not affect each other. Higher MI values indicate stronger dependencies.

The MI between two random variables X and Y is as follows:

$$I(X;Y) = H(X) - H(X|Y) \quad (3)$$

In this equation, *I(X ; Y)* represents the MI for X and Y, H(X) denotes the entropy for X, and H(X | Y) is the conditional entropy of X given Y. The result is measured in bits, with values ranging from zero to one [14].

3. Literature Review

In this section, we discuss and review research studies conducted between 2019 and 2023 on the subject of using ML models in predicting dementia and Alzheimer's disease (AD). These studies utilize the OASIS longitudinal dataset, known as OASIS-2 [15]. The dataset contains 15 columns and 373 rows.

Using both the OASIS-1 and OASIS-2 datasets, Saratxaga et al. [16] proposed the use of deep learning (DL) techniques, as opposed to the traditional image processing and classification methods, in regard to MRI analysis. The study aims to calculate the accuracy and balanced accuracy and compare the results to previous works. Two approaches are used: 2D slice-level and 3D subject-level. Various architectures were used on the OASIS-2 subset, including BrainNet2D, BrainNet3D, and ResNet18. Batch Normalization and Cyclical Learning Rate (CLR) were used to increase the performance.

As for the results, BrainNet2D using CLR triangular achieved an accuracy and baseline accuracy of 92%, and BrainNet3D using Batch Normalization had an accuracy and baseline accuracy of 84%. ResNet18 using CLR triangular and min–max scaling achieved the highest accuracy and baseline accuracy of 93%.

Battineni et al. [17] used Support Vector Machines (SVMs) and achieved 68.75% accuracy and 64.18% precision. These findings substantiate that superior predictive performance in dementia prognosis is obtained with low gamma (1.0×10^{-4}) and high regularized (C = 100) values. Kavitha et al. [18] employed various classifiers, including the Random Forest (RF) technique, with median imputation utilized for addressing the missing values in SES. The RF classifier yielded an accuracy of 86.92%.

Baglat et al. [19], in addition to using both SVM and RF, used Logistic Regression (LR) and Decision Trees (DTs) techniques, with the RF performance achieving the best results. The RF classifier achieved an accuracy of 86.8%. AdaBoost, a boosting algorithm, was implemented to increase the predictive accuracy.

Basheer et al. [20] proposed a different approach, introducing a Modified Capsule Network model, M-CapNet. To increase the performance and accuracy of the model, several optimization techniques were applied. The model achieved 92.39% accuracy.

Dhakal et al. [21] applied various additional ML techniques, including K-Nearest Neighbor (KNN). These techniques were used on features selected using Least Absolute Shrinkage and Selection Operator (LASSO) and the full set of features. Upon comparing the metrics, the SVM model achieved the highest accuracy of 96.77%.

Overall, the reviewed works have shown promising advancements in the utilization of ML models for predicting dementia and Alzheimer's disease. These studies have explored various ML techniques, including deep learning (DL) approaches like BrainNet2D, BrainNet3D, and ResNet18, as well as Support Vector Machines (SVMs), Random Forest (RF), Logistic Regression (LR), Decision Trees (DTs), and Modified Capsule Network

(M-CapNet). In Tables 1 and 2, we present the comparison between all the methods and the advantages and disadvantages of each study, respectively.

Table 1. Comparison of related ML models for dementia prediction in state-of-the-art approaches.

Study	Dataset	Feature Selection Technique	Classifier(s)	Accuracy
Saratxaga et al. [16]	OASIS-2	None	BrainNet2D (CLR triangular)	92%
			BrainNet3D (Batch Norm.)	84%
			ResNet18 (CLR triangular)	93%
Battineni et al. [17]	OASIS-2	Feature Exclusion	SVM	68.75%
Kavitha et al. [18]	OASIS-2	Correlation, Info gain, Chi-Square	Decision Tree	80.46%
			SVM	86.92%
			Random Forest	81.67%
Baglat et al. [19]	OASIS-2	None	Logistic Regression (Imputation)	78.9%
			Logistic Regression (Dropping)	75%
			Decision Tree	81.5%
			Random Forest	86.8%
			SVM	81.5%
Basheer et al. [20]	OASIS-2	None	M-CapNet	92.39%
Dhakal et al. [21]	OASIS-2	None	Decision Tree	89.5%
			K-Nearest Neighbor	95.1%
			Logistic Regression	94.35%
			Random Forest	94.35%
			SVM	96.77%
		LASSO	Decision Tree	89.5%
			K-Nearest Neighbor	95.1%
			Logistic Regression	94.35%
			Random Forest	94.35%
			SVM	94.35%
		Chi-Square	Decision Tree	91.93%
			K-Nearest Neighbor	94.35%
			Logistic Regression	94.35%
			Random Forest	94.35%
			SVM	94.35%

Table 2. Advantages and disadvantages of related ML models for dementia prediction in state-of-the-art approaches.

Study	Advantages	Disadvantages
Saratxaga et al. [16]	High accuracy (up to 93%) using advanced deep learning models like BrainNet2D and ResNet18. Applied optimization techniques such as CLR and Batch Normalization.	No feature selection used, which may include irrelevant data. Requires substantial computational resources.
Battineni et al. [17]	Simple implementation using SVM with moderate accuracy (68.75%).	Lower accuracy compared to other models, limiting effectiveness. Feature exclusion may overlook important predictors.

Table 2. *Cont.*

Study	Advantages	Disadvantages
Kavitha et al. [18]	Good accuracy (up to 86.92%) with Random Forest and effective feature selection methods, such as correlation and chi-square.	Random Forest requires more computational power, and Decision Trees can overfit.
Baglat et al. [19]	Evaluated several classifiers, with Random Forest achieving 86.8% accuracy. Handled missing values through imputation.	Logistic Regression underperformed (75–78.9%), and inconsistent performance across classifiers makes it harder to choose the best model.
Basheer et al. [20]	Achieved high accuracy (92.39%) with M-CapNet, a specialized neural network.	No feature selection used, which may include unnecessary data.
Dhakal et al. [21]	Very high accuracy (up to 96.77%) using SVM and multiple classifiers like KNN, Random Forest, and Logistic Regression.	No specific feature selection was mentioned for SVM, and models could potentially overfit.

4. Materials and Methods

In this section, we will discuss the stages we went through, from collecting, preparing, and processing data to choosing features and models to arrive at our proposed model, which achieves the highest percentage of accuracy.

4.1. Data Collection and Preparation

In this study, we used the OASIS-2 dataset [15], which consists of a longitudinal collection of 150 people aged 60–96 years. Each subject was scanned on two or more visits, separated by at least one year, for a total of 373 imaging sessions. For each subject, 3 or 4 individual T1-weighted MRI scans were included in individual scanning session [22].

All participants in this dataset are right-handed and include both men and women; 72 people were classified as non-demented throughout the study, while another 64 were described as having dementia upon their initial visits, which remained so for subsequent scans, including 51 individuals with mild to moderate Alzheimer's disease [22]. Another 14 subjects were characterized as non-demented at the time of their initial visit and were subsequently characterized as demented at a later visit [22].

This dataset includes 373 instances and includes 10 unique features shown in Table 3 below.

Table 3. Description of key features in the dataset.

Feature	Description
Group	Class label (Demented, Non-demented)
M/F	Male/Female
Age	Ranges from 60 to 96
EDUC	Years of Education
SES	Socioeconomic Status (1–5)
MMSE (Mini Mental State Examination)	A cognitive screening tool that provides a brief objective measure of cognitive function.
CDR (Clinical Dementia Rating)	A numeric scale used to quantify the severity of symptoms of dementia, including cognitive, functional, and social domains in the overall staging.

Table 3. *Cont.*

Feature	Description
eTIV (Estimated total intracranial volume)	Volume of the cranial cavity.
nWBV (Normalized Whole Brain Volume)	Percentage of the intracranial cavity occupied by the brain.
ASF (Atlas Scaling Factor)	Volume-scaling factor required to match each individual to the atlas target.

4.2. Data Analysis

Initially, we employed basic analytical techniques to delineate the properties of each feature, thereby enhancing our understanding of the dataset's characteristics. This foundational analysis is illustrated in Figures 1 and 2a,b.

	Group	M/F	Age	EDUC	SES	MMSE	CDR	eTIV	nWBV	ASF
0	Nondemented	M	87	14	2.0	27.0	0.0	1987	0.696	0.883
1	Nondemented	M	88	14	2.0	30.0	0.0	2004	0.681	0.876
2	Demented	M	75	12	NaN	23.0	0.5	1678	0.736	1.046
3	Demented	M	76	12	NaN	28.0	0.5	1738	0.713	1.010
4	Demented	M	80	12	NaN	22.0	0.5	1698	0.701	1.034

Figure 1. The result of first 5 rows.

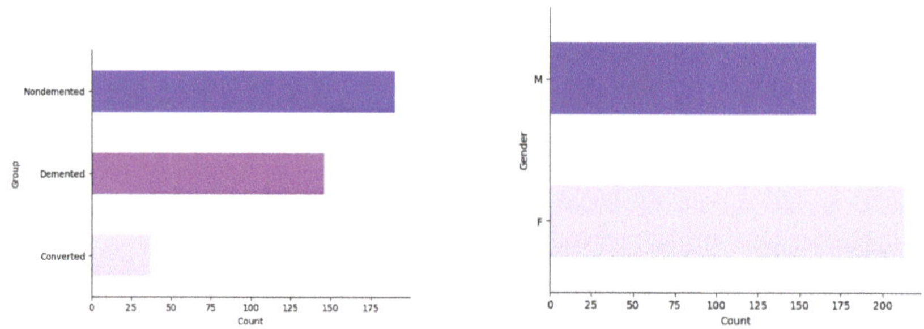

(**a**) (**b**)

Figure 2. Descriptive statistics of the participant demographics within the dataset. (**a**) The distribution of group classifications. (**b**) The gender representation.

All features in the dataset showed numeric properties except "gender" (classified as "M/F") and "Group" (serves as a class label).

4.2.1. Data Transformation

As we can see in Table 4, the "Group" feature, serving as a class label, is divided into three categories. Knowing that the converted column does not contribute to the classification process in the integrated machine learning model, we decided to combine this category with the demented category to achieve a balanced dataset, so the dataset will now contain 190 records. These are classified as non-demented and 183 records are classified as demented.

Table 4. Group descriptions.

	Demented	Non-Demented	Converted
Description	a label given to patients with Alzheimer's disease	a label given to patients without Alzheimer's disease	a label given to patients diagnosed first to not have Alzheimer's disease but were converted to be patients with Alzheimer's disease
Count	146	190	37

4.2.2. Encoding

As we can see in our dataset, we have two columns (group and M/F) that we need to use label encoding to ensure compatibility; the result is shown in Table 5 below:

Table 5. Encoding for group and M/F.

Group	M/F
Demented = 1	Female = 1
Non-demented = 0	Male = 0

4.2.3. Missing Values

When we examined the dataset, we noticed 19 missing values in the "SES" column and two missing values in the "MMSE" column. To choose the appropriate method for dealing with the missing values, we analyzed both columns, considering replacing the missing values with either the mean or the median, as shown in Figure 3.

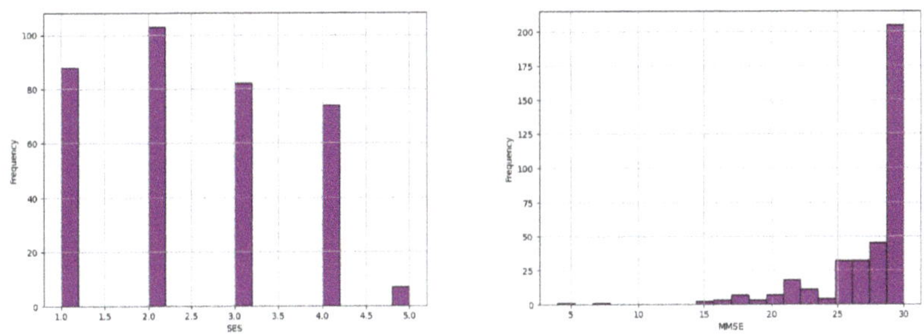

(**a**) (**b**)

Figure 3. Histograms illustrating the distribution of socioeconomic status. (**a**) SES scores among participants. (**b**) MMSE scores among participants.

Given the non-normal distribution observed in both columns, we chose the median as the most appropriate method for dealing with missing values. This choice is supported by the resilience of the median to the disproportionate effect of the outliers compared to the mean, making it a more robust measure of central tendency.

4.2.4. Outlier Analysis

Before determining the appropriate method for dealing with outliers, it is necessary to analyze them, confirm their nature, and determine whether they are real outliers or those arising from human error. In light of this, we analyzed them to identify the nature of the outliers, as shown in Figure 4; it is clear that some bars show outliers. However, our attention was particularly drawn to outliers in the MMSE column, prompting us to further investigate them, as shown in Figure 5.

As shown in Figure 5, the highest value of the MMSE score is 30. A score of 25 or higher is classified as natural. If the score is less than 24, the score usually indicates possible cognitive impairment [23] and is an abnormal result. Therefore, it can be concluded that the outliers observed in the MMSE column are inherent in the distribution of the data and do not arise from human error. As a result, the decision was made to retain these outliers within the dataset.

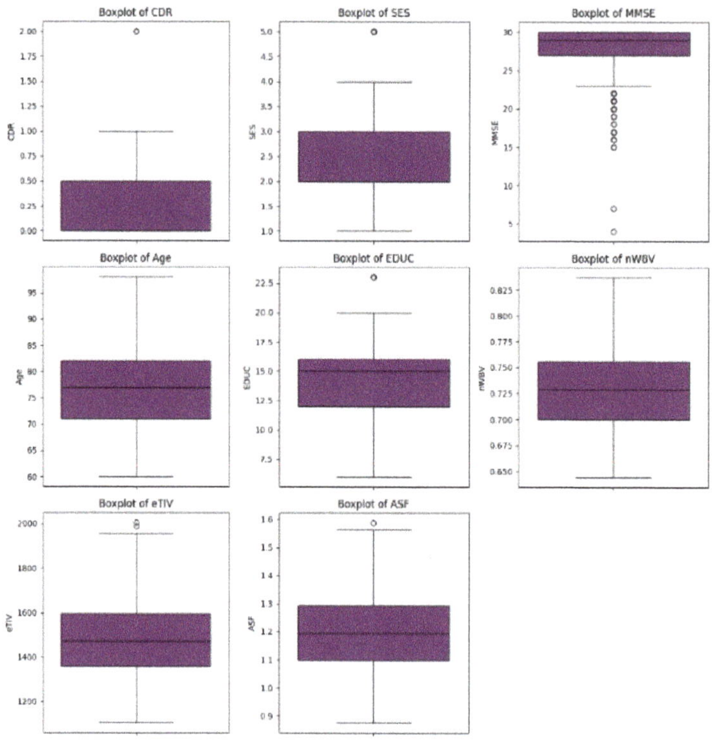

Figure 4. Boxplots representing the distribution and variability of numeric attributes within the dataset, facilitating the identification of potential outliers.

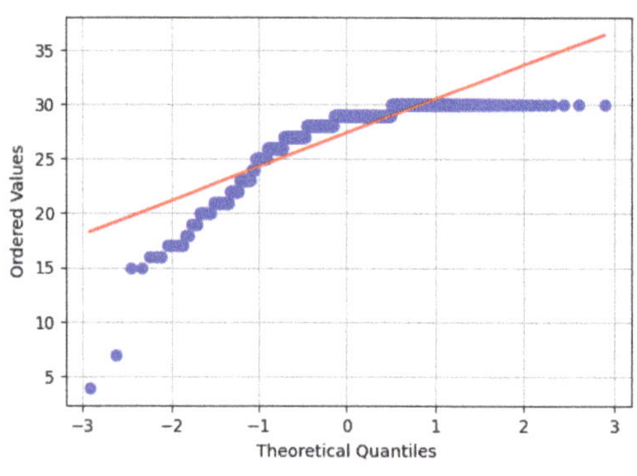

Figure 5. Probability plot depicting the distribution of MMSE scores, highlighting the assessment of cognitive function among participants.

4.3. Feature Selection

In the feature selection process, we used three methodologies for feature selection: mRMR (Minimum Redundancy Maximum Relevance), correlation coefficient analysis, and Mutual Information methodology. Our intent was to select a subset of five features as an initial step in the feature selection process.

The results derived from the application of mRMR and correlation coefficient methodologies are shown in Figures 6 and 7.

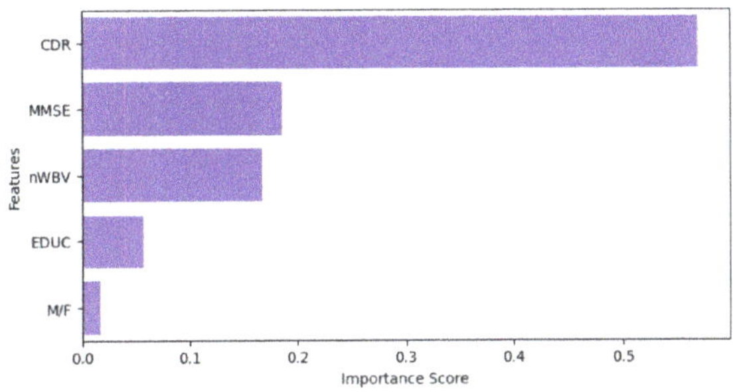

Figure 6. Visualization of feature selection results using mRMR.

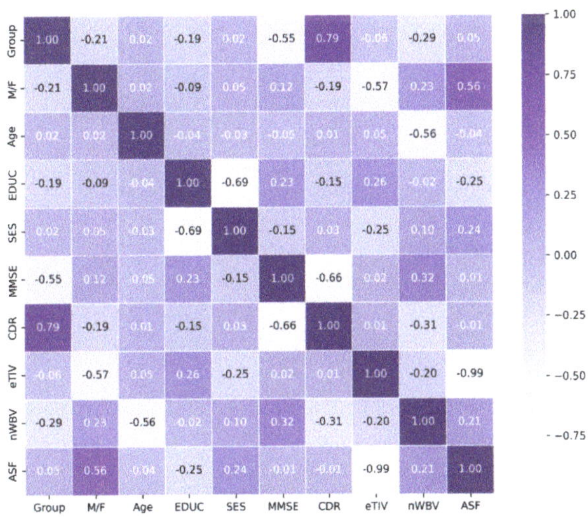

Figure 7. Heat map of correlation coefficient feature selection.

First, we analyzed mRMR and correlation coefficient and found that both methodologies yielded identical results, namely CDR, MMSE, nWBV, EDUC, and M/F. However, these results were not sufficient to satisfy our curiosity, so we decided to use another approach: Mutual Information.

The result after applying Mutual Information is shown in Figure 8.

The result is a little different from the previous methodologies; the first five columns identified through the Mutual Information methodology are CDR, MMSE, EDUC, eTIV, and ASF.

Since mRMR and correlation coefficient yielded identical results, we decided to use mRMR and Mutual Information in our model; the outcomes of feature selection methodologies are shown in Table 6 below.

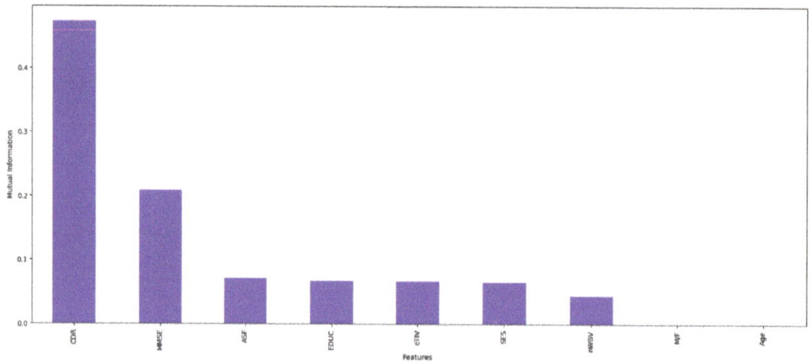

Figure 8. Mutual Information feature selection.

Table 6. Summary of feature selection results.

Feature Selection	Selected Features
mRMR	CDR, MMSE, nWBV, EDUC, M/F
Correlation Coefficient	CDR, MMSE, nWBV, EDUC, M/F
Mutual Information (MI)	CDR, MMSE, EDUC, cTIV, ASF

4.4. *Classifier Models*

We used three classification methods: Support Vector Machine (SVM), Random Forest, and Logistic Regression models. Each model was trained without feature selection, and with mRMR selection and Mutual Information feature selection methodologies. We also tried the cross-validation method, which we will discuss in the Section 5.

5. Results and Analysis

This section presents the findings from the various machine learning models employed in this study, focusing on their predictive performance. Both the hold-out and cross-validation methods were utilized to evaluate the model accuracy, highlighting the effectiveness of the selected algorithms and feature selection techniques in predicting Alzheimer's disease.

5.1. *Hold-Out Method*

In this subsection, we detail the results obtained through the hold-out method, which involves dividing the dataset into training and testing subsets to assess the models' performance. The following results focus on the Support Vector Machine (SVM), Logistic Regression, and Random Forest models, showcasing the predictive capabilities under various feature selection techniques.

5.1.1. SVM

The Support Vector Machine (SVM) model consistently achieved an accuracy of 97.32% when utilizing both the mRMR and MI feature selection methodologies. In contrast, when all the features were considered, the model attained an accuracy of 96.42% (see Figure 9a). Notably, the performance of the SVM model improved significantly following the removal of the outliers, with accuracies reaching 98% across all the methodologies (see Figure 9b).

(**a**) (**b**)

Figure 9. Performance comparison of the SVM model. (**a**) Accuracy before outlier removal. (**b**) Accuracy after outlier removal.

5.1.2. Logistic Regression

Before the removal of the outliers, LR showed similar results to SVM despite their different methods, demonstrating the effectiveness of both models in achieving high accuracy (see Figure 10a).

The results from LR after removing the outliers displayed an overall improvement across all the methodologies, with accuracies ranging from 98% to 99%. This enhancement is evidenced by the highest accuracy achieved at 99% (see Figure 10b).

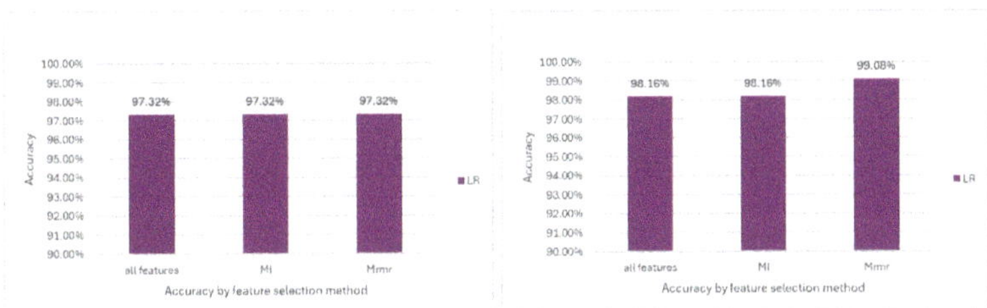

(**a**) (**b**)

Figure 10. Performance comparison of the LR model. (**a**) Accuracy before outlier removal. (**b**) Accuracy after outlier removal.

5.1.3. Random Forest

RF appears to perform slightly lower than SVM and LR. The highest accuracy achieved was recorded at 95.53% for all the feature selection methods; however, it is still lower compared to the accuracy achieved by the other two methodologies (see Figure 11a).

The outcomes derived from RF after removing the outliers exhibited notable enhancements, with accuracies ranging from nearly 96% to 98%, indicating significant advancements in performance (see Figure 11b).

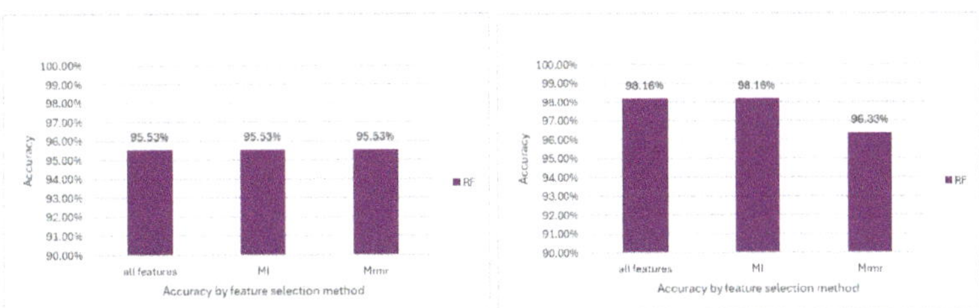

(**a**) (**b**)

Figure 11. Performance comparison of the RF model. (**a**) Accuarcy before outlier removal. (**b**) Accuracy after outlier removal.

5.2. Cross-Validation

To further evaluate the robustness of our models, we employed k-fold cross-validation, a technique that enhances the reliability of accuracy estimates by utilizing multiple subsets of the data. We tested various configurations of K in the k-fold cross-validation process, specifically using K = 10, K = 5, and Leave-One-Out (LOO) strategies.

For the K = 10 cross-validation, we divided the dataset into 10 equal-sized folds such that each fold serves as the validation set once while the remaining folds serve as the training set (see Figure 12).

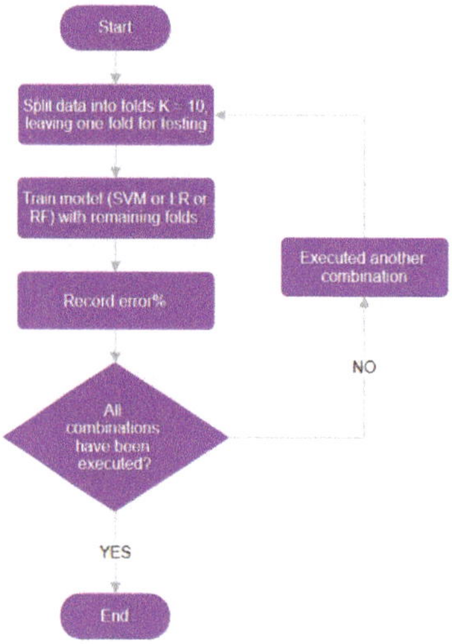

Figure 12. Illustration of the 10-fold cross-validation process, showing the division of the dataset into 10 equal-sized folds for model evaluation.

The K = 5 cross-validation was similar (see Figure 13).

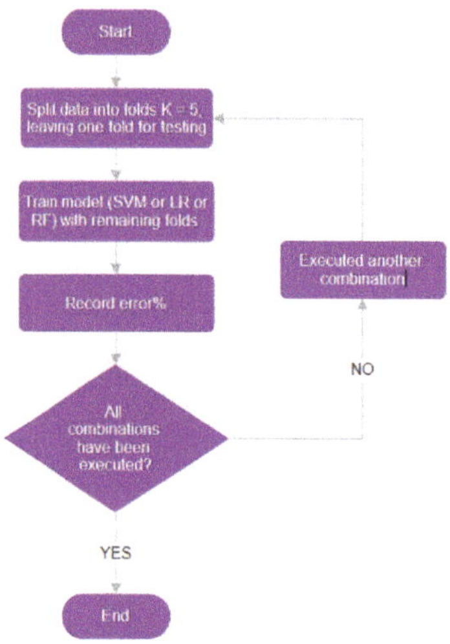

Figure 13. Illustration of the 5-fold cross-validation process, demonstrating the partitioning of the dataset into five subsets for model evaluation.

Finally, the Leave-One-Out method involved creating n folds, where n represents the total number of samples in the dataset, with each sample serving as a validation set once, while the rest of the data formed the training set (Figure 14).

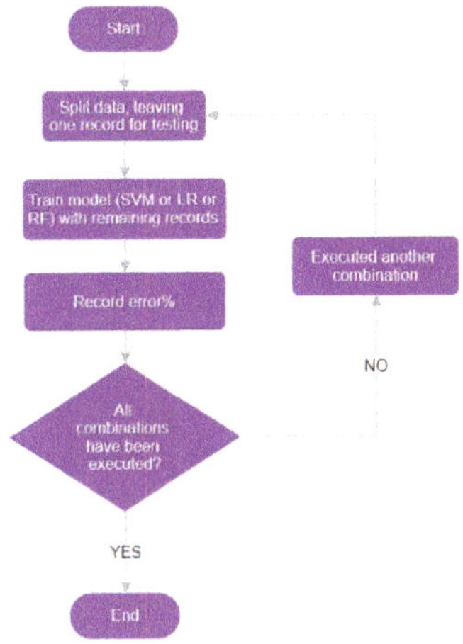

Figure 14. Illustration of the Leave-One-Out (LOO) cross-validation method, depicting the evaluation process where each individual data point serves as a separate validation set.

We applied these techniques to Support Vector Machine (SVM), Random Forest, and Logistic Regression models. First, we used two methods to train our models: training models without feature selection and training models with mRMR feature selection and Mutual Information methods. This process culminated in the production of 27 distinct models. Additionally, we trained each model with and without outliers due to their strong influence on the model results.

We will provide a detailed analysis of the results obtained from each model, including SVM, LR, and RF, focusing on their performance metrics and implications.

5.2.1. Support Vector Machines

SVM showed great performance before removing the outliers and achieved the highest accuracy among all the methods. Specifically, when no feature selection method was applied, in conjunction with K =5 during cross-validation, the SVM achieved an accuracy of 94.90%. Conversely, the lowest accuracy of 93.56% was observed when selecting MI features using K = 5 in the cross-validation process (see Figure 15a). However, after removing the outliers, the results derived from SVM remained largely consistent with those obtained before removing the outliers (see Figure 15b).

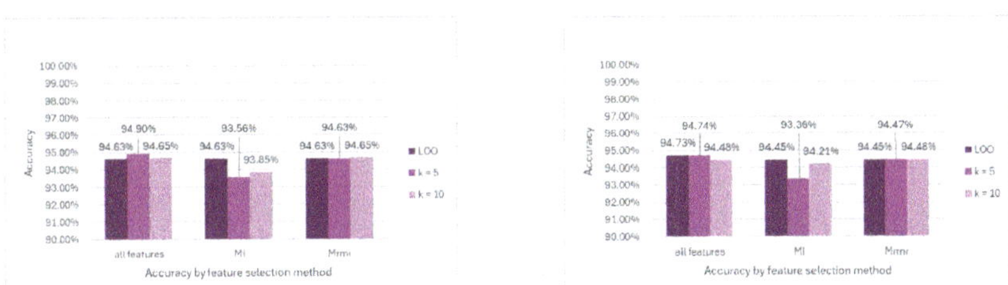

(a) (b)

Figure 15. Comparison of SVM model performance. (**a**) Accuracy prior to outlier removal. (**b**) Accuracy following outlier removal.

5.2.2. Logistic Regression

The results of Logistic Regression before removing the outliers showed remarkable performance, with the highest accuracy of 94.92% achieved by selecting the mRMR feature combined with K = 10 in the cross-validation procedure. In contrast, the lowest accuracy of 91.43% was recorded when no feature selection method was applied, in conjunction with K = 5 during cross-validation (see Figure 16a). After outlier removal, the Logistic Regression results showed no significant changes from the results before outlier removal (see Figure 16b).

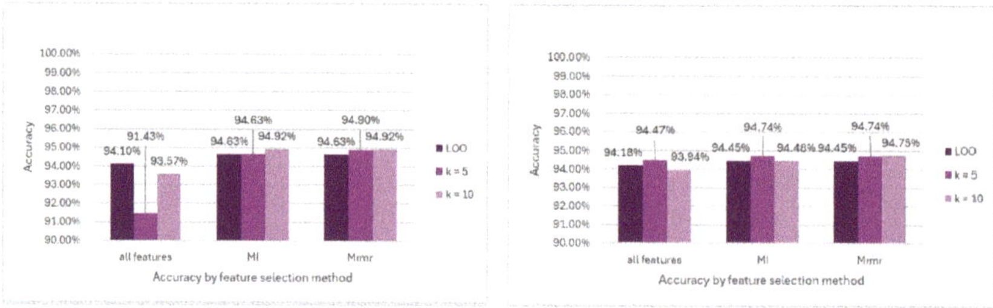

(a) (b)

Figure 16. Comparison of LR model performance. (**a**) Accuracy prior to outlier removal. (**b**) Accuracy following outlier removal.

5.2.3. Random Forest

As for RF before removing the outliers, the results showed a difference in performance from those obtained from the Logistic Regression analysis, where the highest accuracy was also observed at 95.17% by using all the features combined with Leave-One-Out cross-validation. Conversely, the lowest accuracy of 92.75% coincided with those scenarios where the MI was close to K = 10 during the cross-validation (see Figure 17a). For RF with outlier removal, the results showed no changes from the results before the outlier removal (see Figure 17b).

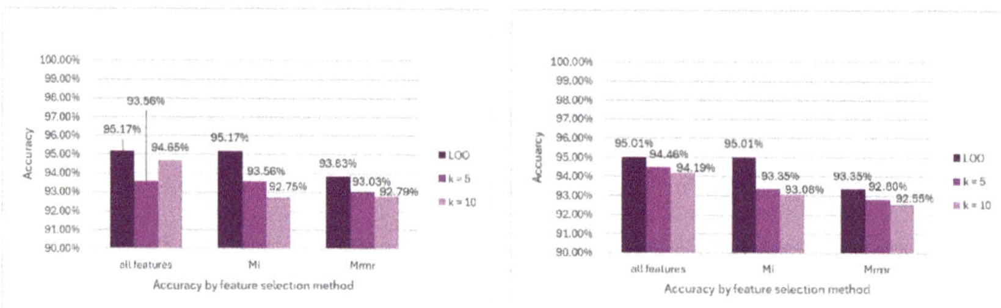

(a) (b)

Figure 17. Comparison of Random Forest model performance. (**a**) Accuracy prior to outlier removal. (**b**) Accuracy following outlier removal.

6. Discussion

Conducting this experiment has sparked inquiries into the sensitivity of the 70–30 split methodology to outliers. While the initial performance was commendable, the removal of the outliers notably improved the performance, contrasting with the minor impact observed in the cross-validation results post-outlier removal. The following tables show all the results we have achieved.

In Table 7, the results obtained from the hold-out method are displayed, including the accuracy and error rates with and without outliers.

Table 7. Results obtained from the hold-out method, showing accuracy and error rates with and without outliers.

Feature Selection	With Outliers						Without Outliers					
	SVM		Logistic Regression		Random Forest		SVM		Logistic Regression		Random Forest	
	Acc.	Error	Acc.	Error	Acc.	Error	Acc.	Error	Acc.	Error	Acc.	Error
mRMR	97.32%	2.68%	97.32%	2.68%	95.53%	4.47%	98.16%	1.84%	99.08%	0.92%	96.33%	3.67%
MI	97.32%	2.68%	97.32%	2.68%	95.53%	4.47%	98.16%	1.84%	98.16%	1.84%	98.16%	1.84%
All Features	96.42%	3.58%	97.32%	2.68%	95.53%	4.47%	98.16%	1.84%	98.16%	1.84%	98.16%	1.84%

As for the cross-validation method, the results we obtained from the cross-validation models did not show significant modifications after applying the outlier removal. In most cases, the accuracy decreased rather than improved after removing the outliers. In Table 8 below, the results obtained from the cross-fold method are displayed, including the accuracy and error rates with and without outliers.

Table 8. Results obtained from the k-fold cross-validation method, showing accuracy and error rates with and without outliers.

	With Outliers						Without Outliers					
	SVM		Logistic Regression		Random Forest		SVM		Logistic Regression		Random Forest	
	Accuracy/Error Rate	Std Dev	Accuracy/Error Rate	Std Dev	Accuracy/Error Rate	Std Dev	Accuracy/Error Rate	Std Dev	Accuracy/Error Rate	Std Dev	Accuracy/Error Rate	Std Dev
mRMR k = 10	94.65%/5.35%	0.039	94.92%/5.08%	0.030	92.79%/7.21%	0.055	94.48%/5.52%	0.040	94.75%/5.25%	0.037	92.55%/7.45%	0.057
mRMR k = 5	94.63%/5.37%	0.028	94.90%/5.10%	0.020	93.03%/6.97%	0.021	94.47%/5.53%	0.028	94.74%/5.26%	0.026	92.80%/7.20%	0.021
mRMR LOO	94.63%/5.37%	0.225	94.63%/5.37%	0.220	93.83%/6.17%	0.240	94.45%/5.55%	0.228	94.45%/5.55%	0.228	93.35%/6.65%	0.249
MI k = 10	93.85%/6.15%	0.050	94.65%/5.35%	0.030	92.75%/7.25%	0.062	94.21%/5.79%	0.050	94.48%/5.52%	0.040	93.08%/6.92%	0.063
MI k = 5	93.56%/6.44%	0.030	94.63%/5.37%	0.020	93.56%/6.44%	0.042	93.36%/6.64%	0.030	94.47%/5.53%	0.026	93.35%/6.65%	0.052
MI LOO	94.63%/5.37%	0.220	94.63%/5.37%	0.220	95.17%/4.83%	0.214	94.45%/5.55%	0.220	94.45%/5.55%	0.220	95.01%/4.99%	0.217
All features k = 10	94.65%/5.35%	0.033	93.57%/6.43%	0.060	94.65%/5.35%	0.033	94.48%/5.52%	0.045	94.48%/5.52%	0.052	94.19%/5.81%	0.035
All features k = 5	94.90%/5.10%	0.033	91.43%/8.57%	0.070	93.56%/6.44%	0.033	94.74%/5.26%	0.026	94.47%/5.53%	0.028	94.46%/5.54%	0.024

Despite the improved outcomes observed following the outlier removal within the hold-out methodology, we think keeping the outliers within the dataset is better. We firmly believe that these outliers hold pivotal significance in accurately diagnosing various conditions that the model may encounter, especially if it is applied in real-world settings, thus underscoring their indispensable role in model robustness and generalization.

Three feature selection methods were used; we can see that the selected features play a crucial role in assessing cognitive impairment, brain atrophy, and cognitive reserve, making them clinically relevant. Integrating these features into clinical decision-making can help clinicians in personalizing patient care, enabling early intervention for those at higher risk.

All three models were trained using the same random seed value to ensure consistency across the experiments. This approach enables a fair comparison of model performance by minimizing the variability introduced by random sampling. In clinical use, Logistic Regression, with its simplicity and easily interpretable coefficients, can offer clear insights for clinicians. Random Forest, while slightly more complex, provides feature importance rankings that can help in identifying the key clinical factors. SVM, although highly accurate, is less interpretable, which may limit its standalone use in clinical settings.

7. Conclusions and Future Work

In this study, we experimented with several feature selection methods and classifiers to predict Alzheimer's disease (AD) based on the OASIS-2 [15] dataset, which consists of a longitudinal cohort of 150 people aged 60 to 96 years. After analyzing the OASIS-2 dataset, our results show that mRMR and the correlation coefficient have the same effect on our classifiers. Mutual Information showed slightly different results from the other two approaches.

According to these results, mRMR and Mutual Information are effective methods for feature selection in predicting Alzheimer's disease using the OASIS-2 dataset. In addition, we used three distinct classification methods: SVM, Logistic Regression, and Random

Forest. Each model was trained without using feature selection and with using the mRMR and Mutual Information feature selection methodologies.

Our results showed that Logistic Regression produced similar results to SVM despite their different methodologies and methods, which essentially confirms the reliability of both models in achieving high accuracy, while the Random Forest model showed slightly lower performance compared to the other models. We also applied two approaches to train the models: a hold-out approach with 70% of the data allocated to training and the remaining portion allocated to testing, and a k-fold cross-validation approach.

In k-fold validation, we used different configurations of k, specifically using k = 10, k = 5, and Leave-One-Out (LOO) strategies. We applied these methods with and without removing outliers. According to our results, the 70–30 split methodology showed sensitivity to outliers. Although it showed commendable results before removing the outliers, removing the outliers significantly improved the performance. In contrast, we observed little impact on the validation results after removing the outliers.

There are several key factors that influenced our work throughout this research. First, feature selection is a major factor that affects the efficiency of the model and the overall quality of the results. In the future, we will focus on applying more advanced feature selection techniques, such as redundant feature elimination (RFE). Another factor is that we relied on the OASIS-2 dataset alone to train the model. In future work, we will focus on training our model on a larger and more comprehensive dataset, Also, in our future work, we will focus on using other types of machine learning algorithms than the ones we used to find out which model performs best, which ultimately improves the accuracy of the final results.

This research aimed to develop models for the early detection of Alzheimer's disease, contributing to advancements in healthcare and education. By improving the accuracy and effectiveness of these models, we hope to ensure their accessibility for those individuals and families affected by Alzheimer's disease. Our findings could lead to significant advancements in diagnostic tools, ultimately improving human well-being and quality of life.

Ultimately, we hope that our findings will have significant implications for the development of more accurate diagnostic tools for the early detection of Alzheimer's disease.

Author Contributions: Conceptualization, H.A.; methodology, H.A., A.A., A.B. and L.A.; validation, H.A., A.A, A.B. and L.A.; formal analysis, A.A. and A.B.; investigation, H.A., A.A., A.B. and L.A.; data curation, H.A.; writing—original draft preparation, A.A., A.B. and L.A.; writing—review and editing, H.A., A.A., A.B. and L.A.; visualization, H.A., A.A. and L.A.; supervision, H.A.; project administration, H.A. All authors have read and agreed to the published version of the manuscript.

Funding: This Research is funded by Research Supporting Project Number (RSPD2024R904), King Saud University, Riyadh, Saudi Arabia.

Institutional Review Board Statement: Not applicable.

Informed Consent Statement: Not applicable.

Data Availability Statement: The original data presented in the study are openly available at Washington University School of Medicine in St. Louis at https://sites.wustl.edu/oasisbrains/home/oasis-2/ (accessed on 20 January 2024).

Acknowledgments: The authors really appreciated the kind support from the Research Supporting Project Number (RSPD2024R904), King Saud University, Riyadh, Saudi Arabia.

Conflicts of Interest: The authors declare no conflicts of interest.

References

1. "Alzheimer's Disease Fact Sheet". National Institute on Aging. Available online: https://www.nia.nih.gov/health/alzheimers-and-dementia/alzheimers-disease-fact-sheet (accessed on 5 May 2024).
2. "What Is Alzheimer's Disease?" Centers for Disease Control and Prevention. Available online: https://www.cdc.gov/aging/aginginfo/alzheimers.htm (accessed on 5 May 2024).

3. IBM. "What Is Machine Learning?" IBM. 2023. Available online: https://www.ibm.com/topics/machine-learning (accessed on 5 May 2024).
4. Solanki, A.; Kumar, S.; Nayyar, A. *Handbook of Research on Emerging Trends and Applications of Machine Learning*; IGI-GLOBAL: Hershey, PA, USA, 2019.
5. Chen, M.; Hao, Y.; Hwang, K.; Wang, L.; Wang, L. Disease Prediction by Machine Learning Over Big Data From Healthcare Communities. *IEEE Access* **2017**, *5*, 8869–8879. [CrossRef]
6. Raza, K. Chapter 8—Improving the prediction accuracy of heart disease with ensemble learning and majority voting rule. *Sci. Direct* **2019**, *1* , 179–196. [CrossRef]
7. Tabsharani, F. What Is Support Vector Machine (SVM)?—Definition from WhatIs.com. WhatIs.com, August 2023. Available online: https://www.techtarget.com/whatis/definition/support-vector-machine-SVM (accessed on 5 May 2024).
8. "Logistic Regression in Machine Learning." GeeksforGeeks. Available online: https://www.geeksforgeeks.org/understanding-logistic-regression/ (accessed on 5 May 2024).
9. "What is Random Forest?". IBM. Available online: https://www.ibm.com/topics/random-forest (accessed on 5 May 2024).
10. Mahendran, N.; Vincent, P.; Srinivasan, K.; Chang, C. Improving the Classification of Alzheimer's Disease Using Hybrid Gene Selection Pipeline and Deep Learning. *Front. Genet.* **2021**, *12*, 784814. [CrossRef]
11. Mazzanti, S. MRMR Explained Exactly How You Wished Someone Explained to You. *Medium*, September 2023. Available online: https://towardsdatascience.com/mrmr-explained-exactly-how-you-wished-someone-explained-to-you-9cf4ed27458b (accessed on 5 May 2024).
12. Fernando, J. The Correlation Coefficient: What It Is and What It Tells Investors. *Investopedia*, February. 2024. Available online: https://www.investopedia.com/terms/c/correlationcoefficient.asp (accessed on 5 May 2024).
13. Admin. Correlation Coefficient Formula for Pearson's, Linear, Sample and Population Correlation Coefficients. *BYJUS*, September. 2020. Available online: https://byjus.com/correlation-coefficient-formula/ (accessed on 5 May 2024).
14. Guhanesvar. Mutual Information Based Feature Selection Based for ML. *Medium*, January 2022. Available online: https://guhanesvar.medium.com/feature-selection-based-on-mutual-information-gain-for-classification-and-regression-d0f86ea5262a (accessed on 5 May 2024).
15. OASIS Brains. OASIS Brains—Open Access Series of Imaging Studies. Available online: https://www.oasis-brains.org/ (accessed on 5 May 2024).
16. Saratxaga, C.L.; Albizuri, A.; Beristain, A.; Remesal, H.; Blesa, J.; Álvarez, I.; Moratal, D. MRI Deep Learning-Based Solution for Alzheimer's Disease Prediction. *J. Pers. Med.* **2021**, *11*, 902. [CrossRef] [PubMed]
17. Battineni, G.; Chintalapudi, N.; Amenta, F. Machine learning in medicine: Performance calculation of dementia prediction by support vector machines (SVM). *Inform. Med. Unlocked* **2019**, *16*, 100200. [CrossRef]
18. Kavitha, C.; Mani, V.; Srividhya, S.R.; Khalaf, O.I.; Romero, C.A.T. Early-stage Alzheimer's disease prediction using machine learning models. *Front. Public Health* **2022**, *10*, 3294. [CrossRef] [PubMed]
19. Baglat, P.; Salehi, A.W.; Gupta, A.; Gupta, G. Multiple machine learning models for detection of Alzheimer's disease using oasis dataset. In *Re-imagining Diffusion and Adoption of Information Technology and Systems: A Continuing Conversation*; Galliers, R.D., Newell, S., Eds.; Springer: Cham, Switzerland, 2020; pp. 614–622._54. [CrossRef]
20. Basheer, S.; Bhatia, S.; Sakri, S.B. Computational Modeling of Dementia Prediction Using Deep Neural Network: Analysis on OASIS Dataset. *IEEE Access* **2021**, *9*, 42449–42462. [CrossRef]
21. Dhakal, S.; Adhikari, S.; Nepal, R.; Rai, P.; Shrestha, S.; Pokharel, S.; Acharya, S. Dementia prediction using machine learning. *Procedia Comput. Sci.* **2023**, *219*, 1297–1308. [CrossRef]
22. Marcus, D.S.; Fotenos, A.F.; Csernansky, J.G.; Morris, J.C.; Buckner, R.L. Open access series of imaging studies: Longitudinal MRI data in nondemented and demented older adults. *J. Cogn. Neurosci.* **2010**, *22*, 2677–2684. [CrossRef] [PubMed]
23. Health Direct. Mini Mental State Examination (MMSE). Available online: https://www.healthdirect.gov.au/mini-mental-state-examination-mmse (accessed on 5 May 2024).

diagnostics

MDPI

Article

Machine Learning-Based Prediction of Clinical Outcomes in Microsurgical Clipping Treatments of Cerebral Aneurysms

Corneliu Toader [1,2], Felix-Mircea Brehar [1,3,*], Mugurel Petrinel Radoi [1,2,*], Razvan-Adrian Covache-Busuioc [1], Luca-Andrei Glavan [1], Matei Grama [4], Antonio-Daniel Corlatescu [1], Horia Petre Costin [1], Bogdan-Gabriel Bratu [1], Andrei Adrian Popa [1], Matei Serban [1] and Alexandru Vladimir Ciurea [1,5]

1 Department of Neurosurgery "Carol Davila", University of Medicine and Pharmacy, 030167 Bucharest, Romania; corneliu.toader@umfcd.ro (C.T.); razvan-adrian.covache-busuioc0720@stud.umfcd.ro (R.-A.C.-B.); luca-andrei.glavan0720@stud.umfcd.ro (L.-A.G.); antonio.corlatescu0920@stud.umfcd.ro (A.-D.C.); horia-petre.costin0720@stud.umfcd.ro (H.P.C.); bogdan.bratu@stud.umfcd.ro (B.-G.B.); andreiadrianpopa@stud.umfcd.ro (A.A.P.); matei.serban2021@stud.umfcd.ro (M.S.); prof.avciurea@gmail.com (A.V.C.)

2 Department of Vascular Neurosurgery, National Institute of Neurology and Neurovascular Diseases, 077160 Bucharest, Romania

3 Department of Neurosurgery, Clinical Emergency Hospital "Bagdasar-Arseni", 041915 Bucharest, Romania

4 Department of Research and Development, Syndical.io, street Icoanei 29A, 020452 Bucharest, Romania; ma-tei.grama@syndical.io

5 Neurosurgery Department, Sanador Clinical Hospital, 010991 Bucharest, Romania

* Correspondence: felixbrehar@yahoo.com (F.-M.B.); petrinel.radoi@umfcd.ro (M.P.R.)

Abstract: Background: This study investigates the application of Machine Learning techniques to predict clinical outcomes in microsurgical clipping treatments of cerebral aneurysms, aiming to enhance healthcare processes through informed clinical decision making. **Methods:** Relying on a dataset of 344 patients' preoperative characteristics, various ML classifiers were trained to predict outcomes measured by the Glasgow Outcome Scale (GOS). The study's results were reported through the means of ROC-AUC scores for outcome prediction and the identification of key predictors using SHAP analysis. **Results:** The trained models achieved ROC-AUC scores of 0.72 ± 0.03 for specific GOS outcome prediction and 0.78 ± 0.02 for binary classification of outcomes. The SHAP explanation analysis identified intubation as the most impactful factor influencing treatment outcomes' predictions for the trained models. **Conclusions**: The study demonstrates the potential of ML for predicting surgical outcomes of ruptured cerebral aneurysm treatments. It acknowledged the need for high-quality datasets and external validation to enhance model accuracy and generalizability.

Keywords: machine learning; ruptured intracranial aneurysm; treatment outcome; microsurgical clipping

1. Introduction

Medical machine learning (ML) has been subject to an exponential growth in the past decade, delivering the promise of improved healthcare delivery and clinical decision making. Traditional approaches to medical data analysis often fall short when trying to capture the subtle relationships between patient characteristics, disease pathophysiology, and treatment outcomes. In contrast, ML techniques offer a powerful framework for extracting meaningful insights from complex medical datasets, paving the way for personalized medicine and precision healthcare. Recent years have seen an alignment among scholars towards a common sense that, along with other technical and scientific advancements in the field, machine learning is going to reshape the future of neuroscience [1]. Intracranial aneurysms present a significant challenge in neurosurgery, particularly when ruptured, as they can lead to life-threatening complications such as subarachnoid hemorrhage (SAH). Understanding their formation, progression, and rupture risk is crucial for effective management and treatment. Surgical clipping remains one of the primary intervention alternatives for both ruptured and unruptured aneurysms, yet its success hinges on a multitude of patient-specific and aneurysm-related factors. In this study, our primary objective is to predict outcomes of surgical clipping interventions for prevalently ruptured aneurysms using machine learning techniques. The topic of intracranial aneurysms has been subject to multiple types of developments, with use-cases ranging from the aneurysm risk forecast [2] and prediction of periprocedural complications related to endovascular treatments [3], to outcome prediction for flow diverter treatments [4]. The general end goal of these ML developments is to improve the treatment through the means of informed decision making. These models can be integrated into clinical workflows to assist neurosurgeons in making data-driven decisions about treatment strategies for intracranial aneurysms. Given the serious implications ruptures have for the outcome of aneurysms' surgical intervention with regard to the morbidity and mortality rates, the study of these cases becomes particularly important. This is why multiple studies have expressly focused on ruptured intracranial aneurysms and the predictive models that can be built concerning these cases, e.g., for rupture risk prediction [5,6] or the predictive factors of rerupture [7]. Our goal is to contribute to the ongoing efforts of the medical AI community by validating the transferability and applicability of existing approaches [8] within our specific scope, namely outcome predictions for surgical clipping interventions of both ruptured and unruptured intracranial aneurysms given the set of preoperative characteristics. Through this research, we aim to not only develop the predictive models but also enhance our understanding of the complex interplay between patient factors and surgical outcomes within this specific scope.

2. Background

In the context of outcome prediction for intracranial aneurysm treatment, the synthesis composed by Velagapudi et al. [8] categorizes the literature on this subject based on the targeted aneurysm type as unruptured, aneurysmatic subarachnoid hemorrhage (aSAH) and combined. The recent years have seen a growing interest in leveraging ML techniques for outcome prediction in the context of unruptured intracranial aneurysms (UIAs). Traditionally, the management of UIAs has relied on various clinical factors to guide treatment decisions. However, the underlying relations between these factors are of great complexity and are thus hindering the chances of accurately predicting patients' individual outcomes. In this context, several studies have explored the feasibility of using ML models for predicting the outcomes following UIA microsurgical treatments. Stroh et al. [9] developed, for instance, an ML-based predictive model for preoperative assessment of UIA surgery outcomes, providing valuable insights into patient-specific risk factors and the potential applicability of these technologies. Similarly, Staartjes et al. [10] conducted a pilot study focusing on the development of ML models for outcome prediction in UIA surgery. Ishankulov et al. [11] also reported a good performance in predicting the outcome in the context of UIA surgical interventions. Taking note of the studies targeting ruptured and aSAH cases, Vin Shen Ban et al. [12] presented the Southwestern Aneurysm Severity Index (SASI), accurately predicting one-year functional outcomes after microsurgical clipping of ruptured aneurysms; Xia et al. [13] demonstrated that a Random Forest model can effectively predict the clinical outcome at discharge for patients with ruptured Anterior Communicating Artery aneurysms; and Zador et al. [14] used a logistic regression model with Bayesian network analysis to identify age and the World Federation of Neurosurgical Societies grade as key predictors of outcomes in acute aSAH. While these studies delivered promising results in improving outcome prediction for this pathology, there remains a need for further investigation and validation of these models across diverse patient populations and treatment settings.

3. Methodology

3.1. Dataset

The dataset was collected from 344 patients that underwent surgical interventions using the microsurgical clipping method for at least one intracranial aneurysm, predominantly saccular or berry-type intracranial aneurysms. It is composed of preoperative characteristics with an established significance in influencing the operatory outcome, such as arterial hypertension and atherosclerosis. Along with the 23 variables representing the input for the ML models, the dataset features the operatory outcome, recorded using the Glasgow Outcome Scale (GOS) for each of the individual surgical interventions. While three of the input variables (age, aneurysm diameter, and aneurysm neck) are numerical, the rest of the variables along with the GOS are categorical. The value distributions for each of those variables is presented in Table 1. A few categorical input features have missing values for an insignificant number of data points and, consequently, have been attributed a separate class to account for the missing information. While some of the input parameters represent the preexisting conditions of each individual, e.g., obesity, diabetes, others are aneurysm- and operatory-specific parameters such as the aneurysm neck and clip shape. In terms of demographics, the dataset covers a wide age range, with a higher prevalence in individuals over 30 years old and notable gender discrepancy, with more cases observed in women.

Table 1. Characteristics of the variables recorded for the study cohort (no. of cases n = 344).

Variable	Description
Age (years), mean $\pm$ σ	55.05 $\pm$ 11.45
Female gender	219 (63.7%)
Aneurysm Diameter (mm), mean $\pm$ σ	7.36 $\pm$ 3.51
Aneurysm Neck, mean $\pm$ σ	3.93 $\pm$ 1.64
Clip Shape, *n* (%)	
Straight	201 (58.4%)
Curved	102 (29.7%)
Angular	21 (6.1%)
Wrap	17 (4.9%)
Arterial Hypertension, *n* (%)	
Grade II	145 (42.2%)
Grade I	67 (19.5%)
Grade III	8 (2.3%)
Atherosclerosis	108 (31.4%)
Diabetes Mellitus (Type 2)	17 (4.9%)
Obesity, *n* (%)	
Not present	256 (74.4%)
Grade 2	53 (15.4%)
Morbid	30 (8.7%)
Grade 1	2 (0.6%)
Preoperative Aphasia	9 (2.6%)
Cognitive Disorder	118 (34.3%)
Intubation	76 (22.1%)
Hemorrhage	311 (90.4%)
Hydrocephalus	124 (36.0%)
Ruptured	308 (89.5%)

Table 1. *Cont.*

Variable	Description
Aneurysm Artery, *n* (%)	
A-Comm	152 (44.2%)
MCA	93 (27.0%)
PCA	35 (10.2%)
Pericallosal	13 (3.8%)
P-Comm	11 (3.2%)
Ophthalmic	11 (3.2%)
Choroidal	7 (2.0%)
ICA	6 (1.7%)
Fetal P-Comm	5 (1.4%)
Basilar	3 (0.9%)
Other (*)	8 (2.3%)
Craniotomy, *n* (%)	
Pterional	178 (51.7%)
Frontobasal	137 (39.8%)
Frontobasal Paramedian	16 (4.7%)
Other	11 (3.2%)
Suboccipital	2 (0.6%)
Vasospasm	101 (29.4%)
Glasgow Coma Scale (GCS), *n* (%)	
GCS 14	219 (63.7%)
GCS 13	57 (16.6%)
GCS 7	33 (9.6%)
GCS 15	29 (8.4%)
GCS 9	18 (5.2%)
GCS 8	16 (4.7%)
GCS 12	15 (4.4%)
Other	59 (17.1%)
Preoperative Hemiparesis, *n* (%)	
5	233 (67.4%)
2	63 (18.3%)
3	26 (7.6%)
1	10 (2.9%)
4	10 (2.9%)
Other	3 (0.9%)
Fisher Scale, *n* (%)	
Grade 3	142 (41.0%)
Grade 4	98 (28.5%)
Grade 2	72 (20.9%)
Grade 1	31 (9.0%)
Grade 0	2 (0.6%)

Table 1. *Cont.*

Variable	Description
Glasgow Outcome Scale (GOS), *n* (%)	
GOS 5	141 (41.0%)
GOS 1	77 (22.4%)
GOS 4	71 (20.6%)
GOS 3	50 (14.5)
GOS 2	5 (1.5%)
σ—standard deviation;	
A-Comm—Anterior Communicating Artery;	
MCA—Middle Cerebral Artery;	
PCA—Posterior Cerebral Artery;	
P-Comm—Posterior Communicating Artery;	
ICA—Internal Carotid Artery;	
Fetal P-Comm—Fetal Posterior Communicating Artery;	
(*) including Carotid Cave, Hypophyseal Artery, Distal Anterior Cerebral Artery, Internal Carotid Artery bifurcation, Posterior Inferior Cerebellar Artery, Vertebral Artery, Anterior Cerebral Artery.	

3.2. *Model Training and Validation*

The proposed ML task consists of predicting the outcome of the surgical intervention based on the preoperative parameters described earlier. This corresponds to predicting the five possible categories of GOS. The task can also be reduced to a binary classification as follows by setting GOS = 5 to correspond to a positive outcome and, conversely, GOS < 5 to a negative outcome. We conducted the experiments for both formulations and report the results in the section below.

Our experiments are constructed upon an extensive evaluation of machine learning classifiers, evaluating different state-of-the-art approaches and models, including Extra Trees (ETs), Random Forest (RF), Support Vector Machines (SVMs), K-Nearest Neighbors (KNNs), Artificial Neural Networks (ANNs), Logistic Regression (LR), Extreme Gradient Boosting (XGB), and Linear Discriminant Analysis (LDA). For each model, specific hyperparameter tuning was conducted for maximizing the outcome of the training, e.g., in the case of ETs, hyperparameters such as the number of estimators in the ensemble and the maximum depth of each tree were optimized through a grid search to enhance the model's predictive performance (Table 2). Key libraries used for data analysis and model development included pandas for data manipulation, numpy for numerical computations, matplotlib and seaborn for data visualization, and scikit-learn for machine learning model implementation.

Techniques such as bootstrap resampling, involving iteratively sampling the dataset with replacement to estimate classifier performance variance accurately, and a stratified sampling for the train–test split are employed to ensure fairness in model evaluation. The stratified test–train split was used to address potential class imbalance issues and ensure representative sampling in both subsets, the training partition consisting of 276 data points and the test partition with 68 data points.

Table 2. Hyperparameter tuning settings for each classifier.

Classifier	Hyperparameters
ANN	hidden_layer_sizes: [(50, 50), (100, 100)], alpha: [0.0001, 0.001, 0.01]
SVM	C: [0.1, 1, 10], kernel: ['linear', 'rbf'], gamma: ['scale', 'auto']
XGB	learning_rate: [0.01, 0.1, 0.2], n_estimators: [100, 200, 300]
RF	n_estimators: [50, 100, 200], max_depth: [None, 10, 20], min_samples_split: [2, 5, 10]
KNN	n_neighbors: [3, 5, 7], weights: ['uniform', 'distance']
LR	C: [0.1, 1, 10], penalty: ['l2', 'none']LDA solver: ['svd', 'lsqr', 'eigen']

3.3. SHAP Analysis

In recent years, SHapley Additive exPlanations (SHAP) has become a significant actor in ML's development due to its ability to provide consistent explanations that help in effective interpretations of the models [15]. Nowadays, it acts as guide for case-specific explanations, model refinement, and feature engineering by essentially quantifying each feature's contribution to predictions. Through the means of this analysis, the input variables are ranked by impact, highlighting key predictors in the model output. In our study, we performed this analysis for the trained models using the KernelExplainer and the TreeExplainer [16] provided within the shap Python package v 0.44.1. The results for the best performing models are discussed in the following sections.

3.4. Ethical Standards Compliance

In line with the principles outlined in the Declaration of Helsinki, informed consent has been obtained for the healthcare data used within this study and the data handling was conducted in compliance with current GDPR regulations. All procedures performed in this study, in accordance with the ethical standards, were reviewed and approved by the Ethics Committee of the National Institute of Neurology and Neurovascular Diseases in Bucharest, Romania (IRB approval name: Ethical Review Board of National Institute of Neurology and Neurovascular Diseases; IRB number: 2/2024).

4. Results

Once the hyperparameter tuning was performed for the eight models included in this study, the achieved performance was assessed making use of bootstrap resampling. The results reported for the task of predicting the specific GOS outcome (Figure 1) indicate an acceptable performance in terms of ROC-AUC, with Extreme Gradient Boosting (XGB) leading with a score of 0.72 $\pm$ 0.03, also indicating relatively stable performance. Coming up next, Extra Trees (ETs) achieved a comparable ROC-AUC of 0.71 $\pm$ 0.03, followed closely by Random Forest (RF) and Support Vector Machines (SVMs) with scores of 0.70 $\pm$ 0.02 and 0.69 $\pm$ 0.02, respectively. Lower performances were observed for Artificial Neural Networks (ANNs) and Linear Discriminant Analysis (LDA), with ROC- AUC scores of 0.67 $\pm$ 0.02 and 0.67 $\pm$ 0.03. K-Nearest Neighbors (KNNs) and Logistic Regression (LR) had the lowest performances in this context, both scoring 0.65 $\pm$ 0.01 and 0.65 $\pm$ 0.03, respectively.

When reducing the task to the binary classification distinguishing between the positive outcome, corresponding to GOS = 5, and the negative outcomes, for GOS < 5, a different model ranking and, as Figure 1 shows in terms of the outcome prediction performance of the different models measured by means of ROC-AUC, a significant increase in the performances were established, with the highest ROC-AUC of 0.78 $\pm$ 0.02retained. being achieved this time by the ET classifier. Following closely behind, KNN achieved a notable ROC-AUC score of 0.75 $\pm$ 0.03. LR and RF also performed well, both attaining ROC-AUC scores of 0.74 $\pm$ 0.04 and 0.74 $\pm$ 0.03, respectively. LDA achieved a moderate performance with a score of 0.72 $\pm$ 0.04. ANN, SVM, and XGB all showed similar performances, each obtaining ROC-AUC scores of 0.70 $\pm$ 0.04, $\pm$ 0.04, and $\pm$ 0.03, respectively. The SHAP analysis we conducted for the best performing models (ET, ANN, XGB, and RF) pointed out the most influential factors affecting the outcome for each of the models (Figure 2). The most notable finding that emerged is the consistent significance of intubation in predicting the GOS value across all models. This feature shows up as the highest-ranking factor in terms of importance. Another interesting observation comes in regard to vasospasm, which consistently shows up among the three most influential parameters. The Hunt and Hess Scale has also consistently ranked high for three out of four of the best-performing models. The observations resulting from our experiments are further discussed in the following section.

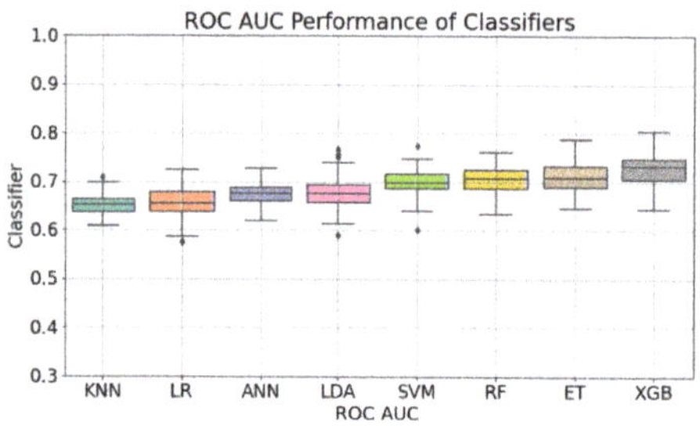

Figure 1. Outcome prediction performance of the different models measured by ROC-AUC.

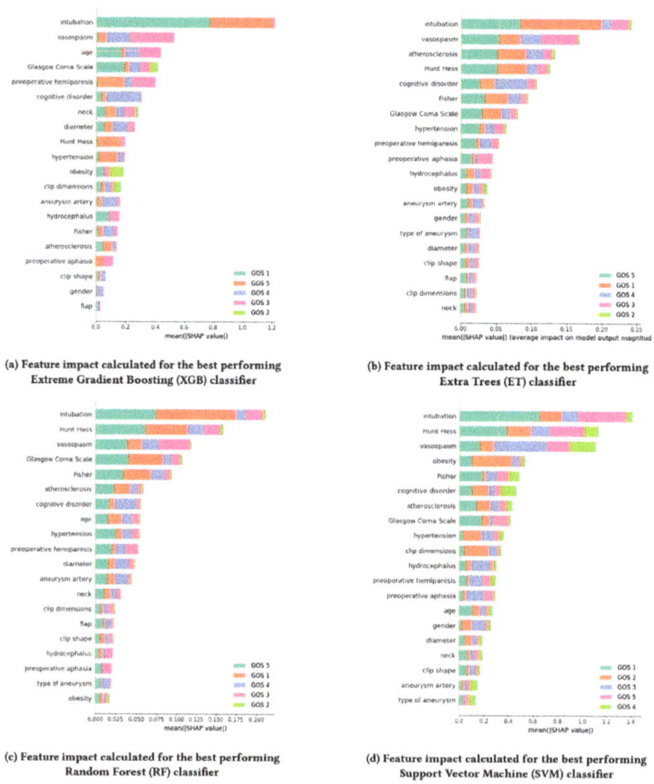

(a) Feature impact calculated for the best performing Extreme Gradient Boosting (XGB) classifier

(b) Feature impact calculated for the best performing Extra Trees (ET) classifier

(c) Feature impact calculated for the best performing Random Forest (RF) classifier

(d) Feature impact calculated for the best performing Support Vector Machine (SVM) classifier

Figure 2. SHAP analysis for the four best performing models (XGB, ET, RF, SVM) consisting of the feature ranking based on the average impact on model output magnitude, with higher values corresponding to a higher impact on GOS prediction.

5. Discussion

In our study, the application of machine learning (ML) models to predict clinical outcomes of microsurgical clipping for cerebral aneurysms shows promising results, achieving good and acceptable performance levels. These findings align with previous research, highlighting the potential of ML to enhance clinical decision making and patient care in neurosurgery. In a similar study conducted by Stroh et al. within the scope of unruptured cerebral aneurysms and the use of ML models for predicting clinical outcomes for microsurgical clipping [9], a good performance for the trained ML models has been reported, indicating a significant potential to enhance prognostic accuracy and customize surgical approaches for individual patients. In their case, for the task of predicting the positive post-operative outcome in terms of GOS, a slightly higher ROC-AUC value has been reported over the test set and lower ROC-AUC for the external validation, 0.79 ± 0.07 and 0.62 ± 0.02, respectively. The models showcased good performance for differentiating the outcomes based on preoperative indicators and clinical parameters. This sort of prognosis can play a part in enhancing nuanced risk stratification and personalized treatment planning, resonating with the ongoing shift towards precision medicine in neurosurgery. However, challenges in improving the generalizability and validity of these predictive models across diverse healthcare settings were noted, highlighted by variations in model performance during external validation. To address these issues, future efforts will need to focus on integrating larger, multi-institutional datasets and continually updating algorithms to reflect current clinical practices, ensuring that ML tools effectively support clinical decision making in the management of cerebral aneurysms. By conducting the training phase across multiple regional datasets, the models' performance can improve and thus increase their applicability in real-time prediction of ruptured aneurysms, enabling an enhanced tracking of the patients' evolution for this pathology. Correctly assessing the outcomes of unstable patients can lead to correct therapeutic conducts, which in turn will lead to better overall care and survival. Moreover, an accurate model can be a helpful tool for neurosurgeons in determining the urgency of surgical interventions and for intensive care specialists to formulate a better image of the outcome of each case.

One of the core strengths of this study lies in its attempt to model clinical outcomes in a real-world setting, where both ruptured and unruptured aneurysms have been routinely encountered. While the majority of the cases in the dataset are ruptured aneurysms, the inclusion of unruptured aneurysms provides a more accurate representation of the treated population in clinical practice subject to this study. In this context, both ruptured and unruptured aneurysms are treated, often within the same clinical environment. Therefore, by including both classes, the model performance is measured over the full spectrum of cases, enabling the findings to be more widely applicable to clinical decision making. Not including both ruptured and unruptured aneurysms would potentially limit its utility in this specific clinical context, and in fact, validating the model on a heterogeneous dataset that includes both types of cases is essential to its generalizability, particularly as ML techniques are applied to increasingly complex patient populations. This aspect is critical for the future development of a versatile and adaptable tool that performs well over the diverse range of neurosurgery cases.

Our study identified several key parameters as being highly influential in predicting the clinical outcomes of microsurgical clipping for cerebral aneurysms, through the means of modern techniques such as explainable machine learning. These relevant parameters, determined through SHAP analysis, may provide valuable insights into the factors that most significantly impact the model prediction outcomes. One of the relevant factors was intubation, as it consistently showed up as the most impactful parameter across the best performing models. The requirement for intubation is often associated with more severe cases and can indicate a higher risk of complications or poorer baseline patient status. Another important parameter was the vasospasm, which is particularly critical as it is associated with delayed cerebral ischemia, a common and severe complication following aneurysmal subarachnoid hemorrhage. Its early prediction can influence treatment decisions and potentially improve outcomes. The Hunt and Hess Scale assesses the initial clinical condition of patients with subarachnoid hemorrhage, and was also highlighted as a key predictor. It reflects the severity of the hemorrhage and the patient's neurological status, which are crucial for outcome prediction. These findings underscore the complexity of cerebral aneurysm management and the value of ML in capturing and analyzing these nuances.

Through the means of such studies, we can explore intricate patterns that are often inaccessible through traditional statistical methods, particularly when dealing with high-dimensional datasets, as demonstrated by the SHAP analysis. For example, the identification of intubation as the most impactful feature was consistent across all models, suggesting a non-obvious link between the preoperative conditions in question and post-surgical outcomes. This level of interpretability allows ML models to highlight parameters previously underappreciated and help refine the understanding of which factors significantly influence patient outcomes in different settings, beyond what clinicians may intuitively assess.

Limitations and Future Directions

The present study encountered several limitations that should inform future directions of research regarding this subject. The dataset displayed imbalances across certain features, which could skew model performance. When low performance levels were observed, this may be an indication of inherent deficiencies within the data, suggesting a need for higher quality training sets to improve model efficacy. In any ML-based analysis, both the data and model selection play crucial roles in determining the outcome prediction. In this study, while models such as Extreme Gradient Boosting (XGB) and Extra Trees (ETs) performed well, the dataset itself introduces limitations that may have influenced these results. For example, the presence of only three basilar aneurysm cases and certain missing points could skew predictions. When there is an imbalance in such critical features, models tend to overfit to the more frequent data points, reducing their ability to generalize to rare events. Nevertheless, in healthcare, the imbalanced nature of data reflects the actual prevalence of pathological characteristics in the population. Manipulating this imbalance, e.g., oversampling rare cases or undersampling common ones, can lead to models that do not generalize well to real-world scenarios due to the sampling biases that might be introduced by the application of such techniques. Handling missing data points can also significantly affect model robustness. Making use of techniques such as imputing missing data by creating separate classes or applying methods like mean imputation that ensure no loss of data may also lead to the same adverse effect of oversimplifying underlying patterns. In many clinical settings, predictions that reflect the true incidence rates of diseases are more meaningful for decision making. Manipulating the data can reduce the utility of a model for clinicians by distorting the probability estimates.

As healthcare moves towards more nuanced outcome measures beyond single-dimensional scales like GOS, current ML models may face limitations. More sophisticated measures could involve a comprehensive view that would combine physical, cognitive, and psychological recovery indicators. Furthermore, the dataset's single origin raises concerns about potential biases and disparities, underscoring the importance of multi-institutional, multi-regional datasets to mitigate these biases and enhance the reliability. Moving forward, future research should prioritize standardizing methodologies, validating findings across varied cohorts, and exploring innovative approaches to enhance both the accuracy and generalizability. Exploring additional predictive features, such as genetic markers and advanced imaging parameters, could further enhance the accuracy and utility of these models. The integration of these predictive models into clinical practice could significantly enhance decision-making processes, potentially leading to better patient outcomes and more personalized treatment plans.

6. Conclusions

Our study shows that the ML models trained for predicting clinical outcomes of microsurgical clipping treatments for intracranial aneurysms achieved acceptable performance levels when trained using the collected patient data. These results come in alignment with similar studies' conclusions reported in the literature, indicating that the use of ML techniques under this scope presents a high potential for enhancing clinical decision making. However, before these models can be deployed in real-world systems with the required levels of generalizability and reliability, further research focusing on diversified patient populations is needed for validation and further refinements. Within the domain of neurosurgery, machine learning holds great potential of enhancing the management of intracranial aneurysms. These vascular anomalies present a formidable challenge to neurosurgeons due to their unpredictable nature and potentially devastating consequences. The development of decision support systems incorporating these predictive models could provide real-time risk assessments and personalized treatment recommendations. Overall, the results contribute to the growing body of evidence supporting the integration of ML techniques into healthcare for better patient management and treatment outcomes.

Author Contributions: Conceptualization, C.T.; methodology, M.G.; software, M.G. and M.S.; formal analysis, M.P.R., H.P.C. and M.S.; investigation, C.T., M.P.R. and R.-A.C.-B.; resources, M.P.R., A.-D.C., B.-G.B. and A.A.P.; data curation, R.-A.C.-B., L.-A.G., B.-G.B. and A.A.P.; writing—original draft, L.-A.G., M.G. and A.-D.C.; writing—review and editing, F.-M.B. and H.P.C.; visualization, F.-M.B. and A.V.C.; supervision, C.T., F.-M.B. and A.V.C.; project administration, A.V.C. All authors have read and agreed to the published version of the manuscript.

Funding: The research received no external funding.

Institutional Review Board Statement: IRB approval name: Ethical Review Board of National Institute of Neurology and Neurovascular Diseases; IRB number: 2/2024. date: 16 January 2024.

Informed Consent Statement: Informed consent was obtained from all subjects involved in the study.

Data Availability Statement: Data requests should be performed according to GDPR laws.

Conflicts of Interest: The authors declare no conflicts of interest.

References

1. Awuah, W.A.; Adebusoye, F.T.; Wellington, J.; David, L.; Salam, A.; Weng Yee, A.L.; Lansiaux, E.; Yarlagadda, R.; Garg, T.; Abdul-Rahman, T.; et al. Recent Outcomes and Challenges of Artificial Intelligence, Machine Learning, and Deep Learning in Neurosurgery. *World Neurosurg.* **2024**, *23*, 100301. [CrossRef] [PubMed]
2. Heo, J.; Park, S.J.; Kang, S.-H.; Oh, C.W.; Bang, J.S.; Kim, T. Prediction of intracranial aneurysm risk using machine learning. *Sci. Rep.* **2020**, *10*, 6921. [CrossRef] [PubMed]
3. Tian, Z.; Li, W.; Feng, X.; Sun, K.; Duan, C. Prediction and analysis of periprocedural complications associated with endovascular treatment for unruptured intracranial aneurysms using machine learning. *Front. Neurol.* **2022**, *13*, 1027557. [CrossRef] [PubMed]
4. Paliwal, N.; Jaiswal, P.; Tutino, V.M.; Shallwani, H.; Davies, J.M.; Siddiqui, A.H.; Rai, R.; Meng, H. Outcome prediction of intracranial aneurysm treatment by flow diverters using machine learning. *Neurosurg. Focus* **2018**, *45*, E7. [CrossRef] [PubMed]
5. Sanchez, S.; Miller, J.M.; Samaniego, E.A. Clinical Scales in Aneurysm Rupture Prediction. Stroke. *Vasc. Interv. Neurol.* **2024**, *4*, e000625. [CrossRef]
6. Zhou, J.; Xia, N.; Li, Q.; Zheng, K.; Jia, X.; Wang, H.; Zhao, B.; Liu, J.; Yang, Y.; Chen, Y. Predicting the rupture status of small middle cerebral artery aneurysms using random forest modeling. *Front. Neurol.* **2022**, *13*, 921404. [CrossRef] [PubMed]
7. Chang, Y.-J.; Liu, C.-K.; Cheng, C.-Y.; Shih, Y.-C.; Wang, S.-C.; Lin, C.-C.; Lin, C.-M. Predictive Factors of Cerebral Aneurysm Rerupture after Clipping. *Front. Neurol.* **2022**, *12*, 296. [CrossRef] [PubMed]
8. Velagapudi, L.; Al Saiegh, F.; Swaminathan, S.; Mouchtouris, N.; Khanna, O.; Sabourin, V.; Gooch, M.R.; Herial, N.; Tjoumakaris, S.; Rosenwasser, R.H.; et al. Machine learning for outcome prediction of neurosurgical aneurysm treatment: Current methods and future directions. *Clin. Neurol. Neurosurg.* **2023**, *224*, 107547. [CrossRef] [PubMed]
9. Stroh, N.; Stefanits, H.; Maletzky, A.; Kaltenleithner, S.; Thumfart, S.; Giretzlehner, M.; Drexler, R.; Ricklefs, F.L.; Dührsen, L.; Aspalter, S.; et al. Machine learning based outcome prediction of microsurgically treated unruptured intracranial aneurysms. *Sci. Rep.* **2023**, *13*, 22641. [CrossRef]
10. Staartjes, V.E.; Sebök, M.; Blum, P.G.; Serra, C.; Germans, M.R.; Krayenbühl, N.; Regli, L.; Esposito, G. Development of machine learning-based preoperative predictive analytics for unruptured intracranial aneurysm surgery: A pilot study. *Acta Neurochir.* **2020**, *162*, 2759–2765. [CrossRef] [PubMed]

11. Ishankulov, T.; Grebenev, F.; Strunina, U.; Shekhtman, O.; Eliava, S.; Danilov, G. The Prediction of Functional Outcome After Microsurgical Treatment of Unruptured Intracranial Aneurysm Based on Machine Learning. *Stud. Health Technol. Inform.* **2022**, *294*, 470–474. [CrossRef] [PubMed]
12. Ban, V.S.; El Ahmadieh, T.Y.; Aoun, S.G.; Plitt, A.R.; Lyon, K.A.; Eddleman, C.; Beecher, J.; McDougall, C.M.; Reisch, J.; Welch, B.G.; et al. Prediction of Outcomes for Ruptured Aneurysm Surgery. *Stroke* **2019**, *50*, 595–601. [CrossRef] [PubMed]
13. Xia, N.; Chen, J.; Zhan, C.; Jia, X.; Xiang, Y.; Chen, Y.; Duan, Y.; Lan, L.; Lin, B.; Chen, C.; et al. Prediction of Clinical Outcome at Discharge After Rupture of Anterior Communicating Artery Aneurysm Using the Random Forest Technique. *Front. Neurol.* **2020**, *11*, 538052. [CrossRef] [PubMed]
14. Zador, Z.; Huang, W.; Sperrin, M.; Lawton, M.T. Multivariable and Bayesian network analysis of outcome predictors in acute aneurysmal subarachnoid hemorrhage: Review of a pure surgical series in the post-International Subarachnoid Aneurysm Trial era. *Oper. Neurosurg.* **2018**, *14*, 603–610. [CrossRef] [PubMed]
15. Lundberg, S.M.; Lee, S.-I. A Unified Approach to Interpreting Model Predictions. In *Advances in Neural Information Processing Systems 30*; Guyon, I., Luxburg, U.V., Bengio, S., Wallach, H., Fergus, R., Vishwanathan, S., Garnett, R., Eds.; Curran Associates, Inc.: Red Hook, NY, USA, 2017; pp. 4765–4774. Available online: http://papers.nips.cc/paper/7062-aunified-approach-to-interpreting-model-predictions.pdf (accessed on 3 July 2024).
16. Lundberg, S.M.; Erion, G.; Chen, H.; DeGrave, A.; Prutkin, J.M.; Nair, B.; Katz, R.; Himmelfarb, J.; Bansal, N.; Lee, S.-I. From local explanations to global understanding with explainable AI for trees. *Nat. Mach. Intell.* **2020**, *2*, 56–67. [CrossRef] [PubMed]

diagnostics

Article

Validating the Accuracy of a Patient-Facing Clinical Decision Support System in Predicting Lumbar Disc Herniation: Diagnostic Accuracy Study

Fatima Badahman [1], Mashael Alsobhi [1], Almaha Alzahrani [2], Mohamed Faisal Chevidikunnan [1], Ziyad Neamatallah [1], Abdullah Alqarni [1], Umar Alabasi [1], Ahmed Abduljabbar [3], Reem Basuodan [4] and Fayaz Khan [1,*]

[1] Department of Physical Therapy, Faculty of Medical Rehabilitation Sciences, King Abdulaziz University, Jeddah 22252, Saudi Arabia; fatimahassanb@gmail.com (F.B.); mgalsobhi@kau.edu.sa (M.A.); mfaisal@kau.edu.sa (M.F.C.); zneamatallah@kau.edu.sa (Z.N.); amalgarni@kau.edu.sa (A.A.); ualabasi@kau.edu.sa (U.A.)
[2] Department of Physical Therapy, King Faisal Hospital, Makkah 24236, Saudi Arabia; aahmedalzahrani0035@stu.kau.edu.sa
[3] Department of Radiology, Faculty of Medicine, King Abdulaziz University, Jeddah 22252, Saudi Arabia; aabduljabar@kau.edu.sa
[4] Department of Rehabilitation Sciences, College of Health and Rehabilitation Sciences, Princess Nourah Bint Abdulrahman University, Riyadh 11671, Saudi Arabia; rmbasoudan@pnu.edu.sa
* Correspondence: rkhan2@kau.edu.sa

Abstract: Background: Low back pain (LBP) is a major cause of disability globally, and the diagnosis of LBP is challenging for clinicians. Objective: Using new software called Therapha, this study aimed to assess the accuracy level of artificial intelligence as a Clinical Decision Support System (CDSS) compared to MRI in predicting lumbar disc herniated patients. Methods: One hundred low back pain patients aged ≥18 years old were included in the study. The study was conducted in three stages. Firstly, a case series was conducted by matching MRI and Therapha diagnosis for 10 patients. Subsequently, Delphi methodology was employed to establish a clinical consensus. Finally, to determine the accuracy of the newly developed software, a cross-sectional study was undertaken involving 100 patients. Results: The software showed a significant diagnostic accuracy with the area under the curve in the ROC analysis determined as 0.84 with a sensitivity of 88% and a specificity of 80%. Conclusions: The study's findings revealed that CDSS using Therapha has a reasonable level of efficacy, and this can be utilized clinically to acquire a faster and more accurate screening of patients with lumbar disc herniation.

Keywords: artificial intelligence; machine learning; back pain; clinical decision support system; lumbar; disc herniation

1. Introduction

Low back pain (LBP) is a highly prevalent condition globally [1], with a prevalence rate of approximately 53.5% in Saudi Arabia [2]. LBP and its related disabilities are influenced by a combination of psychological, social, and biological factors, making it a complex disorder [3]. LBP can appear in different clinical and pathognomonic manifestations such as lumbar disc herniation and radiculopathy which are the two primary factors responsible for causing LBP [4]. However, lumbar disc herniation can be the cause of different types of pain which may lead to the overlap of dermatome and myotome, resulting in misdiagnosis of nerve root involvement [5]. Therefore, numerous clinical examinations have been conducted to determine the most accurate diagnostic test for detecting the likelihood and presence of radiculopathy caused by lumbar disc herniation. These examinations include straight leg raising, crossed straight leg raising, tendon reflexes, atrophy, and sensory deficits tests [6].

In addition, radiographic examination has traditionally been employed in primary healthcare settings and is broadly undertaken as a valid reference standard for identifying lumbar herniated discs [7]. Magnetic resonance imaging (MRI) and computed tomography (CT) have a considerable degree of accuracy in terms of specificity and sensitivity. The specificity of CT in detecting lumbar herniated discs with nerve root involvement ranges from 0.7 to 0.87, while its sensitivity ranges from 0.62 to 0.9. On the other hand, MRI has a specificity ranging from 0.43 to 0.97 and a sensitivity ranging from 0.6 to 1.0 in the same diagnosis [7]. However, the diagnosis of LBP is challenging for clinicians and requires a high level of skill, experience, and understanding of the complex physiological and anatomical structure of the human body [8].

A study done by Vining R et al. reported that only around 15% of surveyed LBP patients had accurate diagnoses with some degree of certainty [9]. Clinicians often gain diagnostic experience through an arduous, time-consuming process [8]. For example, understanding the human body's complicated physiological and anatomical structure [8], a large patient load, and human limitations such as not being up-to-date on manual errors and the latest advancements [10]. Thus, the quality of medical care is frequently compromised, particularly in developing countries [11]. Therefore, this existing dilemma might be resolved by using CDSS to support the diagnosis of LBP and improve timely knowledge access [8].

Artificial intelligence (AI) will assist healthcare practitioners by providing reliable and expedient medical consultation services to improve the quality of treatment. It has the ability to supplement and enhance decision-making in various contexts of high importance [12–14]. CDSS can process patient information intelligently and present it to clinicians [15] or physiotherapists to support decision-making on the administration and assessment of various medical conditions such as assessing gait deviation in anterior cruciate ligament (ACL) injury [16], diagnosing and classifying peripheral neuropathy [17], predicting elderly quality of life [18], detecting coronary heart disease by using fuzzy rule-based system with multi-objective genetic algorithm [19] and helping clinicians in cancer treatment [20].

In the current study, researchers conducted a diagnostic accuracy study to assess the accuracy of an AI-enabled platform and an algorithm such as CDSS. The study examined a new web-based software called Therapha, which incorporates CDSS technology. This software aims to assist clinicians in diagnosing LBP by providing them with real-time support and guidance. Therapha is a triaging software that guides the clinician through a comprehensive patient history consisting of medical history, social history, family medical history, self-reported pain evaluation, functional deficits, psychological issues, sleep patterns, and posture. Upon concluding the patient history, it generates the most likely differential diagnosis related to the LBP. This emerging technology has the potential to serve as a valuable instrument for medical healthcare professionals in the process of making clinical decisions [21,22]. Consequently, it has the potential to enhance both the accuracy of diagnoses and the promptness of accessing knowledge [8]. The objective of this study was to assess the diagnostic precision of AI as a CDSS in comparison to MRI for diagnosing patients with lumbar disc herniation.

2. Materials and Methods

2.1. Study Design

The study was designed in three stages: (i) a case series was conducted to generate hypotheses and examine the effectiveness of a novel patient-facing assessment; (ii) the Delphi method was used to validate the prediction accuracy of Therapha and evaluate the capability of the CDSS in comparison to experts' opinions; (iii) a diagnostic accuracy study was conducted to assess the prediction accuracy of Therapha web-based software in contrast to MRI in diagnosing lumbar radiculopathy.

2.2. Participants

The participants were male and female patients over the age of 18 complaining of LBP with radiating symptoms extending down to the leg. Patients who were suffering from any other LBP pathology, history of cancer, previous spine surgery and failed back syndrome were excluded. Before obtaining the patients' clinical and personal background, all participants provided their signature on a written informed consent document. The ethical approval was obtained from the research ethics committee at King Abdulaziz University Hospital (KAUH) (HA-02-J-008), Jeddah, Saudi Arabia.

2.3. Case Series

The current study began with a preliminary case series to determine the Therapha web software's accuracy in diagnosing patients with herniation of the lumbar disc. This study comprised 10 patients with LBP who met the inclusion criteria. All patients underwent an MRI, and the patients' history was collected by a blind assessor using Therapha web software on the same day before the MRI procedure. Therapha analyzed the patient data and generated the most likely differential diagnosis of the LBP. A study assistant with 7 years' experience of working in the field of radiology matched the software's final report to the MRI report and tabulated the data (Therapha and MRI final diagnosis) for each patient (Figure 1).

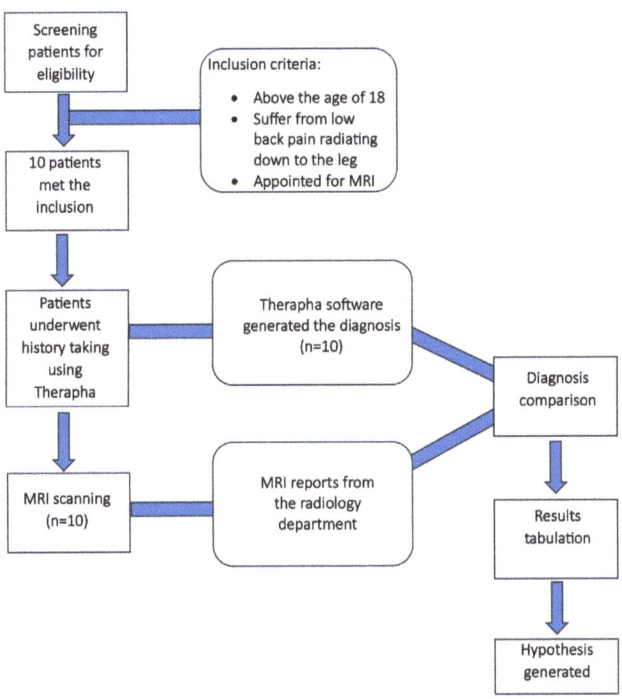

Figure 1. Design of case series to generate the study hypothesis.

2.4. Delphi Method

The Delphi method has been used in various medical and healthcare settings [23,24] to determine expert consensus in situations where there is insufficient evidence available [25]. The Delphi method is an effective process that uses multiple rounds of expert voting, in which a questionnaire is distributed to a group of experts who respond anonymously to the questions. The survey results are then arranged in a tabular form and reported back to the experts, and each panel is requested to complete the questionnaire once again [24] (Figure 2).

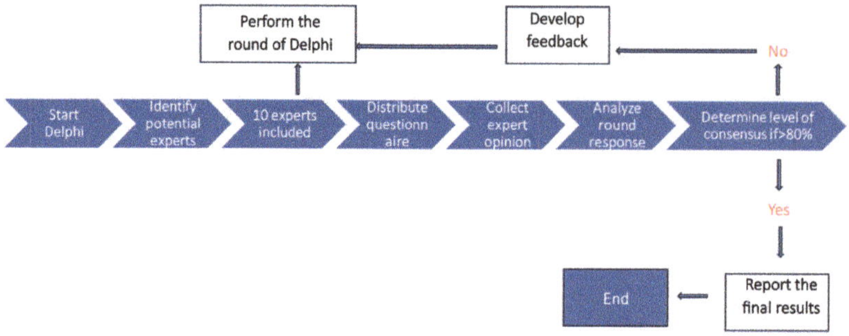

Figure 2. Delphi study steps for Therapha software validation on patients with low back pain.

2.4.1. Panel Selection

A total of ten experts have participated in this study to validate Therapha software. The panel was chosen based on their research and clinical experience in assessing and treating patients with LBP. Experts were from multiple countries including Saudi Arabia, India, the United Arab Emirates, the United Kingdom, and the United States.

2.4.2. Round 1

In the first round, all experts were emailed a set of statements, instructions for rating, and an explanation of the study aims. Experts were asked to read the patient clinical history summaries that were automated by the Therapha software from subjective reporting and were asked to give a final diagnosis for each patient. Then, a rating was made on the experts' final diagnosis (agree/disagree with the Therapha diagnosis) on a scale of 1 to 9, where a score of 1 to 3 score was classified as "inappropriate", 4 to 6 as "uncertain", and those in the range of 7 to 9 as "appropriate" [24]. Panelists were asked to justify their vote if they chose to vote for "disagreement".

Therapha prediction was consolidated with an 80% agreement among the experts (8 out of 10 panelists) to accept or remove a statement during the construction of the final diagnosis. Based on Lynn's et. al. findings, 80% was determined as an adequate cut-off [26]. Statements that did not receive 80% agreement were amended based on feedback from the expert panel and circulated to panelists in the second round.

2.5. Diagnostic Accuracy

2.5.1. Sample Size

The sample size was calculated by using G*power software (version 3.1.9.7, Dusseldorf, Germany), with an alpha of 0.05, a power of 0.8 and an effect size of 0.5. Based on the software calculation, the minimum required number was 90.

2.5.2. Assessment

Upon receiving the patient's consent, their personal and clinical history were taken using Therapha software on the same day before the patient underwent an MRI. Each software-assisted patient evaluation session lasted around 10–15 min. The history encompassed initial demographic details and questions about the patient's medical history, family history, social history, functional deficits, the severity of the back pain, nature of the pain, duration, onset, location, easing and aggravating factors, abnormal sensations, and the mechanism of injury and its relationship to the emergence of pain. Also, patients were asked about symptoms that may indicate systemic disease infection, fracture, or neurological impairments.

2.5.3. Procedure

Patients were referred to the Radiology department from orthopedic, neurology, and spine surgery clinics. The Therapha software is only available in English, hence requiring the use of a blind assessor. An experienced physical therapist, who was blinded to the patient's information, utilized Therapha software to gather the patient's personal and clinical history before the MRI. The final report received from the software was compared to the gold standard (MRI reports) by the study assistant and excluded any other pathology such as infection, fracture or tumor (Figure 3).

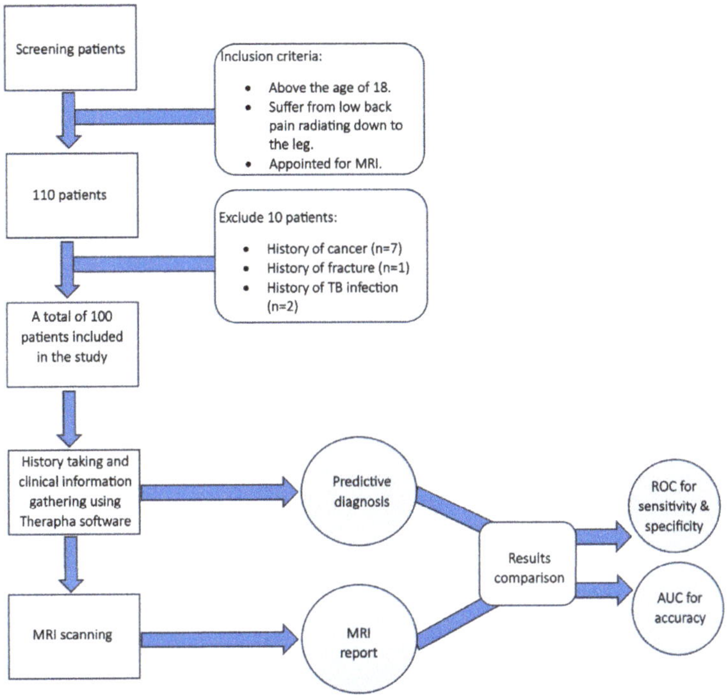

Figure 3. Flowchart demonstrating the diagnostic accuracy study process of the Therapha software.

2.6. Outcome Measures

2.6.1. Therapha Software

Therapha (v1.3.5) web software is a CDSS that is considered a comprehensive screening tool and triaging system for spine pathology which supports self-service by the patient or clinician. This system utilizes the power of information technology to identify medically established patterns by incorporating over 4000 independent variables, which is evidently superior to the clinical reasoning by a human clinician. This algorithmic approach enables the tool to predict the possible differential diagnoses in an order of probability percentile. The system divides spine disorders into subgroups besides conventional classification. The supervised AI model identifies patterns in the same manner as a clinician conducts an evaluation, and the Therapha system is capable of modifying their algorithm in the light of newer research findings. This improves the solution's predictive accuracy over the period through an iterative process. CDSS uses data from the clinician or patient self-administered questionnaire to refine clinical reasoning and predictions.

The information gathered from the patient generates a 'digital twin', which serves as the primary input to this clinical decision support solution. The other main component is the knowledge repository, which utilizes the data captured by the chatbot to generate the top three hypothetical diagnoses. The algorithm is the software that embodies AI, which is the system's most innovative component.

2.6.2. MRI

In this study, all patients were scanned with MRI, using 3 Tesla Verio and Skyra systems (Siemens, Erlangen, Germany). A slice thickness of 3 mm disc by disc was used with an interslice gap of 1–0.5 mm.

2.7. Statistical Analysis

Data were statistically analyzed by using SPSS for Windows (version 23.0; SPSS Inc, Chicago, IL, USA) to evaluate the predictive accuracy of Therapha web software with MRI reports. Data were analyzed for normality using the Shapiro–Wilk and Kolmogorov–Smirnov tests. Anthropometric data were analyzed using descriptive analysis. The accuracy of the software performance was measured using the Receiver Operating Curve (ROC) to find the Sensitivity and Specificity. The ROC curve was plotted by True Positive Rate (TPR)/False Positive Rate (FPR), where TPR or Sensitivity is expressed as TPR = TP/(TP + FN), where TP is the number of true positive instances, and FN is the number of false negative instances. The FPR or Specificity is calculated as FP/FP + TN, where FP is the number of false positives and TN is the number of true negatives. p-value < 0.05 was considered statistically significant.

3. Results

3.1. Case Series

Lumbar disc herniation was verified by MRI for all patients (n = 10) except for cases # (4 and 6) who had a different spine diagnosis by AI software. Patients' Therapha software inferences and MRI reports are illustrated in Table 1.

Table 1. MRI findings and Therapha software inferences.

Patient No.	Age	BMI	MRI Diagnosis	THERAPHA Prediction
1	55	38	Mild diffuse disc bulge	Lumbar radiculopathy/Lumbar disc herniation/Lumbago with Sciatica
2	67	26	At the level of T12-L1 there is mild central disc bulge	Lumbar facet joint pain/lumbar zygapophyseal/non-specific low back pain
3	59	22	Mild diffuse disc bulge	Lumbar radiculopathy/lumbar disc herniation/lumbago with sciatica
4	40	21	No significant disc disease	Lumbar radiculopathy/lumbar disc herniation/lumbago with sciatica
5	33	25	Diffuse posterior bulge	Lumbar radiculopathy/lumbar disc herniation/lumbago with sciatica
6	38	25	Extradural meningeal cyst	Lumbar radiculopathy/lumbar disc herniation/lumbago with sciatica
7	65	32	Minimal diffuse posterior disc bulges	Lumbar radiculopathy/lumbar disc herniation/lumbago with sciatica
8	61	34	Lumbar disc bulge L4-5, L5-S1	Lumbar radiculopathy/lumbar disc herniation/lumbago with sciatica
9	40	29	Lumbar disc bulge L4-5, L5-S1	Lumbar radiculopathy/lumbar disc herniation/lumbago with sciatica
10	69	25	Lumbar disc bulge L4-5, L5-S1	Lumbar radiculopathy/lumbar disc herniation/lumbago with sciatica

MRI findings for case# 4 showed "no significant disc disease"; however, Therapha identified it as "Lumbar Radiculopathy/Lumbar Disc Herniation/Lumbago with Sciatica". This can be explained by the nature of the pain. As the patient fed into the chatbot, they reported dull and radiating pain from the back spreading to the L3-L4 dermatomal regions, along with tingling and numbness covering the lateral aspect of the thigh and lower leg corresponding to the L4-S1 dermatomes. There was also tenderness in the midline corresponding to the L4 segment. Furthermore, the patient mentioned trouble with functional activities such as squatting, climbing stairs, and walking, and also reported decreased intolerance to activities like bending forward, standing, and sitting. As a result, patients complain of restriction in daily living activities and notice guarding and limited movement of the involved areas.

Regarding case# 6, the MRI report was "extradural meningeal cyst", while Therapha predicted the patient as "Lumbar Radiculopathy/Lumbar Disc Herniation/Lumbago with Sciatica". Researchers can interpret this from the patient's medical history since the patient reported his/her pain as a dull ache, radiating in nature, and marked the corresponding area to T12-L1 dermatomal distribution. The patient also had abnormal sensations of tingling and numbness in the feet in L_4-S_1 dermatomal regions and the presence of tender spots. In addition, the patient's pain worsened in the morning and got better as the day progressed. Also, the pain was exacerbated by deep breathing, laughing, sneezing or coughing. The patient reported difficulty in moving the affected area, as well as performing functional activities like squatting, climbing stairs, and walking. The complaints noted by the patient were aggravated with standing, walking, and sustained positions. Moreover, the patient reported a decreased tolerance in activities such as coming up from a bent-over position and from sitting to standing positions, bending forward, standing, and sitting. As a consequence, the patient complained of restriction in daily living activities and noticed guarding or limiting movement of the involved area. Furthermore, the patient had a history of falling. The patient reported an easing of their symptoms with rest in the supine position and medications.

3.2. Delphi Method

Following the completion of the case series, the assessor collected and tabulated the experts' rating sheets. In the first round, all respondents agreed (n = 10) that the patients' medical histories were congruent with Therapha's prediction, thereby achieving consensus. As all 10 cases had a consensus of above 80%, there was no need to re-circulate rating sheets for a second round.

Regarding cases# (2, 3, 4, 5, 8, 9, and 10), there was 100% consensus that the patients' medical histories were consistent with Therapha predictions. However, regarding cases #1, #6, and #7, there was a consensus of 90% among all 10 experts.

3.3. Diagnostic Predictive Accuracy

The demographic characteristics of patients who participated in this study are displayed in Table 2. The detailed clinical and MRI data of the 100 patients who were included in the diagnostic study are described in Supplementary Material Table S1. The accuracy of predictions from the software was determined by using ROC. The curve was up toward the left with an area under the curve of 0.84 (p = 0.001, 95% CI; 0.6 to 1.0) (Figure 4). Therapha had a sensitivity of 88%, specificity of 80%, positive predictive value of 99%, negative predictive value of 27%, positive likelihood ratio of 4.4, and negative likelihood ratio of 0.15.

Table 2. Descriptive statistics (n = 100).

	Female		Male	
	Mean ± SD	**Median (Min–Max)**	**Mean ± SD**	**Median (Min–Max)**
Age (Years)	48 ± 14	48	48.5 ± 13.3	50
Height (cm)	158.8 ± 5	159	173 ± 6	170.5
Weight (kg)	75.3 ± 15	72	82 ± 12.6	80
BMI	29.7 ± 6	29	27.6 ± 3.9	26

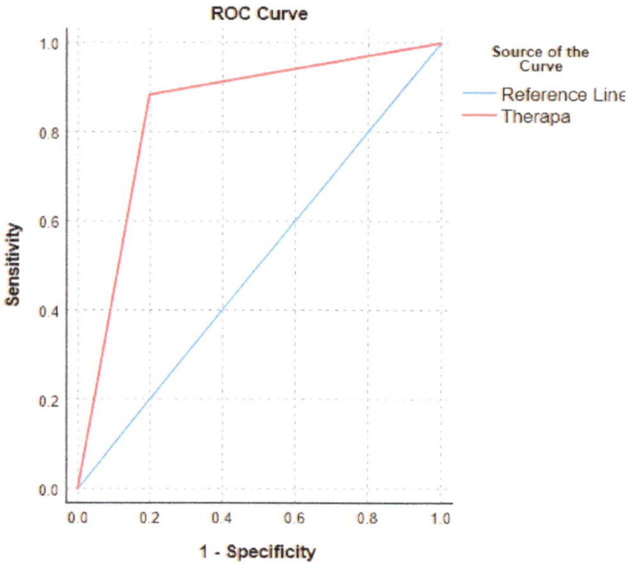

Figure 4. Receiver Operating Curve (ROC).

4. Discussion

In this study, a novel web software has been proposed for predicting lumbar disc herniation. The results of this study showed that Therapha software can be used for predicting and assisting with the diagnosis of lumbar disc herniation with a high level of accuracy. Findings revealed that Therapha technology has a satisfactory level of effectiveness, making it suitable for expedited and more effective screening of patients with lumbar disc herniation. To the best of our knowledge, this is the first study to examine the efficacy of an AI-enabled tool compared to the gold standard (MRI) for diagnosing patients with lumbar disc herniation.

4.1. Delphi Method

The Delphi method was used to reach an agreement in diagnosing patients with lumbar herniated discs. In the present study, experts agreed on statements representing the final diagnosis of patients that were generated by Therapha software which can be used as a guideline for clinical practice to help in decision-making process. This reinstates the significance of the tool's potential to be used as an adjunct to provide care that is backed by the power of technology.

Authors began with the patient history collection, ensuring a comprehensive and unbiased approach, and subsequently moved to analyzing the MRI results in detail. This methodological sequence, which kept the assessor blind as to whether patients had lumbar disc prolapse or a different pathology, was designed to mitigate any potential biases from the assessor during the initial phase (patient history-taking).

Each patient's signs and symptoms were in match to their MRI findings. Since most patients had the same history of lumbar herniated disc, patient compliance with LBP accompanied by radiculopathy and abnormal sensations like numbness and tingling were reported. Most patients in this study showed an increase in their symptoms with flexion activities which eased with extension-based movements. The majority of patients complained of pain during prolonged activities such as standing, sitting, or even walking for a long time. Also, patients reported that sneezing, deep breathing, or coughing exacerbated their symptoms. In addition, patients had the same type of pain "dull, aching, radiating to the lateral side down to the leg and foot". Furthermore, muscle spasms and weakness of either the back or thigh muscles were reported.

Concerning cases # (4 and 6), the Therapha predictions aligned with the expert diagnostic findings. However, the experts' agreements were not entirely consistent with the results obtained from the MRI scans. They had a consensus of 100% and 90% for cases # (4 and 6), respectively. Based on the findings, it could be stated that Therapha can identify patients with lumbar disc herniation efficiently which could help clinicians in their decision-making process by serving as a cognitive aid.

Therapha software is an innovative tool that can assist healthcare professionals in analyzing patient history (screening and diagnostic questions) and provide the most likely differential diagnosis without disturbing the clinical workflow. The solution accomplishes this in an innovative way by replacing the traditional patient intake forms with a conversational chatbot. The entire interaction is automated into an organized clinical summary and provides an actionable, insightful report that includes probable differential diagnoses. Screening questions help to identify patients who require special or alternative care pathways [23] while diagnostic questions help to identify the proper pathology, allowing for suitable treatment selections [23].

The use of proper diagnostic and screening questions can improve patient care by successfully identifying problems and discovering diseases correctly [23]. History-taking alone has been proven to accurately identify clinical problems 56% to 82.5% of the time [27,28]. Therapha, however, has an accuracy of 84% in predicting lumbar disc herniation pathology through history-taking incorporated with AI-based technology.

However, this procedure does not diminish the significance of physical examination. To confirm a diagnosis, a physical examination should be done in conjunction with a history. Physical examinations can sometimes reveal unexpected diagnoses or contribute to previously obtained data [29]. Investigations can also help in LBP diagnosis; medical imaging advances, in particular, have offered healthcare professionals new, non-invasive approaches to improve patient care [30].

4.2. Diagnostic Predictive Accuracy

Various systems have been developed to assist in diagnosing different pathologies. However, there are limited systems in the literature that are specifically designed to help in the diagnosis and treatment of LBP [31–33]. Although AI-based solutions have been developed in a variety of medical disciplines, research incorporating AI in LBP is still in its embryonic stage [34].

The result of this study was in line with Kadhim et al. who studied a fuzzy expert system for diagnosing LBP patients depending on medical observation symptoms detected using the fuzzy system and other factors that may influence back pain diagnoses such as body mass index, gender, age, and patient history. Thereafter, depending on the information provided, this system offered the most likely spine pathology and treatment. The system showed an accuracy of 90% when compared to other specialist diagnoses [35].

Furthermore, Lin et al. examined a web-based decision support system that used production for knowledge representation to evaluate a patient's information and suggest a diagnosis for LBP. The system infers multi-part diagnoses rapidly using a mini-Bayesian approach. It includes two interfaces for knowledge update and convenient system access, an inference engine, a knowledge base and a case repository. The CDSS shows performance equal to that of human experts [8]. However, this web-based decision support system needs to expand on its knowledge replenishment support. A clinician's diagnostic knowledge is likely to accumulate over time, making the system's knowledge replenishment support important. On the other hand, Therapha web software is currently built on a hybrid algorithm that utilizes over 4000 independent variables and is periodically updated depending on feedback from clinicians on the predictions made. They verify and discuss the feedback from clinicians and then make the required changes to improve the algorithmic power. The vision is to create a learning system as the system absorbs large quantities of clinical data over the course of time.

Sari et al. examined two expert systems (adaptive neuro-fuzzy inference system and artificial neural network) to objectively assess the level of pain intensity in low back pain patients. Researchers used two input variables including skin resistance, a visual analog scale, and one output variable which was LBP intensity [36]. They found higher accuracy in predicting LBP when the two systems were used in combination. Sari et al. proposed an objective and subjective method that can be used to scale a patient's level of pain [36]. However, this cannot be used as a diagnostic tool, rather it can be used as a screening tool. On the other hand, the current study presents a clinical decision support tool (Therapha) which is part of its algorithm to determine LBP intensity subjectively by using a visual analog scale and detailed history related to pain.

In recent years, the area of AI has been growing rapidly, with multiple studies demonstrating high levels of accuracy in various AI applications. Indeed, a growing body of literature has reported the high level of accuracy of AI systems in various medical fields, especially in medical imaging. Lee et al. converted lumbar CT scans into axial T2-weighted MRI slices by using generative adversarial networks. Then, they compared them to each patient's actual lumbar MRI images [14]. The results show an 80.2% similarity to the actual lumbar MRI images. Moreover, Ramirez et al. used AI techniques such as decision trees, support vector machines, and a logistic regression classifier to identify scoliosis curve changes. They found that the support vector machine generated the best results on 141 radiological spine scans, with an accuracy of 86%, and applied AI on MRI scans to estimate the surgical level of patients having disc decompression surgery [37,38]. Harada et al. recently developed an algorithm that identified patients at risk of re-herniation after microdiscectomy. After validation, the prediction method showed a recall of 80%, an area under the ROC of 72%, and an accuracy of 70% [39].

The above-mentioned studies showed high accuracy and performance, as Therapha software has shown. However, Therapha is a triaging system assessing physical problems related to the musculoskeletal system, but it does not assess and read the MRI itself. Image processing and computer vision are gaining momentum as computational resources such as powerful graphics processing units become more readily accessible. Indeed, current spine research studies using AI methods are focused on medical imaging. However, AI research in other areas such as spine pathology diagnosis is lacking, and more studies and research are required.

A systematic review examined the effectiveness of neurological tests, including sensory, motor, reflex, and neuro-dynamic assessments, in diagnosing lumbo-sacral radiculopathy by comparing them to MRI, electro-diagnostic tests, or intra-operative findings. The findings indicated that the diagnostic accuracy of most clinical neurological tests is generally low to moderate. There appears to be a lack of consensus among researchers regarding the diagnostic value of lower limb neuro-dynamic tests [40]. It remains unclear whether these tests can reliably identify disc herniation and subsequent nerve root compression, or if they are more indicative of hypersensitivity in the sacral and femoral plexus due to mechanical irritation [40].

Further, a systematic review and meta-analysis assessed the diagnostic effectiveness of sensory, motor, and reflex tests for identifying disc herniation and radiculopathy. The review compared neurological test results with surgical and radiological reference standards for disc herniation, and uniquely examined the connection between neurological findings and specific levels of herniation. The overall findings indicated that the diagnostic accuracy of all components of the neurological examination in detecting disc herniation in patients with suspected radiculopathy was limited. This was consistent regardless of the testing procedure or the specific level of herniation. The tests demonstrated poor diagnostic accuracy, with low sensitivity, moderate specificity, and limited overall accuracy, independent of the reference standard or herniation level [5]. Conversely, Therapha showed high sensitivity compared to the gold standard MRI. Therapha's approach not only assesses patient signs and symptoms without aggravating pain but also includes a comprehensive patient history, covering medical history, social history, family medical history, self-reported pain evaluation, functional deficits, psychological issues, sleep patterns, and posture [5].

This study had several limitations. The software was only used for the lumbar area to identify patients with radiculopathy, excluding other conditions or lesions associated with back pain. Additionally, although the software was intended to detect various pathologies connected to lower back pain, it did not encompass all possible conditions. Furthermore, the study exhibited a generalization bias by exclusively recruiting individuals with symptoms related to the lower back and lower limbs, which accounts for the lower negative predictive value (NPV) of 27%. Also, the age group of the patients in the study is worth mentioning; it is noteworthy to state the age group of the participants involved in the study as a limitation to generalizing the findings.

In addition, the study had no control over the MRI devices, radiology technicians, and radiology consultants. Another limitation of the study was that the software is only available in the English language. As mentioned before, Therapha software supports patient self-service. However, the majority of the patients in this study did not understand English, so translation to other languages is important in order to increase access to the patient self-service features in the software. In this study, a patient's history was taken and entered into the software by the clinician. Further, this study had several strengths, the software is a cost-effective and quick screening tool that may aid patients who are contraindicated to MRI such as claustrophobic individuals. For future research, it is recommended to investigate and study other spinal conditions such as cervical, thoracic, sacroiliac, and other conditions related to the lumbar region.

5. Conclusions

Given the increasing expenses associated with diagnosing and treating LBP, Therapha could be a highly cost-efficient tool to improve diagnostic accuracy and deliver effective care. Therapha's patient-centered approach utilizing AI to augment decision-making will lead to enhanced clinical outcomes in spine care. This study provides new insights for healthcare researchers, practitioners, and policymakers into the potential of implementing emerging technologies, especially AI, in healthcare delivery and diagnosis. As evidenced in the study, CDSS can be an efficient cognitive support for clinicians to render optimal patient care. Future studies are required to investigate additional spinal illnesses such as sacroiliac, facet joint disorders, etc., not limited to the lumbar region but to the thoracic, and cervical segments as well.

Supplementary Materials: The following supporting information can be downloaded at: https://www.mdpi.com/article/10.3390/diagnostics14171870/s1, Table S1: MRI findings and Therapha software inferences of patients in the Diagnostic Accuracy study.

Author Contributions: Conceptualization, F.B., A.A. (Almaha Alzahrani), M.F.C., Z.N., U.A., A.A. (Abdullah Alqarni) and F.K.; methodology, F.B., M.A., A.A. (Almaha Alzahrani), A.A. (Ahmed Abduljabbar), Z.N., M.F.C. and F.K.; validation, Z.N., U.A., A.A. (Abdullah Alqarni), A.A. (Ahmed Abduljabbar), R.B. and F.K.; formal analysis, F.B., M.A., M.F.C., R.B. and F.K.; investigation, A.A. (Almaha Alzahrani), Z.N., U.A., A.A. (Abdullah Alqarni), A.A. (Ahmed Abduljabbar) and F.K.; resources, M.A., Z.N., A.A. (Abdullah Alqarni), A.A. (Ahmed Abduljabbar) and F.K.; data curation, F.B., M.A., A.A. (Almaha Alzahrani), R.B. and M.F.C.; writing—original draft preparation, F.B., M.A., A.A. (Almaha Alzahrani), M.F.C. and R.B.; writing—review and editing, Z.N., A.A. (Abdullah Alqarni), U.A., A.A. (Ahmed Abduljabbar) and F.K.; visualization, Z.N., M.F.C., A.A. (Ahmed Abduljabbar) and F.K.; supervision, M.A., U.A., A.A. (Abdullah Alqarni) and F.K.; project administration, M.A., A.A. (Abdullah Alqarni), R.B. and F.K.; funding acquisition, R.B. All authors have read and agreed to the published version of the manuscript.

Funding: This research was funded by the Princess Nourah Bint Abdulrahman University researchers supporting project number (PNURSP2024R267), Princess Nourah Bint Abdulrahman University, Riyadh, Saudi Arabia.

Institutional Review Board Statement: The study was conducted in accordance with the Declaration of Helsinki, and approved by the Institutional Review Board of Research Ethics Committee at King Abdulaziz University Hospital (KAUH) (No. 125-22; HA-02-J-008; 14 April 2022), Jeddah, Saudi Arabia.

Informed Consent Statement: Informed consent was obtained from all subjects involved in the study.

Data Availability Statement: The data presented in this study are available on request from the corresponding author. The data are not publicly available due to privacy restrictions.

Acknowledgments: The authors would like to extend their sincere appreciation to the Princess Nourah Bint Abdulrahman University researchers supporting project number (PNURSP2024R267), Princess Nourah Bint Abdulrahman University, Riyadh, Saudi Arabia.

Conflicts of Interest: The authors declare no conflicts of interest.

References

1. Walker, B.F. The Prevalence of Low Back Pain: A Systematic Review of the Literature from 1966 to 1998. *J. Spinal Disord.* **2000**, *13*, 205–217. [CrossRef] [PubMed]
2. Alotaibi, M.A.; Alfaifi, R.M.; Alhowimel, A.S.; Alodaibi, F.A.; Alzahrani, H.; Alenazi, A.M.; Alqahtani, B.A.; Elnaggar, R.K. The Key Determinants of Low Back Pain among Lifestyle Behaviors in Adolescents: A Cross-Sectional Study from Saudi Arabia. *Medicine* **2024**, *103*, e37669. [CrossRef] [PubMed]
3. Hartvigsen, J.; Hancock, M.J.; Kongsted, A.; Louw, Q.; Ferreira, M.L.; Genevay, S.; Hoy, D.; Karppinen, J.; Pransky, G.; Sieper, J.; et al. What Low Back Pain Is and Why We Need to Pay Attention. *Lancet* **2018**, *391*, 2356–2367. [CrossRef]
4. Scaia, V.; Baxter, D.; Cook, C. The Pain Provocation-Based Straight Leg Raise Test for Diagnosis of Lumbar Disc Herniation, Lumbar Radiculopathy, and/or Sciatica: A Systematic Review of Clinical Utility. *J. Back Musculoskelet. Rehabil.* **2012**, *25*, 215–223. [CrossRef]
5. Al Nezari, N.H.; Schneiders, A.G.; Hendrick, P.A. Neurological Examination of the Peripheral Nervous System to Diagnose Lumbar Spinal Disc Herniation with Suspected Radiculopathy: A Systematic Review and Meta-Analysis. *Spine J.* **2013**, *13*, 657–674. [CrossRef]
6. Rebain, R.; Baxter, G.D.; McDonough, S. A Systematic Review of the Passive Straight Leg Raising Test as a Diagnostic Aid for Low Back Pain (1989 to 2000). *Spine* **2002**, *27*, E388–E395. [CrossRef]
7. Jarvik, J.G.; Deyo, R.A. Diagnostic Evaluation of Low Back Pain with Emphasis on Imaging. *Ann. Intern. Med.* **2002**, *137*, 586–597. [CrossRef]
8. Lin, L.; Hu, P.J.-H.; Liu Sheng, O.R. A Decision Support System for Lower Back Pain Diagnosis: Uncertainty Management and Clinical Evaluations. *Decis. Support Syst.* **2006**, *42*, 1152–1169. [CrossRef]
9. Vining, R.; Potocki, E.; Seidman, M.; Morgenthal, A.P. An Evidence-Based Diagnostic Classification System for Low Back Pain. *J. Can. Chiropr. Assoc.* **2013**, *57*, 189–204.
10. Byyny, R.L. The Data Deluge: The Information Explosion in Medicine and Science. *Pharos Alpha Omega Alpha Honor Med. Soc.* **2012**, *75*, 2–5.
11. Aggarwal, L.M. Advances in Medical Technology and Its Impact on Health Care in Developing Countries. *Int. J. Radiol. Radiat. Ther.* **2017**, *2*, 55–56. [CrossRef]
12. Cai, C.J.; Reif, E.; Hegde, N.; Hipp, J.; Kim, B.; Smilkov, D.; Wattenberg, M.; Viegas, F.; Corrado, G.S.; Stumpe, M.C.; et al. Human-Centered Tools for Coping with Imperfect Algorithms During Medical Decision-Making. In Proceedings of the 2019 CHI Conference on Human Factors in Computing Systems, Scotland, UK, 4–9 May 2019; pp. 1–14. [CrossRef]
13. Cai, C.J.; Winter, S.; Steiner, D.; Wilcox, L.; Terry, M. "Hello AI": Uncovering the Onboarding Needs of Medical Practitioners for Human-AI Collaborative Decision-Making. *Proc. ACM Hum. Comput. Interact.* **2019**, *3*, 104:1–104:24. [CrossRef]
14. Lee, M.H.; Siewiorek, D.; Smailagic, A.; Bernardino, A.; Badia, S. Opportunities of a Machine Learning-Based Decision Support System for Stroke Rehabilitation Assessment. *arXiv* **2020**, arXiv:2002.12261.
15. Musen, M.A.; Middleton, B.; Greenes, R. Clinical Decision-Support Systems. In *Biomedical Informatics*; Springer: London, UK, 2014; pp. 643–674. ISBN 978-1-4471-4473-1.
16. Park, J.H.; Choi, M.-H.; Lee, J.; Han, H.-S.; Lee, M.C.; Ro, D.H. Gait Deviations of Patients with Ruptured Anterior Cruciate Ligament: A Cross-Sectional Gait Analysis Study on Male Patients. *Knee Surg. Relat. Res.* **2021**, *33*, 45. [CrossRef] [PubMed]
17. Kunhimangalam, R.; Ovallath, S.; Joseph, P.K. A Clinical Decision Support System with an Integrated EMR for Diagnosis of Peripheral Neuropathy. *J. Med. Syst.* **2014**, *38*, 38. [CrossRef]
18. Ahmadi, M.; Nopour, R. Clinical Decision Support System for Quality of Life among the Elderly: An Approach Using Artificial Neural Network. *BMC Med. Inform. Decis. Mak.* **2022**, *22*, 293. [CrossRef] [PubMed]
19. Lahsasna, A.; Ainon, R.N.; Zainuddin, R.; Bulgiba, A. Design of a Fuzzy-Based Decision Support System for Coronary Heart Disease Diagnosis. *J. Med. Syst.* **2012**, *36*, 3293–3306. [CrossRef] [PubMed]

20. Emani, S.; Rui, A.; Rocha, H.A.L.; Rizvi, R.F.; Juaçaba, S.F.; Jackson, G.P.; Bates, D.W. Physicians' Perceptions of and Satisfaction With Artificial Intelligence in Cancer Treatment: A Clinical Decision Support System Experience and Implications for Low-Middle–Income Countries. *JMIR Cancer* **2022**, *8*, e31461. [CrossRef]
21. Azimi, P.; Yazdanian, T.; Benzel, E.C.; Aghaei, H.N.; Azhari, S.; Sadeghi, S.; Montazeri, A. A Review on the Use of Artificial Intelligence in Spinal Diseases. *Asian Spine J.* **2020**, *14*, 543–571. [CrossRef]
22. Azimi, P.; Mohammadi, H.R.; Benzel, E.C.; Shahzadi, S.; Azhari, S.; Montazeri, A. Artificial Neural Networks in Neurosurgery. *J. Neurol. Neurosurg. Psychiatry* **2015**, *86*, 251–256. [CrossRef]
23. Eubank, B.H.; Mohtadi, N.G.; Lafave, M.R.; Wiley, J.P.; Bois, A.J.; Boorman, R.S.; Sheps, D.M. Using the Modified Delphi Method to Establish Clinical Consensus for the Diagnosis and Treatment of Patients with Rotator Cuff Pathology. *BMC Med. Res. Methodol.* **2016**, *16*, 56. [CrossRef]
24. Fitch, K.; Bernstein, S.J.; Aguilar, M.D.; Burnand, B.; LaCalle, J.R.; Lázaro, P.; van het Loo, M.; McDonnell, J.; Vader, J.P.; Kahan, J.P. *The Rand/UCLA Appropriateness Method User's Manual*; Fitch, K., Ed.; MR/Rand DG-XII/RE; Rand: Santa Monica, CA, USA, 2001; ISBN 978-0-8330-2918-8.
25. Meshkat, B.; Cowman, S.; Gethin, G.; Ryan, K.; Wiley, M.; Brick, A.; Clarke, E.; Mulligan, E. Using an E-Delphi Technique in Achieving Consensus across Disciplines for Developing Best Practice in Day Surgery in Ireland. *J. Hosp. Adm.* **2014**, 3, p1. [CrossRef]
26. Lynn, M.R. Determination and Quantification of Content Validity. *Nurs. Res.* **1986**, *35*, 382–385. [CrossRef]
27. Peterson, M.C.; Holbrook, J.H.; Von Hales, D.; Smith, N.L.; Staker, L.V. Contributions of the History, Physical Examination, and Laboratory Investigation in Making Medical Diagnoses. *West. J. Med.* **1992**, *156*, 163–165. [CrossRef]
28. Hampton, J.R.; Harrison, M.J.; Mitchell, J.R.; Prichard, J.S.; Seymour, C. Relative Contributions of History-Taking, Physical Examination, and Laboratory Investigation to Diagnosis and Management of Medical Outpatients. *Br. Med. J.* **1975**, *2*, 486–489. [CrossRef] [PubMed]
29. Davis, J.L.; Murray, J.F. History and Physical Examination. In *Murray and Nadel's Textbook of Respiratory Medicine*; Elsevier: Amsterdam, The Netherlands, 2016; pp. 263–277.e2. [CrossRef]
30. Chou, R.; Qaseem, A.; Owens, D.K.; Shekelle, P.; Clinical Guidelines Committee of the American College of Physicians. Diagnostic Imaging for Low Back Pain: Advice for High-Value Health Care from the American College of Physicians. *Ann. Intern. Med.* **2011**, *154*, 181–189. [CrossRef]
31. Iqbal, K.; Yin, X.-C.; Hao, H.-W.; Ilyas, Q.M.; Ali, H. An Overview of Bayesian Network Applications in Uncertain Domains. *IJCTE Int. J. Comput. Theory Eng.* **2015**, *7*, 416–427. [CrossRef]
32. Matar, H.E.; Navalkissoor, S.; Berovic, M.; Shetty, R.; Garlick, N.; Casey, A.T.H.; Quigley, A.-M. Is Hybrid Imaging (SPECT/CT) a Useful Adjunct in the Management of Suspected Facet Joints Arthropathy? *Int. Orthop.* **2013**, *37*, 865–870. [CrossRef]
33. Hurri, H.; Karppinen, J. Discogenic Pain. *Pain* **2004**, *112*, 225–228. [CrossRef]
34. D'Antoni, F.; Russo, F.; Ambrosio, L.; Vollero, L.; Vadalà, G.; Merone, M.; Papalia, R.; Denaro, V. Artificial Intelligence and Computer Vision in Low Back Pain: A Systematic Review. *Int. J. Environ. Res. Public Health* **2021**, *18*, 10909. [CrossRef]
35. Kadhim, M.A.; Alam, M.A.; Kaur, H. Design and Implementation of Fuzzy Expert System for Back Pain Diagnosis. *Int. J. Innov. Technol. Creat. Eng.* **2011**, *1*, 16–22.
36. Sari, M.; Gulbandilar, E.; Cimbiz, A. Prediction of Low Back Pain with Two Expert Systems. *J. Med. Syst.* **2012**, *36*, 1523–1527. [CrossRef]
37. Cabitza, F.; Locoro, A.; Banfi, G. Machine Learning in Orthopedics: A Literature Review. *Front. Bioeng. Biotechnol.* **2018**, *6*, 75. [CrossRef] [PubMed]
38. Ramirez, L.; Durdle, N.G.; Raso, V.J.; Hill, D.L. A Support Vector Machines Classifier to Assess the Severity of Idiopathic Scoliosis from Surface Topography. *IEEE Trans. Inf. Technol. Biomed.* **2006**, *10*, 84–91. [CrossRef] [PubMed]
39. Harada, G.K.; Siyaji, Z.K.; Mallow, G.M.; Hornung, A.L.; Hassan, F.; Basques, B.A.; Mohammed, H.A.; Sayari, A.J.; Samartzis, D.; An, H.S. Artificial Intelligence Predicts Disk Re-Herniation Following Lumbar Microdiscectomy: Development of the "RAD" Risk Profile. *Eur. Spine J.* **2021**, *30*, 2167–2175. [CrossRef] [PubMed]
40. Tawa, N.; Rhoda, A.; Diener, I. Accuracy of Clinical Neurological Examination in Diagnosing Lumbo-Sacral Radiculopathy: A Systematic Literature Review. *BMC Musculoskelet. Disord.* **2017**, *18*, 93. [CrossRef]

diagnostics

MDPI

Article

Gastric Cancer Detection with Ensemble Learning on Digital Pathology: Use Case of Gastric Cancer on GasHisSDB Dataset

Govind Rajesh Mudavadkar [1], Mo Deng [1], Salah Mohammed Awad Al-Heejawi [1], Isha Hemant Arora [2], Anne Breggia [3], Bilal Ahmad [4], Robert Christman [4], Stephen T. Ryan [4] and Saeed Amal [5,*]

1 College of Engineering, Northeastern University, Boston, MA 02115, USA; mudavadkar.g@northeastern.edu (G.R.M.); m.deng@northeastern.edu (M.D.); s.al-heejawi@northeastern.edu (S.M.A.A.-H.)
2 Khoury College of Computer Sciences, Northeastern University, Boston, MA 02115, USA; arora.isha@northeastern.edu
3 MaineHealth Institute for Research, Scarborough, ME 04074, USA
4 Maine Medical Center, Portland, ME 04102, USA; bilal.ahmad@spectrumhcp.com (B.A.); robert.christman@spectrumhcp.com (R.C.); stephen.ryan@mainehealth.org (S.T.R.)
5 The Roux Institute, Department of Bioengineering, College of Engineering at Northeastern University, Boston, MA 02115, USA
* Correspondence: s.amal@northeastern.edu

Simple Summary: Gastric cancer is a major worldwide health concern, underscoring the importance of early detection to enhance patient outcomes. Traditional histological analysis, while considered the gold standard, is labour intensive and manual. Deep learning (DL) is a potential approach, but existing models fail to extract all of the visual data required for successful categorization. This work overcomes these constraints by using ensemble models that mix different deep-learning architectures to improve classification performance for stomach cancer diagnosis. Using the Gastric Histopathology Sub-Size Images Database, the ensemble models obtained an average accuracy of more than 99% at various resolutions. ResNet50, VGGNet, and ResNet34 performed better than EfficientNet and VitNet, with the ensemble model continuously delivering higher accuracy. These findings show that ensemble models may accurately detect important characteristics from smaller picture patches, allowing pathologists to diagnose stomach cancer early and increasing patient survival rates.

Abstract: Gastric cancer has become a serious worldwide health concern, emphasizing the crucial importance of early diagnosis measures to improve patient outcomes. While traditional histological image analysis is regarded as the clinical gold standard, it is labour intensive and manual. In recognition of this problem, there has been a rise in interest in the use of computer-aided diagnostic tools to help pathologists with their diagnostic efforts. In particular, deep learning (DL) has emerged as a promising solution in this sector. However, current DL models are still restricted in their ability to extract extensive visual characteristics for correct categorization. To address this limitation, this study proposes the use of ensemble models, which incorporate the capabilities of several deep-learning architectures and use aggregate knowledge of many models to improve classification performance, allowing for more accurate and efficient gastric cancer detection. To determine how well these proposed models performed, this study compared them with other works, all of which were based on the Gastric Histopathology Sub-Size Images Database, a publicly available dataset for gastric cancer. This research demonstrates that the ensemble models achieved a high detection accuracy across all sub-databases, with an average accuracy exceeding 99%. Specifically, ResNet50, VGGNet, and ResNet34 performed better than EfficientNet and VitNet. For the 80 × 80-pixel sub-database, ResNet34 exhibited an accuracy of approximately 93%, VGGNet achieved 94%, and the ensemble model excelled with 99%. In the 120 × 120-pixel sub-database, the ensemble model showed 99% accuracy, VGGNet 97%, and ResNet50 approximately 97%. For the 160 × 160-pixel sub-database, the ensemble model again achieved 99% accuracy, VGGNet 98%, ResNet50 98%, and EfficientNet 92%, highlighting the ensemble model's superior performance across all resolutions. Overall, the ensemble model consistently provided an accuracy of 99% across the three sub-pixel categories. These findings show that ensemble models may successfully detect critical characteristics from smaller patches and achieve high performance. The findings will help pathologists diagnose gastric cancer using histopathological images, leading to earlier identification and higher patient survival rates.

Keywords: cancer detection; machine learning; gastrointestinal cancer; deep learning; histopathology

1. Introduction

The GI tract, spanning 25 feet from the oral cavity to the anus, transports ingested substances. The digestive process begins with the oesophagus and continues through the stomach and small intestines, extracting important nutrients. Waste is then eliminated through the colon and rectum [1,2]. Tumours in these organs often result from aberrant cell growth caused by DNA changes [3]. Mutations can be caused by a variety of reasons, including health conditions, genetics, or lifestyle. The uncontrolled proliferation of malignant cells in the gastrointestinal system is caused by genetic, environmental, and lifestyle factors that interact. Common gastrointestinal (GI) cancers include oesophageal cancer, colorectal cancer, gastric cancer, bile duct cancer, anal cancer, colon cancer, gallbladder cancer, pancreatic cancer, gastrointestinal stromal tumours, liver cancer, rectal cancer, gastric cancer, and small intestine cancer. In 2020, gastric cancer was one of the top three most common cancers in 19 nations, with around 1.1 million cases reported (720,000 men, 370,000 females). Early identification of GI cancer aids in cancer treatment and reduces health-related complications.

Traditional approaches to the identification of cancer include the estimation of body fat percentage and its subsequent correlation with cancer. Other methods include identifying common microbes associated with cancer in food. Another way is the use of Indocyanine Green (ICG) in gastrointestinal surgery, which is gaining popularity, particularly for lymph node diagnosis and operative field imaging [4,5].

However, pathologists must physically assess tissue samples, which is a tough, time-consuming, and subjective procedure. Moreover, different pathologists may provide different results, making the analysis susceptible to errors. The accuracy of histopathological analysis is heavily dependent on the pathologists' experience and knowledge, making the manual process susceptible to mistakes such as incorrect detection and diagnosis. Furthermore, a scarcity of pathologists creates significant delays in examining patient cases, potentially leading to late cancer discovery [6,7].

Various computer-aided detection (CAD) strategies have been investigated for the diagnosis of gastric cancer utilizing histopathological imaging. For more than 30 years, researchers have studied computer-aided diagnosis in gastroenterology, creating datasets from endoscopic images using various methodologies. The most widely researched issue is the identification of aberrant pathological signs in a specific location of the GI tract, notably polyps. There has also been research on the detection and categorization of disorders throughout the GI system, which has included clinical findings, anatomical markers, and therapies [8]. Figure 1 shows the steps of extraction of the histopathological image from slides.

2. Materials and Methods

2.1. Literature Reviews

Various computer-aided detection (CAD) strategies have been investigated for the diagnosis of gastric cancer utilizing histopathological imaging. For more than 30 years, researchers have studied computer-aided diagnosis in gastroenterology, creating datasets from endoscopic images using various methodologies. The most widely researched issue is the identification of aberrant pathological signs in a specific location of the GI tract, notably polyps. There has also been research on the detection and categorization of disorders throughout the GI system, which has included clinical findings, anatomical markers, and therapies [8]. During the first 20 years of development, image processing required the extraction of features using various approaches before their categorization using statistical methods [9]. These characteristics are divided into three categories: spatial, frequency, and high-level descriptors. Spatial characteristics are retrieved using pixel-based and histogram techniques, whilst frequency information is collected using Fourier and wavelet transform algorithms. High-level characteristics are retrieved with edge and region-based methods. Statistical machine learning approaches were frequently employed to categorize these characteristics.

Later, machine learning (ML) was to become commonly used in CAD to diagnose gastric cancer by extracting handmade elements such as colour, texture, and form. For this purpose, support vector machines (SVM), random forests, and Adaboost are among the most frequently employed machine learning classifiers. In recent research, deep learning has been utilized to automate the feature selection process. Several studies have shown that deep convolutional neural networks (CNN) excel at tasks such as recognizing and segmenting histopathological images related to cancer, metastasis, and genetic mutation analysis. Some investigations have indicated that these networks perform similarly to human pathologists [10,11].

Deep learning techniques have advanced significantly over the past decade, notably with the CNN architecture. This design allows for the extraction and categorization of spatial as well as high-level features, making it a major area of study for academia. Several solutions have been proposed, including hybrid approaches based on CNN attributes, transfer learning, the development of novel CNN models, and research into other deep learning networks [12].

Gastroenterology CAD research has a long history and covers a wide range of topics. For studies involving the classification of the KvasirV2 and HayperKvasir datasets, which were employed in the studies conducted [9,12], Melaku et al. (2019) utilized VGGNet and InceptionV3 to classify the Hyper KVASIR dataset with 98% accuracy using SVM. M Hmoud et al. (2020) evaluated GoogLeNet, ResNet-50, and AlexNet on the KVASIR dataset, with AlexNet outperforming the others and achieving 97% accuracy. Yogapriya et al. (2021) used VGG16, ResNet-18, and GoogLeNet on the KVASIR v2 dataset, with VGG16 leading the way with an accuracy of about 96.33%. Furthermore, Zenebe et al. (2022) proposed a unique deep convolutional neural network (CNN) with a spatial attention mechanism for categorizing gastrointestinal (GI) illnesses [13,14]. When evaluated on a dataset of 12,147 GI images, the model demonstrated an impressive accuracy of 92.84%. This study emphasizes the importance of using pre-trained models in the correct diagnosis of gastrointestinal disorders, showcasing numerous methodologies, and achieving significant advances in this sector.

The study by Weiming Hua, Chen Lia, et al., [15] centres on enhancing deep learning models for cancer detection. This research utilizes various deep learning architectures and training methodologies to boost model accuracy, achieving notable results in cancer detection. Ensemble learning, which integrates multiple models to enhance performance, has seen increasing popularity in the field of medical imaging. Previous studies, including those by Hua et al. and others, have demonstrated that ensemble methods often surpass the performance of individual models. However, the specific combination of models and their application to particular datasets, like gastric histopathology images, have not been extensively investigated. Furthermore, the referenced paper does not deeply address data preprocessing techniques, cross-validation methodologies, the selection of the best epochs, or detailed performance metrics. These areas present opportunities for further exploration to enhance the robustness and generalizability of deep learning models in medical image analysis.

Several studies on medical image analysis have investigated the possibility of deep learning models for reliable diagnosis. Varun Sapra, Luxmi Sapra, and Akashdeep Bhardwaj, et al. [16] have proposed strategies that include pruning, quantization, and knowledge distillation to minimize ensemble model size and computing costs while maintaining accuracy. While their research produced important theoretical insights, it required a thorough technical review of real datasets.

To assess how well the proposed models worked, this research evaluated them using the recently available Gastric Histopathology Sub-Size Image Database (GasHisSDB) [17]. The main contributions of this study are the development of effective deep ensemble learning models for detecting gastric cancer that outperform current other models included in the research on the GasHisSDB dataset and the development of their ability to successfully identify gastric histology images with lower resolution, perhaps resulting in a reduction of the digital scanners, data storage, and computer servers needed for histopathology activities. This may increase the chance of the early detection of gastric cancer and enhance the rates of patient survival [18,19].

2.2. Dataset Description

The Gastric Histopathology Sub-Size Image Database (GasHisSDB), with a link to the data available in the Supplementary Material, consists of 600 pictures of stomach cancer pathology obtained from a patient's pathological slides (see Figure 1) obtained from a specific section of their gastrointestinal tract, each measuring 2048 × 2048 pixels. The images were obtained by scanning with a new USB camera at a magnification of 20. Four pathologists from Longhua Hospital at the Shanghai University of Traditional Chinese Medicine then provided tissue-specific labelling. In conjunction with five Northeastern University biomedical experts, the photos were cropped into 245,196 sub-sized gastric cancer pathology images. Two qualified pathologists from Liaoning Cancer Hospital and Institute calibrated these pictures. The dataset was then classed as aberrant or normal, with photographs reduced to three different sizes (160 × 160, 120 × 120, and 80 × 80) for each group. The dataset of gastrointestinal images is separated into three sizes: 80 × 80, 120 × 120, and 160 × 160. Each size category is further classified into two groups: abnormal and normal (see Figure 2b–d). The 80 × 80 size category has 81,390 photos, with 34,350 classed as abnormal and 47,040 labelled as normal. With regard to the 120 × 120 size category, there are 65,260 photos altogether, with 24,800 rated as abnormal and 40,460 as normal. Finally, in the 160 × 160 size group, there are 33,272 photos, with 13,117 classified as abnormal and 20155 as normal (as described in see Figure 3). In this project, the dataset of each class was divided into a five-fold cross validation.

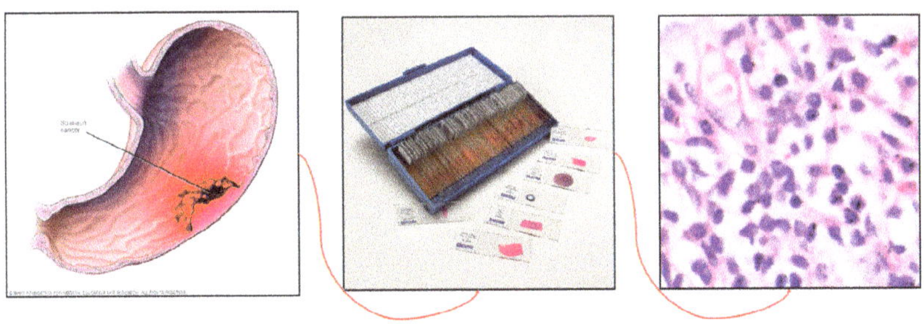

Figure 1. Process of extraction of the histopathology image.

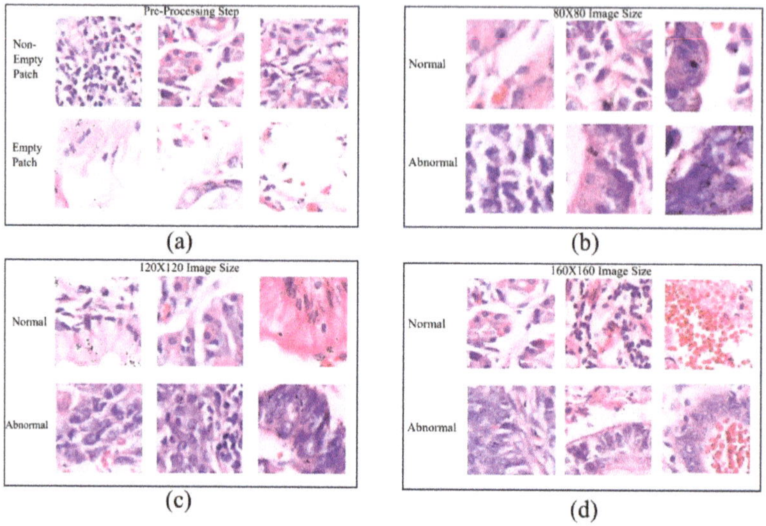

Figure 2. Examples of histopathological gastric images. (**a**) The pre-processing step and (**b–d**) examples of different image sizes.

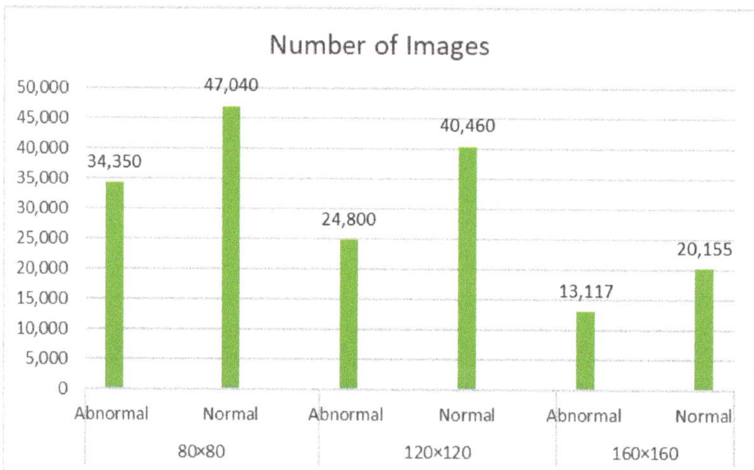

Figure 3. Summaries of the total number of images for every subclass in an experiment setup.

2.3. Methodology Overview

In this work, CNN architectures utilizing transfer learning and ensemble approaches were demonstrated to detect gastric cancer patches. The process is divided into four major steps: (1) creating the dataset by eliminating empty patches (see Figure 2a) and augmenting, (2) tailoring pre-trained networks or base models, (3) choosing the most effective base models to form ensemble models, and (4) assessing and presenting the models using different metrics and the class activation map.

To enhance the model performance, data preprocessing was conducted to create a more balanced dataset by removing non-informative empty patches. This was undertaken because the presence of these empty patches would bias the training process and therefore jeopardize the model performance. After empty patches are removed, data augmentation was employed to increase the dataset size for training.

2.4. Empty Patch Removal Process

Empty patches are defined as those where more than half of the pixels have an RGB intensity value greater than 230 across all channels. The following is an overview of the empty patch removal procedure, including the proportion of empty patches deleted. In the 120 × 120 resolution dataset, 15.92% of patches were removed from the abnormal subclass, while 45.01% were removed from the normal subclass. For the 160 × 160 resolution dataset, 14.77% of patches were removed from the abnormal subclass, and 44.23% were removed from the normal subclass. In the 80 × 80 resolution dataset, 17.45% of patches were removed from the abnormal subclass and 45.54% were removed from the normal subclass. After eliminating the empty patches, the remaining patches were utilized to train the model. This preprocessing phase guaranteed that the dataset contained only relevant images, allowing the model to extract detailed visual characteristics for precise categorization. Figure 4 illustrates the count differences between empty patches, which are mostly backgrounds, and non-empty patches, which represent tissues in the histopathological images.

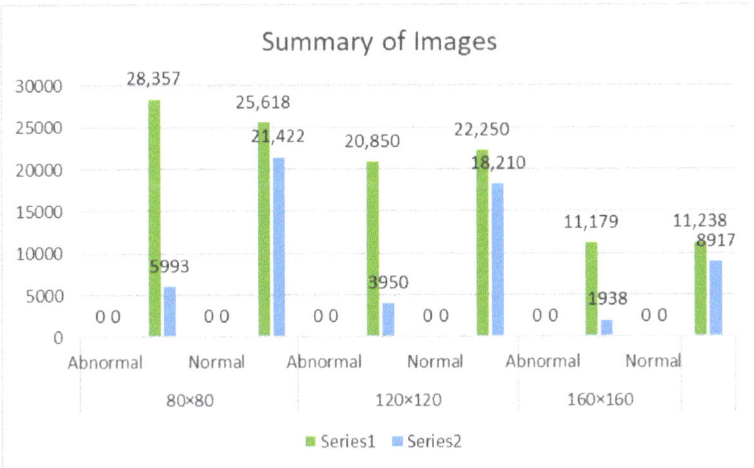

Figure 4. Summarises the total number of the empty and non-empty patch images in the dataset.

2.5. Pretrained Networks as Base Models

Convolutional neural networks (CNNs) have played an important role in numerous applications since deep learning's inception, this is due to continual advances in strength, efficiency, and adaptability. CNNs are an excellent illustration of this breakthrough, as they are particularly built for computer vision problems and use convolutional layers inspired by natural visual processes. Multiple CNN structures have evolved throughout time, each improving accuracy, speed, and overall performance, and are usually compared with the ImageNet project, a massive visual database that promotes advancements in computer vision. Historically, training CNNs from scratch required significant computer resources and time. Transfer learning (TL) provides a practical shortcut by exploiting prior information from trained models to accelerate optimization and perhaps improve classification accuracy. TL entails the transfer of weights from pre-trained models, using insights acquired from varied datasets, and increasing the speed of training processes to improve model accuracy, particularly in complicated architectures [20,21].

2.5.1. ResNet34 Architecture

ResNet34 is a member of the residual networks (ResNet) family, which was introduced by He et al. in 2015. ResNet34 uses residual learning to address the issue of vanishing gradients, which is common in deep neural networks. This design is made up of 34 layers, with shortcut connections that allow gradients to flow straight across the network. These residual connections enable the training of very deep networks by overcoming the degradation issue. ResNet34 strikes a balance between depth and computational efficiency, beating shallower networks while maintaining manageable computational costs.

ResNet34 is a variation of ResNet, a CNN architecture created by Microsoft Research. ResNet34 includes 34 layers and uses residual connections to overcome the vanishing gradient problem, making training more efficient. ResNet models are popular due to their efficacy in a variety of applications.

The design starts with an input layer that processes a two-dimensional picture, followed by a 7×7 convolutional layer with 64 filters and a stride of 2, which includes padding to retain spatial dimensions (see Figure 5). A max pooling layer with a stride of two decreases the spatial dimensions. The network is divided into multiple stages of residual blocks. The first stage consists of three residual blocks, each with two 3×3 convolutional layers with 64 filters and shortcut connections. The second stage consists of four residual blocks, each with two 3×3 convolutional layers and 128 filters. The first block down-samples the feature maps with a stride of two, and the shortcut connection adjusts dimensions. The third stage consists of six residual blocks, each with two 3×3 convolutional layers and 256 filters. The first block downscales with a stride of 2 and the shortcut connection adjusts appropriately. The last stage is made up of three residual blocks, each with two 3×3 convolutional layers and 512 filters. The first block down-samples with a stride of two, and the shortcut connection does the same. An average pooling layer then reduces spatial dimensions to 1×1, and the pooled feature maps are fed via a fully connected layer with 1000 output units, which correspond with the number of classes in the ImageNet dataset.

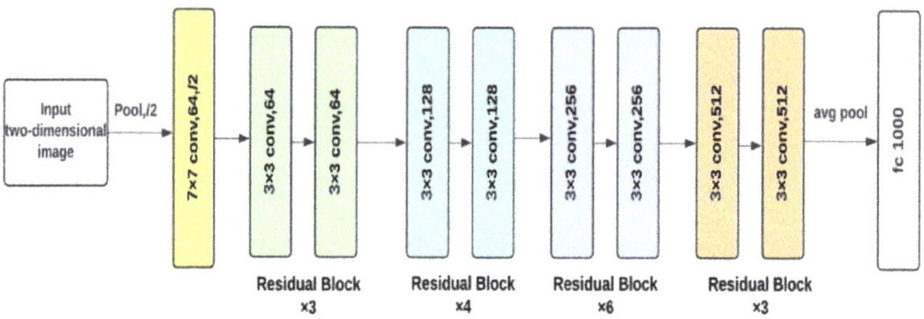

Figure 5. ResNet34 model architecture.

2.5.2. ResNet50 Architecture

ResNet50 is a 50-layer version of the ResNet architecture, deeper than ResNet34. This depth can improve performance on certain tasks, but it also necessitates more computational resources for training.

ResNet50's architecture handles two-dimensional input data using a succession of layers geared for effective feature extraction. It starts with zero-padding, then adds a convolutional layer, batch normalization, ReLU activation, and max-pooling to minimize spatial dimensions (see Figure 6). The network is organized into five phases, each of which contains convolutional and identity blocks. Stage 1 prepares the data, whereas Stages 2–5 contain convolutional blocks for down-sampling and identity blocks for deep feature extraction. Following these steps, an average pooling layer reduces the spatial dimensions to one by one, followed by a flattening layer and a fully linked layer for classification. The inclusion of identity blocks allows for the direct addition of input to output, which alleviates the vanishing gradient problem and enables deeper network training. This design facilitates the effective learning of complicated characteristics, hence improving image recognition performance.

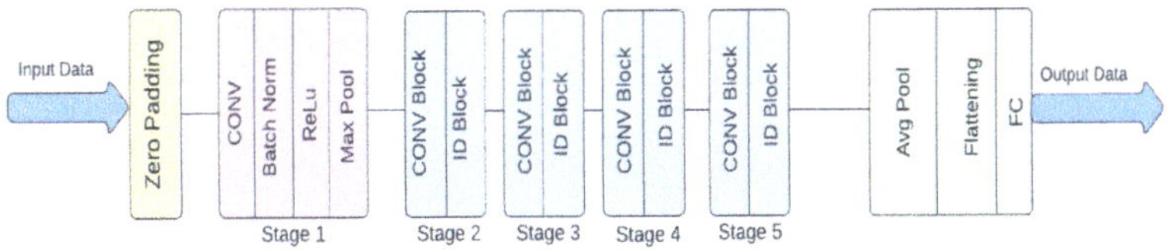

Figure 6. ResNet50 model architecture.

2.5.3. VGGNet16 Architecture

The Visual Geometry Group (VGG) at the University of Oxford developed VGGNet16, a deep CNN architecture. It is well known for being both simple and successful in picture classification applications. VGGNet16 is made up of 16 layers, the first 13 of which are convolutional, followed by three fully connected layers. Each convolutional layer has a 3×3 kernel, and max-pooling layers are used after a sequence of convolutional layers to minimize spatial dimensions. Although VGGNet16 has a large number of parameters, which makes it computationally costly, it achieves good accuracy on benchmark datasets thanks to its deep architecture and consistent layer design.

VGGNet, created by the Visual Geometry Group at the University of Oxford, is renowned for its simplicity. It is mostly composed of layered convolutional layers with 3×3 kernels and max-pooling layers. VGGNet includes several variations, including VGG16 and VGG19, which differ in the number of layers.

The VGGNet16 architecture (see Figure 7) runs an input picture through a sequence of convolutional layers, each followed by ReLU activation. It starts with two convolutional layers of 64 filters each, followed by max-pooling (see Figure 7). This pattern is repeated with increasing numbers of filters (128, 256, and eventually 512). To minimize spatial dimensions, max-pooling is applied after each convolutional layer step. Three fully connected layers follow the convolutional layers, with the last layer serving as a softmax layer for 1000-class classification. This deep design, which makes regular use of modest 3×3 filters, enables good feature extraction while keeping reasonable computing cost, making VGGNet16 a strong model for image recognition applications.

Figure 7. VGGNet16 model architecture.

2.5.4. EfficientNet Architecture

EfficientNet is a class of CNN models developed by Google AI that outperforms earlier models using fewer parameters and FLOPs. They employ a novel scaling technique to improve network depth, breadth, and resolution for improved resource management.

The EfficientNet design starts with a 3×3 convolutional layer, then numerous mobile inverted bottleneck (MBConv) blocks with different kernel sizes and expansion factors (as shown in Figure 8). The design employs a mix of 3×3 and 5×5 MBConv blocks to efficiently record spatial characteristics at various sizes. Each MBConv block is composed of a depth wise convolution followed by a pointwise convolution, allowing the model to capture complicated patterns while being computationally efficient. The network gradually raises the number of filters and the resolution of feature maps using these MBConv blocks, resulting in a highly detailed and rich feature map. This architecture strikes a compromise between depth, breadth, and resolution, producing an efficient and strong model for picture categorization tasks.

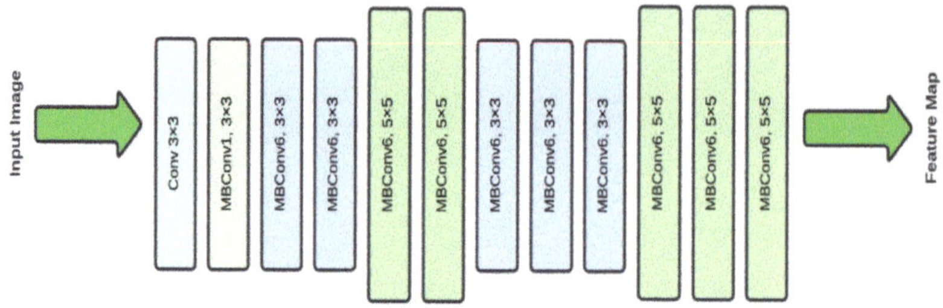

Figure 8. EfficientNet model architecture.

2.5.5. VitNet Architecture

The vision transformer is a novel CNN architecture created by Google Research that employs self-attention techniques seen in transformer topologies. Rather than employing convolutional layers like typical CNNs, ViT employs a transformer encoder, which allows it to recognize relationships across large distances in pictures. This method has shown outstanding results in several computer vision tasks, particularly when trained on large datasets.

The vision transformer (ViT) architecture (as shown in Figure 9) divides an input picture into fixed-size, non-overlapping patches, which are then flattened into a 1D vector (as shown in Figure 9). The vectors are then projected linearly onto a lower-dimensional space. Position embeddings are applied to these vectors in order to keep the positioning information of the patches inside the original picture. The generated sequence is routed through a transformer encoder, which consists of many layers of multi-head self-attention mechanisms and feed-forward neural networks. This encoder captures both the global environment and the interactions between patches. The output of the transformer encoder is then passed into a multi-layer perceptron (MLP) head, which computes the final predicted probability for each class. This design enables ViT to harness the benefits of transformers in collecting long-range relationships and global picture context.

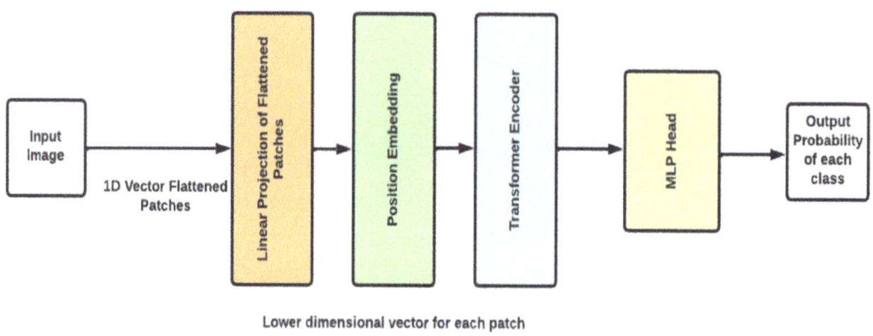

Figure 9. VITNet model architecture.

2.5.6. Ensemble Architecture

Convolutional neural networks (CNNs) and transfer learning have significantly increased neural network performance, but there is still potential for improvement. This article proposes the use of ensemble methods to improve the effectiveness of the three models that have been pre-trained. Ensemble learning, a machine learning and statistics-based approach combines the skills of many algorithms to extract relevant insights from data [19]. During this analysis, stacking was determined to be the most appropriate method. This requires training various ML algorithms on the information before merging them to create a composite algorithm capable of successfully combining their input.

In this research, ensemble model architecture was used using ResNet34 and VGGNet16 as basis models. Figure 10 shows the proposed ensemble, which comprises both ResNet34 and VGGNet16. Initially, each model was trained independently to determine its unique performance. We then selected the best-performing epochs for each model based on validation accuracy. The best-performing epochs were then used to build the ensemble model. To improve the robustness and generalizability of the ensemble techniques, cross-validation was used. During each fold of the cross-validation approach, the research utilized the best weights from the previous folds to train the ensemble. For example, if cross-validation was undertaken on fold 1, the best epochs from folds 2, 3, 4, and 5 of ResNet34 and VGGNet16 were used to establish the ensemble model. The ensemble model architecture was created to leverage the complementary characteristics of ResNet34 and VGGNet16. ResNet34, with its residual connections, successfully mitigates the vanishing gradient problem, allowing for deeper network training. In contrast, VGGNet16, noted for its simplicity and constant layer architecture, excels in capturing fine-grained characteristics. By integrating both models, the ensemble makes use of VGGNet16's comprehensive feature extraction and ResNet34's depth-wise learning capabilities. Training the ensemble model entailed freezing the early layers of both base models to preserve their pre-trained feature extraction capabilities while fine-tuning the subsequent layers to fit them to the unique dataset. This hybrid strategy enabled the ensemble model to outperform individual models, as indicated by the improved accuracy and resilience across many validation criteria. In summary, the proposed ensemble model, which includes ResNet34 and VGGNet16, showed considerable performance increases. The strategic use of cross-validation and the incorporation of complementary model architectures demonstrate the effectiveness of ensemble techniques in deep learning applications.

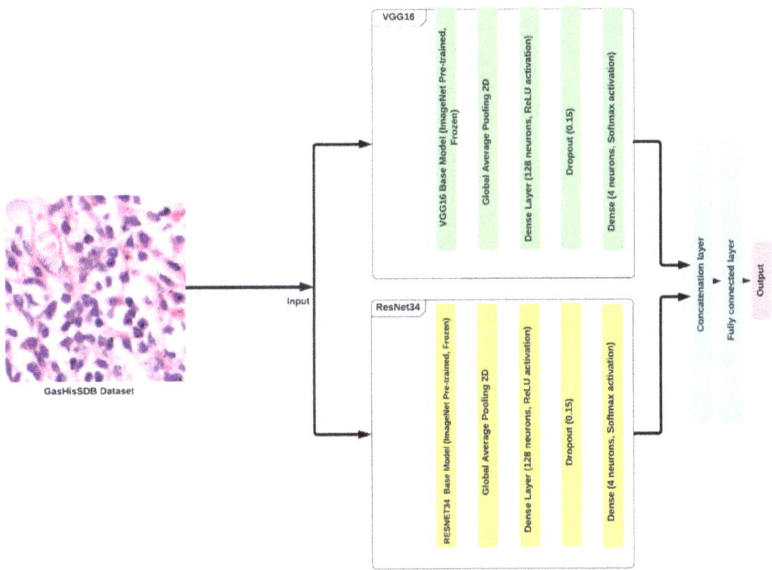

Figure 10. Ensemble model detailed architecture.

2.6. Interpretability of the Ensemble Model

Understanding how ensemble models make conclusions is critical for gaining acceptance and confidence in medical diagnoses. Our work makes use of class activation mapping (CAM) tools to display and comprehend the ensemble models' decision-making processes. CAM identifies the key areas in histopathological pictures that have the most impact on the model's predictions, offering insights into the characteristics examined by the models. During the training phase, we divided the dataset into a subset of validation pictures in order to determine which areas the models focused on when discriminating between normal and diseased tissues. Our cooperating pathologists then inspected the produced maps to confirm that the highlighted locations matched significant histological findings. This technique not only confirmed the model's focus areas, but also helped to better grasp its decision-making framework.

Furthermore, we used ensemble learning with a stacking technique, in which each of the base models (ResNet34, VGGNet16, and so on) were trained separately. These models' outputs were integrated using a meta-classifier, which pooled their predictions. This multi-model decision-making technique improves resilience and minimizes the danger of incorrect classifications by combining the capabilities of many architectures. By adding these interpretability methodologies, we may offer a clear picture of the model's underlying workings, increasing the trustworthiness and dependability of diagnostic results.

2.7. Experimental Setting

The data were divided into training and validation sets. Each network was trained for 20 epochs using 5-fold cross-validation to create the model. The weights from the epoch with the best validation accuracy were chosen as the final representations for each model. Various metrics were then employed to assess accuracy, followed by many objective assessment factors to determine overall performance. The dataset was rigorously separated into training and validation sets to ensure a thorough evaluation of the model's performance. The training approach used 5-fold cross-validation across 20 epochs for each network. This strategy provides a thorough evaluation by cycling through many train–test divides, lowering the danger of overfitting and guaranteeing that the model generalizes well to new data. For example, using an 80 × 80 dataset, we split it into five equal sections. During the initial fold, the model was trained on the first four components and verified on the fifth. In the second fold, the model was trained on parts two, three, four, and five, with validation on the first component. The third fold featured training on parts three, four, five, and one, followed by validation on part two. This method was repeated for the remaining two folds, ensuring that each portion was only used as a validation set once. After running the 20 epochs for each fold, we chose the weights from the epoch with the best validation accuracy to represent the final model for that fold. This strategy ensured that the model parameters that were picked functioned best on data that were not known, hence increasing the model's dependability. Various metrics were then employed to assess accuracy, followed by many objective assessment factors to determine overall performance [22].

3. Results

The performance evaluation criteria used include accuracy, sensitivity, specificity, Jaccard index, and area under the curve (AUC). Positive samples are those that include abnormal or malignant patches, whereas negative samples contain normal or healthy patches. The phrases true positive (TP), false positive (FP), true negative (TN), and false negative (FN) are used to describe the various prediction results.

1. Accuracy: Accuracy is the ratio of properly identified samples to the total number of samples. This is computed as follows:

$$\text{Accuracy} = \frac{(TP + TN)}{(TP + TN + FP + FN)} \tag{1}$$

2. Sensitivity (recall): Sensitivity, also known as recall, is the proportion of real positive samples that the model properly identifies. This is provided as follows:

$$\text{Sensitivity} = \frac{(TP)}{(TP + FN)} \tag{2}$$

3. Specificity refers to the fraction of real negative samples properly detected by the model. This is computed as follows:

$$\text{Specificity} = \frac{(TN)}{(TN + FP)} \quad (3)$$

4. The Jaccard index, commonly known as the intersection over union (IoU), assesses the similarity between expected and observed positive samples. This is provided as follows:

$$\text{Jaccard index} = \frac{(TP)}{(TP + FP + FN)} \quad (4)$$

5. Area under the curve (AUC): The area under the receiver operating characteristic (ROC) curve, or AUC, relates the true positive rate (sensitivity) to the false positive rate (1-specificity). A higher AUC implies improved model performance.

By examining these measures, a thorough picture of the model's performance can be acquired, particularly when discriminating between aberrant (positive) and normal (negative) data.

When analysing the success of machine learning models, performance measures must be considered. These metrics provide values that indicate the overall performance of a statistical or machine learning technique. In classification tasks, performance measures evaluate the model's capacity to accurately categorize data points as well as its consistency in producing the right classifications. Classification accuracy and F1 score are both measures of classification task accuracy, whereas AUC reflects a model's overall ability to forecast accurately. The study's findings, which were obtained by examining these performance metrics, are shown in the Tables 1–3 below.

Table 1. The effectiveness of the several deep learning models was assessed using an 80-pixel sub-database, displayed above.

Model	Fold	Train Accuracy	Train Loss	Val Accuracy	Val Loss	Jaccard Index	AUC	Specificity	Sensitivity
Resnet34	1	98.7508	0.0331	93.6131	0.2175	0.7447	0.9719	0.9604	0.8494
	2	97.8192	0.0604	93.7206	0.2092	0.7472	0.9717	0.9638	0.8427
	3	98.8546	0.0329	93.8111	0.2210	0.7384	0.9760	0.9699	0.8204
	4	**97.7773**	**0.0596**	**93.8226**	**0.1724**	**0.7466**	**0.9777**	**0.9625**	**0.8491**
	5	98.1767	0.0487	93.7910	0.2399	0.7640	0.9754	0.9518	0.8900
ResNet50	1	98.4818	0.0409	93.7408	0.1936	0.8763	0.8979	0.9682	0.8277
	2	**96.2419**	**0.0958**	**94.2439**	**0.1880**	**0.8860**	**0.9069**	**0.9703**	**0.8435**
	3	98.4410	0.0413	93.4971	0.2024	0.8742	0.9042	0.9577	0.8508
	4	97.0654	0.0769	92.7688	0.2016	0.8601	0.9071	0.9431	0.8711
	5	93.5411	0.1673	93.8459	0.1869	0.8772	0.9064	0.9648	0.8480
VitNet	1	83.6243	0.3631	83.7591	0.3555	0.6945	0.7132	0.9347	0.4916
	2	82.9835	0.3816	84.2836	0.3490	0.6969	0.7076	0.9493	0.4660
	3	78.4974	0.4755	78.8656	0.4560	0.5946	0.5054	0.9988	0.0121
	4	**82.5612**	**0.3898**	**84.2841**	**0.3505**	**0.6986**	**0.7015**	**0.9488**	**0.4542**
	5	78.3217	0.5231	77.4105	0.5353	0.5674	0.5	1	0
VggNet	1	98.7371	0.0366	93.7591	0.2579	0.8763	0.9746	0.9747	0.8053
	2	**96.2693**	**0.0969**	**94.3341**	**0.1785**	**0.8868**	**0.9757**	**0.9738**	**0.8353**
	3	98.5683	0.0417	93.9220	0.2120	0.8763	0.9751	0.9720	0.8178
	4	94.3224	0.1519	93.7681	0.1621	0.8756	0.9733	0.9685	0.8245
	5	98.3636	0.0487	93.7910	0.2329	0.8768	0.9724	0.9693	0.8302
EfficientNet	1	74.9474	0.5035	75.4057	0.5142	0.6094	0.7478	0.7172	0.7784
	2	82.4435	0.3866	83.0474	0.3685	0.6829	0.6726	0.9421	0.4030
	3	81.0359	0.4139	82.3709	0.3855	0.6704	0.6346	0.9463	0.3229
	4	**82.3379**	**0.3879**	**84.2231**	**0.3531**	**0.6975**	**0.6725**	**0.9419**	**0.4030**
	5	82.6571	0.3805	83.5210	0.3651	0.6944	0.6816	0.9401	0.4232
Ensemble	**1**	**98.3587**	**0.0436**	**99.3430**	**0.0252**	**0.9867**	**0.9904**	**0.9957**	**0.9850**
	2	98.5187	0.0410	99.2421	0.0211	0.9839	0.9875	0.9962	0.9787
	3	98.6001	0.0384	99.2056	0.0237	0.9836	0.9898	0.9936	0.9861
	4	97.9234	0.0627	97.8197	0.0651	0.9562	0.9642	0.9886	0.9398
	5	98.5642	0.0390	99.0869	0.0221	0.9823	0.9866	0.9943	0.9789

The most interesting results are shown in bold.

Table 2. The effectiveness of the several deep learning models was assessed using a 120-pixel sub-database, displayed above.

Model	Fold	Train Accuracy	Train Loss	Val Accuracy	Val Loss	Jaccard Index	AUC	Specificity	Sensitivity
Resnet34	1	98.8627	0.0327	96.8167	0.1097	0.8347	0.9879	0.9812	0.9075
	2	**99.2847**	**0.0187**	**97.0297**	**0.1327**	**0.8444**	**0.9815**	**0.9884**	**0.8884**
	3	98.9145	0.0336	96.9364	0.1025	0.8392	0.9907	0.9901	0.8762
	4	99.1640	0.0234	96.4475	0.1208	0.8175	0.9879	0.9836	0.8778
	5	99.2957	0.0214	96.6893	0.1074	0.8149	0.9897	0.9877	0.8642
ResNet50	1	99.5552	0.0125	96.7257	0.1120	0.9316	0.9518	0.9756	0.9281
	2	99.2449	0.0213	96.9856	0.1156	0.9372	0.9254	0.9951	0.8557
	3	**99.7229**	**0.0089**	**97.0930**	**0.1198**	**0.9388**	**0.9493**	**0.9833**	**0.9154**
	4	99.5007	0.0142	96.8166	0.1253	0.9339	0.9379	0.9853	0.8905
	5	99.5042	0.0147	97.0748	0.0995	0.9379	0.9406	0.9860	0.8951
VitNet	1	86.6963	0.3109	87.4715	0.2875	0.7518	0.7395	0.9488	0.5301
	2	84.6948	0.3495	86.1826	0.3185	0.7254	0.7123	0.9470	0.4775
	3	82.2817	0.4486	81.7531	0.4287	0.6310	0.5	1	0
	4	**85.7447**	**0.3298**	**87.9123**	**0.2929**	**0.7581**	**0.7607**	**0.9464**	**0.5750**
	5	84.2084	0.3465	86.3265	0.3085	0.7231	0.6767	0.9582	0.3951
VggNet	1	76.6185	0.4764	75.6162	0.5053	0.6152	0.7657	0.7776	0.7538
	2	**88.3740**	**0.2758**	**89.1768**	**0.2611**	**0.7889**	**0.7476**	**0.9592**	**0.5361**
	3	88.1351	0.2822	88.3388	0.2732	0.7756	0.7372	0.9605	0.5139
	4	86.6576	0.3140	88.2826	0.2813	0.7673	0.6979	0.9591	0.4368
	5	88.3702	0.2745	88.4198	0.2781	0.7779	0.7460	0.9594	0.5325
EfficientNet	1	99.1949	0.0265	96.8167	0.1372	0.9340	0.9878	0.9867	0.8818
	2	99.8978	0.0054	96.9416	0.1716	0.9363	0.9886	0.9873	0.8884
	3	**98.4339**	**0.0464**	**97.1824**	**0.1130**	**0.9403**	**0.9883**	**0.9901**	**0.8897**
	4	96.4544	0.1011	96.0322	0.1098	0.9176	0.9860	0.9757	0.8905
	5	99.8366	0.0061	96.9614	0.1410	0.9338	0.9902	0.98363	0.9005
Ensemble	1	97.2581	0.0728	97.6125	0.0646	0.9501	0.9593	0.9853	0.9332
	2	97.7973	0.0622	98.1738	0.0533	0.9627	0.9647	0.9913	0.9381
	3	97.7894	0.0603	97.8756	0.0618	0.9531	0.9636	0.9874	0.9399
	4	97.8457	0.0625	97.5778	0.0685	0.9486	0.9549	0.9876	0.9223
	5	**98.9735**	**0.0316**	**99.4250**	**0.0226**	**0.9872**	**0.9895**	**0.9965**	**0.9826**

The most interesting results are shown in bold.

Table 3. The effectiveness of the several deep learning models was assessed using a 160-pixel sub-database, displayed above.

Model	Fold	Train Accuracy	Train Loss	Val Accuracy	Val Loss	Jaccard Index	AUC	Specificity	Sensitivity
Resnet34	1	99.6311	0.0132	97.9807	0.0648	0.8949	0.9961	0.9887	0.9398
	2	99.5496	0.0148	98.1776	0.0656	0.9022	0.9913	0.9922	0.9341
	3	99.7337	0.0083	98.1038	0.0535	0.8928	0.9976	0.9912	0.9308
	4	97.1184	0.0896	97.2005	0.0821	0.8665	0.9931	0.9806	0.9361
	5	**99.5437**	**0.0213**	**98.4187**	**0.0738**	**0.9105**	**0.9929**	**0.9862**	**0.9739**
ResNet50	**1**	**99.3660**	**0.0288**	**98.4396**	**0.0579**	**0.9678**	**0.9739**	**0.9904**	**0.9573**
	2	99.5842	0.0170	98.3143	0.0592	0.9628	0.9719	0.9894	0.9544
	3	99.8263	0.0066	98.1038	0.0776	0.9599	0.9684	0.9874	0.9494
	4	99.9538	0.0018	97.9348	0.0743	0.9574	0.9620	0.9903	0.9338
	5	99.7718	0.0112	98.3708	0.0625	0.9648	0.9832	0.9839	0.9826
VitNet	1	82.2614	0.4526	81.6888	0.4532	0.6295	0.5	1	0
	2	84.4919	0.3426	86.2870	0.3141	0.7307	0.6980	0.9555	0.4405
	3	81.9212	0.4638	83.0248	0.4527	0.6543	0.5	1	0
	4	84.3706	0.3502	84.6718	0.3393	0.6977	0.6841	0.9498	0.4184
	5	**82.0483**	**0.3951**	**86.3919**	**0.3069**	**0.7274**	**0.6337**	**0.9776**	**0.2898**
VggNet	1	77.3246	0.4784	76.5826	0.4820	0.6291	0.7732	0.7643	0.7822
	2	90.0875	0.2296	93.0702	0.1838	0.8607	0.7924	0.9604	0.6244
	3	**91.0623**	**0.2155**	**92.8018**	**0.1911**	**0.8572**	**0.8139**	**0.9641**	**0.6636**
	4	91.0069	0.2178	93.0925	0.1655	0.8589	0.8168	0.9628	0.6709
	5	91.4476	0.2077	92.0146	0.1945	0.8444	0.8186	0.9659	0.6712
EfficientNet	**1**	**99.9769**	**0.0014**	**98.2101**	**0.1121**	**0.9623**	**0.9946**	**0.9915**	**0.9398**
	2	99.6882	0.0098	97.9498	0.0795	0.9584	0.9944	0.9952	0.9088
	3	98.8773	0.0347	97.9683	0.0896	0.9579	0.9921	0.9934	0.9122
	4	99.2738	0.0244	97.2005	0.0940	0.9464	0.9912	0.9834	0.9243
	5	99.6692	0.0091	98.1312	0.0814	0.9586	0.9948	0.9896	0.9391

Table 3. *Cont.*

Model	Fold	Train Accuracy	Train Loss	Val Accuracy	Val Loss	Jaccard Index	AUC	Specificity	Sensitivity
Ensemble	1	98.4785	0.0412	98.8067	0.0313	0.9751	0.9829	0.9910	0.9749
	2	99.0877	0.0268	98.9066	0.0325	0.9753	0.9775	0.9955	0.9594
	3	99.1203	0.0298	99.1873	0.0260	0.9830	0.9866	0.9945	0.9787
	4	**99.4813**	**0.0150**	**99.7246**	**0.0079**	**0.9942**	**0.9964**	**0.9977**	**0.9952**
	5	98.9165	0.0319	99.3291	0.0245	0.9836	0.9901	0.9948	0.9855

The most stunning results are shown in bold.

4. Discussion

In this work, the proposed model was trained on datasets with varied picture sizes, such as 80 × 80 and 120 × 120 and 160 × 160. Findings highlights the critical significance of ongoing innovation and exploration in increasing medical machine learning and hence improving healthcare practices. Notably, the findings show that the top five ensemble models had high detection accuracy across all sub-databases. The overall ensemble model demonstrated the highest accuracy, (see Tables 1–3) surpassing the performance of VGGNet, ResNet34, and ResNet50, which also outperformed VitNet and EfficientNet. The only exception was in the 160 × 160 sub-database, where EfficientNet achieved an accuracy of 92%. By combining multidisciplinary techniques and technology breakthroughs, one can pave the way towards a future in which the early and precise identification of gastrointestinal disorders is not only achievable but also common practice in protecting human health and wellbeing.

This deliberate method was used to assist the model build a solid knowledge of multiple picture dimensions, making it more flexible to a variety of real-world scenarios. Nonetheless, it is worth noting that the proposed model's knowledge may have been expanded much more with greater computing resources. Greater computational capability might have enabled a more in-depth examination of the dataset's intricacies, as well as the discovery of insights beyond existing capabilities.

These advanced methodologies offer various structures and improvement tactics that may increase the efficacy of the model. Time restrictions precluded the incorporation of these algorithms into the current system; nonetheless, their use holds great promise for improving cancer detection techniques.

This research prioritizes in innovation and placing people at the heart of the research. The objective is to make research applications as user-friendly as possible while also emphasizing how they might benefit medical practitioners. This research strives to integrate the most recent research work with practical applications, avoiding plagiarism and crafting a research story that is honest and true to the research's commitment to expanding scientific understanding.

Experimental Setting

In the future, the focus will be on two major goals: developing an easy-to-use web app for cancer detection and expanding current algorithms to include new forms of cancer. The primary goal of this research is to develop a model that will work for all forms of cancer, not just specific forms. This will be accomplished by utilizing many data sources and innovative algorithms in order to obtain a thorough understanding of cancer, hence assisting healthcare professionals with diagnosis.

By integrating multidisciplinary methodologies and capitalizing on technological developments, we hope to contribute towards a future in which the early and precise identification of GI problems is not only possible, but is also the standard for protecting human health and wellbeing.

5. Conclusions

Detecting and diagnosing gastrointestinal (GI) illnesses is critical for human health, yet it can be challenging owing to limited medical competence and expensive expenses.

The use of machine learning, particularly deep learning techniques, has the potential to increase the speed and accuracy of GI illness identification. This research study investigated the efficiency of ensemble approaches using five pre-trained models on the Gastric Histopathology Sub-size Image Database (GasHisSDB), which comprises a diverse set of pictures in various pixel sizes and categories. Considerable boost in prediction accuracy can be seen when utilizing ensemble learning, which combines the predictive skills of several models, as opposed to using individual models and in contrast with previous studies that used comparable datasets. This demonstrates the potential of ensemble approaches to improve the capabilities of medical machine learning systems, resulting in more effective and precise diagnoses. The current approach is based on transfer learning, a technique that improves model learning by leveraging knowledge from previously trained models. Moreover, we employed an ensemble strategy to improve performance by merging various classifiers. Following a rigorous review, the current approach revealed a good accuracy for the test dataset, exceeding current evaluation techniques. This demonstrates how deep learning may help alleviate the pressure on healthcare systems while also improving human health outcomes [22–25]. In this study, advanced deep ensemble learning models were created that used transfer learning from multiple pre-trained networks, including VitNet, EfficientNet, VGGNet, ResNet34, and ResNet50, to improve stomach cancer diagnosis. The study found that using base models in ensemble learning resulted in high identification accuracy (97.57% to 99.72%) for histopathology pictures with resolutions ranging from 80 × 80 pixels to 160 × 160 pixels. The experimental results demonstrate that ensemble models may extract key information even from smaller picture patches while retaining good performance. This improvement implies the possibility of adopting digital scanners with lower specifications, as well as reduced data storage and computing needs for histopathology operations. This, in turn, might increase the speed of stomach cancer identification and perhaps increase survival rates. Continued work in these areas is intended to push the boundaries of medical image processing and enhance clinical results.

However, it is critical to recognize the limits of the current research. The usage of a restricted dataset emphasizes the need to have access to larger, higher-quality datasets to enhance and confirm the approaches. Furthermore, computational restrictions may have influenced the scope of the results. Future studies might seek to introduce new preprocessing methods and to optimize algorithms to boost performance. Furthermore, the field of medical image retrieval offers several chances for continuous research, including the ability to use multiple deep learning approaches and models for complete examination.

Supplementary Materials: The GasHisSDB dataset is openly available at this Link: https://paperswithcode.com/dataset/gashissdb (Access Date: 11 March 2024).

Author Contributions: Conceptualization, S.M.A.A.-H., M.D. and S.A.; Methodology, M.D., G.R.M., S.M.A.A.-H. and S.A.; Software, G.R.M., I.H.A., M.D., S.M.A.A.-H. and S.A.; Validation, I.H.A., G.R.M., S.M.A.A.-H., A.B., B.A., R.C., S.T.R. and S.A.; Formal analysis, S.A.; Investigation, S.M.A.A.-H., A.B., B.A., R.C., S.T.R. and S.A.; Writing—original draft, G.R.M.; Writing—review & editing, S.M.A.A.-H. and S.A.; Visualization, S.A.; Supervision, S.A. All authors have read and agreed to the published version of the manuscript.

Funding: This research received no external funding.

Institutional Review Board Statement: Not applicable.

Informed Consent Statement: Informed consent was obtained from all subjects involved in the study.

Data Availability Statement: The data presented in this study are available in this article.

Acknowledgments: We would like to express our deepest gratitude to the Roux Institute, the IEAI and the Alfond Foundation for their invaluable support and contributions.

Conflicts of Interest: The authors declare no conflicts of interest.

References

1. Rai, H.M. Cancer detection and segmentation using machine learning and deep learning techniques: A review. *Multimed. Tools Appl.* **2024**, *83*, 27001–27035. [CrossRef]
2. Kuntz, S.; Krieghoff-Henning, E.; Kather, J.N.; Jutzi, T.; Höhn, J.; Kiehl, L.; Hekler, A.; Alwers, E.; von Kalle, C.; Fröhling, S.; et al. Gastrointestinal cancer classification and prognostication from histology using deep learning: Systematic review. *Eur. J. Cancer* **2021**, *155*, 200–215. [CrossRef] [PubMed]
3. Suzuki, H.; Yoshitaka, T.; Yoshio, T.; Tada, T. Artificial intelligence for cancer detection of the upper gastrointestinal tract. *Dis. Esophagus* **2021**, *33*, 254–262. [CrossRef] [PubMed]
4. Chao, W.L.; Manickavasagan, H.; Krishna, S.G. Application of Artificial Intelligence in the Detection and Differentiation of Colon Polyps: A Technical Review for Physicians. *Diagnostics* **2019**, *9*, 99. [CrossRef]
5. Li, B.; Meng, M.Q.H. Automatic polyp detection for wireless capsule endoscopy images. *Expert. Syst. Appl.* **2012**, *39*, 10952–10958. [CrossRef]
6. Guo, L.; Gong, H.; Wang, Q.; Zhang, Q.; Tong, H.; Li, J.; Lei, X.; Xiao, X.; Li, C.; Jiang, J.; et al. Detection of multiple lesions of gastrointestinal tract for endoscopy using artificial intelligence model: A pilot study. *Surg. Endosc.* **2021**, *35*, 6532–6538. [CrossRef] [PubMed]
7. Charfi, S.; El Ansari, M.; Balasingham, I. Computer-aided diagnosis system for ulcer detection in wireless capsule endoscopy images. *IET Image Process.* **2019**, *13*, 1023–1030. [CrossRef]
8. Mirza, O.M.; Alsobhi, A.; Hasanin, T.; Ishak, M.K.; Karim, F.K.; Mostafa, S.M. Computer Aided Diagnosis for Gastrointestinal Cancer Classification Using Hybrid Rice Optimization With Deep Learning. *IEEE Access* **2023**, *11*, 76321–76329. [CrossRef]
9. Liedlgruber, M.; Uhl, A. Computer-aided decision support systems for endoscopy in the gastrointestinal tract: A review. *IEEE Rev. Biomed. Eng.* **2011**, *4*, 73–88. [CrossRef]
10. Sivari, E.; Bostanci, E.; Guzel, M.S.; Acici, K.; Asuroglu, T.; Ercelebi Ayyildiz, T. A New Approach for Gastrointestinal Tract Findings Detection and Classification: Deep Learning-Based Hybrid Stacking Ensemble Models. *Diagnostics* **2023**, *13*, 720. [CrossRef]
11. Gunasekaran, H.; Ramalakshmi, K.; Swaminathan, D.K.; J, A.; Mazzara, M. GIT-Net: An Ensemble Deep Learning-Based GI Tract Classification of Endoscopic Images. *Bioengineering* **2023**, *10*, 809. [CrossRef] [PubMed] [PubMed Central]
12. Naz, J.; Sharif, M.; Yasmin, M.; Raza, M.; Khan, M.A. Detection and Classification of Gastrointestinal Diseases using Machine Learning. *Curr. Med. Imaging Former. Curr. Med. Imaging Rev.* **2020**, *17*, 479–490. [CrossRef]
13. Wang, X.; Qian, H.; Ciaccio, E.J.; Lewis, S.K.; Bhagat, G.; Green, P.H.; Xu, S.; Huang, L.; Gao, R.; Liu, Y. Celiac disease diagnosis from videocapsule endoscopy images with residual learning and deep feature extraction. *Comput. Methods Programs Biomed.* **2020**, *187*, 105236. [CrossRef]
14. Renna, F.; Martins, M.; Neto, A.; Cunha, A.; Libânio, D.; Dinis-Ribeiro, M.; Coimbra, M. Artificial Intelligence for Upper Gastrointestinal Endoscopy: A Roadmap from Technology Development to Clinical Practice. *Diagnostics* **2022**, *12*, 1278. [CrossRef] [PubMed]
15. Hu, W.; Li, C.; Li, X.; Rahaman, M.M.; Ma, J.; Zhang, Y.; Chen, H.; Liu, W.; Sun, C.; Yao, Y.; et al. GasHisSDB: A new gastric histopathology image dataset for computer aided diagnosis of gastric cancer. *Comput. Biol. Med.* **2022**, *142*, 105207. [CrossRef]
16. Sapra, V.; Sapra, L.; Bhardwaj, A.; Bharany, S.; Saxena, A.; Karim, F.K.; Ghorashi, S.; Mohamed, A.W. Integrated approach using deep neural network and CBR for detecting severity of coronary artery disease. *Alex. Eng. J.* **2023**, *68*, 709–720. [CrossRef]
17. Yong, M.P.; Hum, Y.C.; Lai, K.W.; Lee, Y.L.; Goh, C.H.; Yap, W.S.; Tee, Y.K. Histopathological Gastric Cancer Detection on GasHisSDB Dataset Using Deep Ensemble Learning. *Diagnostics* **2023**, *13*, 1793. [CrossRef] [PubMed] [PubMed Central]
18. Su, Q.; Wang, F.; Chen, D.; Chen, G.; Li, C.; Wei, L. Deep convolutional neural networks with ensemble learning and transfer learning for automated detection of gastrointestinal diseases. *Comput. Biol. Med.* **2022**, *150*, 106054. [CrossRef] [PubMed]
19. Almanifi, O.R.A.; Razman, M.A.M.; Khairuddin, I.M.; Abdullah, M.A.; Majeed, A.P.P.A. Automated Gastrointestinal Tract Classification Via Deep Learning and The Ensemble Method. In Proceedings of the 2021 21st International Conference on Control, Automation and Systems (ICCAS), Jeju, Republic of Korea, 12–15 October 2021; pp. 602–606. [CrossRef]
20. Haile, M.B.; Salau, A.O.; Enyew, B.; Belay, A.J.; Jin, Z. Detection and classification of gastrointestinal disease using convolutional neural network and SVM. *Cogent Eng.* **2022**, *9*, 2084878. [CrossRef]
21. Billah, M.; Waheed, S. Gastrointestinal polyp detection in endoscopic images using an improved feature extraction method. *Biomed. Eng. Lett.* **2018**, *8*, 69–75. [CrossRef]
22. Abin, D.; Mudavadkar, G.; Minase, R.; Handoo, M.; Kumawat, R. MRI Image Enhancement for Brain Tumor Detection using Hybridization of Contrast Enhancement Techniques. In Proceedings of the 2022 International Conference on Knowledge Engineering and Communication Systems (ICKES), Chickballapur, India, 28–29 December 2022; pp. 1–5. [CrossRef]
23. Ghanzouri, I.; Amal, S.; Ho, V.; Safarnejad, L.; Cabot, J.; Brown-Johnson, C.G.; Leeper, N.; Asch, S.; Shah, N.H.; Ross, E.G. Performance and Usability Testing of an Automated Tool for Detection of Peripheral Artery Disease Using Electronic Health Records. *Sci. Rep.* **2022**, *12*, 13364. [CrossRef]

24. Singh, A.; Wan, M.; Harrison, L.; Breggia, A.; Christman, R.; Winslow, R.L.; Amal, S. Visualizing Decisions and Analytics of Artificial Intelligence Based Cancer Diagnosis and Grading of Specimen Digitized Biopsy: Case Study for Prostate Cancer. In Proceedings of the Companion Proceedings of the 28th International Conference on Intelligent User Interfaces, Sydney, Australia, 27–31 March 2023; Association for Computing Machinery: New York, NY, USA, 2023; pp. 166–170.
25. Yasmin, F.; Hassan, M.M.; Hasan, M.; Zaman, S.; Bairagi, A.K.; El-Shafai, W.; Fouad, H.; Chun, Y.C. GastroNet: Gastrointestinal Polyp and Abnormal Feature Detection and Classification with Deep Learning Approach. *IEEE Access* **2023**, *11*, 97605–97624. [CrossRef]

diagnostics

Article

Brain and Ventricle Volume Alterations in Idiopathic Normal Pressure Hydrocephalus Determined by Artificial Intelligence-Based MRI Volumetry

Zeynep Bendella [1,2], Veronika Purrer [2,3], Robert Haase [1,2], Stefan Zülow [1], Christine Kindler [2,3], Valerie Borger [4], Mohammed Banat [4], Franziska Dorn [1], Ullrich Wüllner [2,3], Alexander Radbruch [1,2] and Frederic Carsten Schmeel [1,2,*]

1 Department of Neuroradiology, Faculty of Medicine, University Hospital Bonn, Rheinische Friedrich-Wilhelms-Universität Bonn, 53127 Bonn, Germany; zeynep.bendella@ukbonn.de (Z.B.); robert.haase@ukbonn.de (R.H.); stefan.zuelow@ukbonn.de (S.Z.); franziska.dorn@ukbonn.de (F.D.); alexander.radbruch@ukbonn.de (A.R.)
2 German Center of Neurodegenerative Diseases (DZNE), 53127 Bonn, Germany; veronika.purrer@ukbonn.de (V.P.); christine.kindler@ukbonn.de (C.K.); ullrich.wuellner@ukbonn.de (U.W.)
3 Department of Neurodegenerative Diseases, University Hospital Bonn, Rheinische Friedrich-Wilhelms-Universität Bonn, 53127 Bonn, Germany
4 Department of Neurosurgery, University Hospital Bonn, Rheinische Friedrich-Wilhelms-Universität Bonn, 53127 Bonn, Germany; valerie.borger@ukbonn.de (V.B.); mohammed.banat@ukbonn.de (M.B.)
* Correspondence: carsten.schmeel@ukbonn.de; Tel.: +49-0228-28716507; Fax: +49-0228-28714321

Abstract: The aim of this study was to employ artificial intelligence (AI)-based magnetic resonance imaging (MRI) brain volumetry to potentially distinguish between idiopathic normal pressure hydrocephalus (iNPH), Alzheimer's disease (AD), and age- and sex-matched healthy controls (CG) by evaluating cortical, subcortical, and ventricular volumes. Additionally, correlations between the measured brain and ventricle volumes and two established semi-quantitative radiologic markers for iNPH were examined. An IRB-approved retrospective analysis was conducted on 123 age- and sex-matched subjects (41 iNPH, 41 AD, and 41 controls), with all of the iNPH patients undergoing routine clinical brain MRI prior to ventriculoperitoneal shunt implantation. Automated AI-based determination of different cortical and subcortical brain and ventricular volumes in mL, as well as calculation of population-based normalized percentiles according to an embedded database, was performed; the CE-certified software mdbrain v4.4.1 or above was used with a standardized T1-weighted 3D magnetization-prepared rapid gradient echo (MPRAGE) sequence. Measured brain volumes and percentiles were analyzed for between-group differences and correlated with semi-quantitative measurements of the Evans' index and corpus callosal angle: iNPH patients exhibited ventricular enlargement and changes in gray and white matter compared to AD patients and controls, with the most significant differences observed in total ventricular volume (+67%) and the lateral (+68%), third (+38%), and fourth (+31%) ventricles compared to controls. Global ventriculomegaly and marked white matter reduction with concomitant preservation of gray matter compared to AD and CG were characteristic of iNPH, whereas global and frontoparietally accentuated gray matter reductions were characteristic of AD. Evans' index and corpus callosal angle differed significantly between the three groups and moderately correlated with the lateral ventricular volumes in iNPH patients [Evans' index ($r > 0.83$, $p \leq 0.001$), corpus callosal angle ($r < -0.74$, $p \leq 0.001$)]. AI-based MRI volumetry in iNPH patients revealed global ventricular enlargement and focal brain atrophy, which, in contrast to healthy controls and AD patients, primarily involved the supratentorial white matter and was marked temporomesially and in the midbrain, while largely preserving gray matter. Integrating AI volumetry in conjunction with traditional radiologic measures could enhance iNPH identification and differentiation, potentially improving patient management and therapy response assessment.

Keywords: normal pressure hydrocephalus; brain atrophy; quantitative neuroimaging; automated volumetrization

1. Introduction

Diagnosing and monitoring neurodegenerative diseases often involve assessing volume loss in one or multiple brain regions through brain imaging studies. Beyond traditional methods like manual volume measurements through classical segmentation or employing scoring systems, machine learning advancements have led to a rise in automated software solutions available to radiologists. One such solution is artificial intelligence (AI)-driven software for automated brain volumetry for patients suspected of having neurodegenerative diseases, probably complementing standard diagnostic procedures.

Idiopathic normal pressure hydrocephalus (iNPH) poses diagnostic and therapeutic challenges in the realm of neurodegenerative disorders. Characterized by gait disturbance, cognitive decline, and urinary incontinence, iNPH presents with seemingly paradoxical cerebral ventricular enlargement despite normal cerebrospinal fluid (CSF) pressure [1]. Predominantly affecting the elderly, iNPH is often misdiagnosed due to symptom overlap with age-related conditions like Alzheimer's disease (AD), Parkinson's disease, or age-related cognitive decline [2]. It is estimated that 1.6–5.4% of individuals with dementia are impacted by iNPH [3]. In fact, iNPH is frequently misidentified as Alzheimer's disease (AD) due to shared pathophysiological features and comparable imaging characteristics, such as ventricular enlargement and increased white matter lesions (WMLs). The extent to which any patient exhibits reversible symptoms largely depends on the clinician's ability to accurately differentiate iNPH from other neurodegenerative diseases, such as AD. In cases of severe iNPH, there is a 75% likelihood of overlapping characteristics with AD, complicating the differential diagnosis, especially with standard imaging techniques, and leading to varied outcomes in patient management [4,5]. While several reviews focus on the use of various MRI techniques to detect AD and iNPH separately and assess their clinical utility, almost none comprehensively address the advantages and disadvantages of these techniques in distinguishing between AD and iNPH or their ability to predict shunt responsiveness [6]. It was recently shown that the variations in the extent and distribution of WMLs in iNPH and AD, particularly the predominance of deep WMLs over periventricular lesions in iNPH, were indicative of reduced fluid and amyloid-beta (Aβ) clearance [7]. However, currently, no imaging studies have been able to definitively and reliably diagnose or differentiate between AD and iNPH.

Understanding iNPH pathophysiology necessitates exploring alterations in brain and ventricle volumes. Modern imaging techniques have revealed specific radiological signs aiding iNPH diagnosis and elucidating underlying structural brain changes. Notably, ventriculomegaly in iNPH reflects dynamic CSF flow and cerebral blood flow dynamics, potentially impacting surrounding brain tissue [8]. In the realm of radiological markers, the "Evans' index" stands out as an emblem of ventriculomegaly, where a ratio greater than 0.3 indicates enlargement of the frontal horns relative to the internal diameter of the skull [9]. Further features include the "tight high convexity", reflecting narrowing subarachnoid spaces over the brain's high convexity, and the "disproportionately enlarged subarachnoid-space hydrocephalus (DESH)" pattern, pointing to a combination of ventricular expansion and enlarged Sylvian fissures [10]. Another pivotal sign is the corpus callosum angle. Found at the junction of the corpus callosum and the septum pellucidum in sagittal MRI scans, a diminished or flattened angle in iNPH patients suggests ventricle enlargement's effects on neighboring cerebral structures [11]. Ishii et al.'s research emphasizes the angle's clinical relevance, asserting its capacity to differentiate iNPH from other neurodegenerative conditions [12]. In clinical practice, however, it is often necessary if not mandatory, according to recent guideline recommendations, to combine several of the above-mentioned radiological markers in order to achieve an acceptably high accuracy for the diagnosis of iNPH. This emphasizes the importance and complexity of comprehensive imaging analysis for the accurate diagnosis of iNPH.

Yet, no dedicated quantitative analysis of brain and ventricular volume differences has been conducted for iNPH, as imaging studies have focused primarily on visually assessable and semi-quantitative features. Consequently, it remains unclear whether structural brain changes exist alongside apparent ventricular enlargement and if the quantitative extent of volumetric differences in brain and ventricular size in iNPH can be adequately assessed with currently applied semi-quantitative methods. Objective quantitative imaging markers, together with clinical presentation, could provide a holistic view and pave the way for targeted therapeutic interventions and improved patient outcomes. Therefore, this exploratory retrospective study aims to investigate brain and ventricular volume distributions in clinically confirmed iNPH patients compared to those with AD and age- and sex-matched healthy controls, using AI-based brain magnetic resonance imaging (MRI) volumetry to analyze different cortical, subcortical, and ventricular volumes and to compare these findings with two established semi-quantitative measurement techniques commonly employed in clinical settings.

2. Materials and Methods

2.1. Study Population

The research received approval from the Ethics Committee for Clinical Trials on Humans and Epidemiological Research with Personal Data at the Faculty of Medicine of the Rheinische Friedrich-Wilhelms-Universität Bonn (reference No. 118/22). Due to the retrospective nature of this study, written informed consent was not required. Adult patients who underwent brain MR imaging in our department in 2000 or later and met the specified inclusion criteria were identified and included in this study through a review of our internal radiologic information system.

The criteria for inclusion encompassed (a) brain MR imaging that featured an unenhanced, three-dimensional (3D), T1-weighted (T1w) sequence; (b) a verified clinical diagnosis of idiopathic normal pressure hydrocephalus (iNPH) with benefit of firstly a corticospinal fluid (CSF) tap test and secondly a response to a following ventricular shunting; and (c) documentation of imaging and clinical pathology related to iNPH diagnosis in the written report. The clinical diagnosis of iNPH was based on the following criteria:

1. A record of walking difficulties, advancing cognitive decline, and a sense of urgency or loss of control over urination;
2. Imaging characteristics of hydrocephalus with disproportionated enlargement of the ventricles in comparison to the outer subarachnoid spaces, whereby an Evans' index >0.30 was indicative of iNPH on computed tomography (CT) or magnetic resonance (MR) imaging;
3. A CSF opening pressure (appropriately measured) of <24 cm of water.

The Evans' index is calculated as the ratio between the maximum width of the frontal horns of the lateral ventricles and the greatest internal diameter of the skull on the same axial plane, as observed in CT and MRI images. The CSF tap test involves the assessment of walking and cognitive functions both prior to and following the removal of 40–50 mL of lumbar CSF. It is the sole technique capable of momentarily replicating the impact of a permanent shunt, probably allowing it to forecast not just the results of the surgical procedure but also the extent of recovery [13,14].

The International Guidelines have recommended the following key imaging features for the diagnosis of iNPH and the selection of shunt-responsive patients [15,16]:

1. Expansion of the ventricles that cannot be fully ascribed to cerebral shrinkage or congenital enlargement (Evans' index > 0.3).
2. Absence of any visible blockage impeding the flow of CSF.
3. Presence of at least one of the following indicative characteristics:

- Enlargement of the temporal horns of the lateral ventricles not entirely attributable to hippocampus atrophy;
- Callosal angle of <90°;
- Evidence of altered brain water content, including periventricular signal changes on CT and MRI not attributable to microvascular ischemic changes or demyelination;
- An aqueductal or fourth ventricular flow void on MRI.

Forty-one patients met these criteria and were grouped into the iNPH cohort. An additional group of 41 patients with clinically diagnosed Alzheimer's disease (AD) and a healthy control group of 41 patients, with each group being matched for age and gender, were added for group comparisons. The control group was formed from patients who had undergone elective brain MRI for assessment of non-specific headache but exhibited entirely normal brain MR imaging results according to written clinical reports of board-certified radiologists. The general demographic characteristics of the patients are detailed in Table 1.

Table 1. Demographic characteristics of the study participants (n = 82).

	NPH (n = 41)	Control Group (n = 41)	Alzheimer's Disease (n = 41)	*p*-Value †
Age—yr	79.4 ± 7.5	79.0 ± 4.9	81.6 ± 8.4	0.766
Male sex—No. (%)	28 (68%)	27 (66%)	21 (51%)	0.817

† a *p*-value < 0.05 was considered statistically significant. Group differences were calculated using the Kruskal–Wallis test.

2.2. Magnetic Resonance Imaging

All iNPH patients, AD patients, and healthy controls underwent a standardized brain MR imaging study performed at a clinical 1.5 T or 3 T clinical whole-body MRI system (both Achieva TX, Philips Healthcare, Best, the Netherlands), each equipped with an 8-channel head coil and consistent scanning protocols. Exemplary protocol details regarding the MRI scanning parameters for the 3 T MRI scanner are provided in Table 2.

Table 2. Sequence parameters.

Sequence	Pulse Type	Orientation	TR (ms)	TE (ms)	Reconstructed Voxel Size (mm)	Matrix (mm)	Slices
T2w	Turbo spin echo	axial	13.257	90	0.94 × 0.94 × 1	240 × 174	140
SWI	3D fast field echo	axial	31	0	0.6 × 0.6 × 2	384 × 316	145
DWI	*b* values (0, 500, 1000 s/mm^2)	axial	2725	41	1 × 1 × 5	128 × 127	24
T1w	MPRAGE	sagittal	7.3	3.9	1 × 1 × 1	256 × 256	180
FLAIR	3D gradient echo	sagittal	4800	275.776	1.12 × 1.12 × 1.12	240 × 240	321

DWI, diffusion-weighted imaging; MPRAGE, magnetization-prepared rapid gradient echo; SWI, susceptibility-weighted imaging; T1w, T1-weighted; T2w, T2-weighted; TE, echo time; TR, repetition time.

2.3. Image Analysis

Board-certified radiologists with >5 years of neuroradiological experience visually examined the MRI datasets of all subjects for acute cerebral pathology and the presence of the imaging inclusion criteria. An experienced neuroradiologist (Z.B. with nine years' experience in neuroimaging) verified all the documented qualitative radiological signs of iNPH and measured two semi-quantitative scores, namely the Evans' index and the corpus callosal angle. The interpretation of the Evans' index is as follows: 0.20–0.25, normal; 0.25–0.30, possible or early ventriculomegaly; and >0.30, ventriculomegaly [17]. The ratio varies with age and sex; therefore, the ratio can be higher in a normal healthy elderly individual [18]. The corpus callosal angle should be measured on a coronal image perpendicular to the anterior commissure–posterior commissure (AC-PC) plane at the level of the posterior commissure [16,19]. INPH patients have smaller angles than those with ventriculomegaly from atrophy or normal controls: a normal value is between 100 and 120°; in patients with iNPH, the value is <90°. Examples of both radiological markers are visually shown in Figure 1.

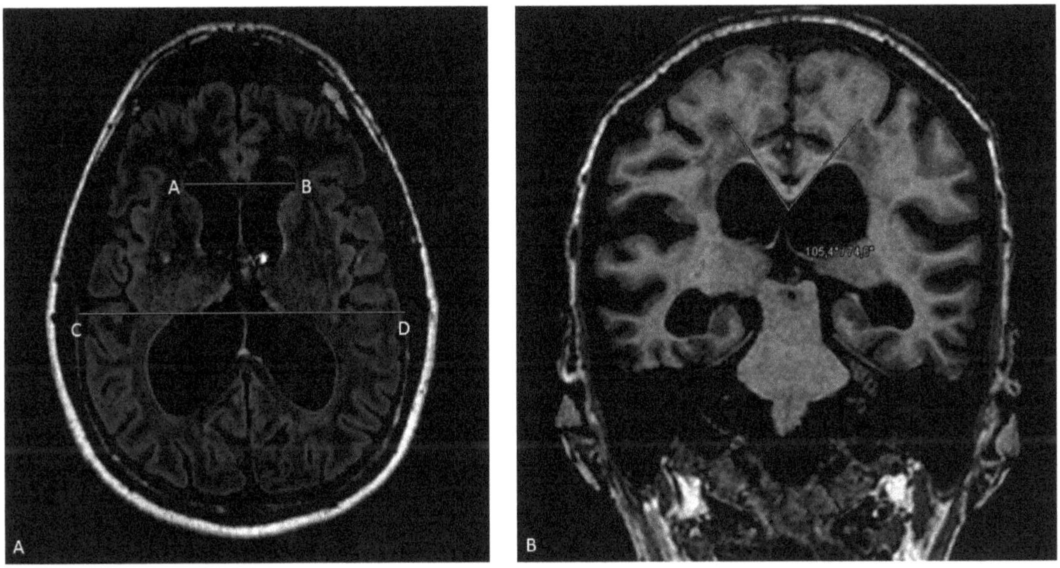

Figure 1. Manual measurements of the Evans' index: (**A**) The ratio between the maximum width of the frontal horns of the lateral ventricles (line from A to B) and the greatest internal diameter of the skull (line from C to D) on the same axial plane, in an MRI image. The corpus callosal angle (**B**) measured on a coronal image perpendicular to the anterior commissure–posterior commissure (AC-PC) plane at the level of the posterior commissure.

2.4. Post-Processing and Artificial Intelligence (AI)-Based Volumetry

An automated AI-driven software tool was employed to perform quantitative volume analyses of various brain regions, measured in mL. The analysis was supplemented by age- and gender-adjusted percentiles derived from an internal reference group of age- and gender-matched healthy individuals embedded in the software. This MRI post-processing software, "mdbrain", is a product of Mediaire GmbH, Berlin, Germany, a certified manufacturer under the European Medical Device Directive 93/42/EEC, and also aligns with the DIN EN ISO 13485:2016 standards [20]. The "mdbrain" software, holding a CE mark, facilitates automatic brain volumetry of numerous brain sections using 3D T1-weighted sequences, leveraging a vast normative database from the general population. Importantly, this software's algorithm and its encompassed normative database are refined based on a broad dataset of multi-center imaging data, surpassing the confines of a single institution, and was endorsed for its precision [21,22].

The 3D T1w MPRAGE sequence was uploaded from the clinical PACS into the mdbrain software, version 4.4.1 or subsequent versions, to facilitate automatic volume determination and percentile assessment. Afterward, volumes and percentiles for a comprehensive list of structures—from the entire brain to individual ventricles—were automatically generated, stored, and then verified for their credibility, as shown in Figure 2.

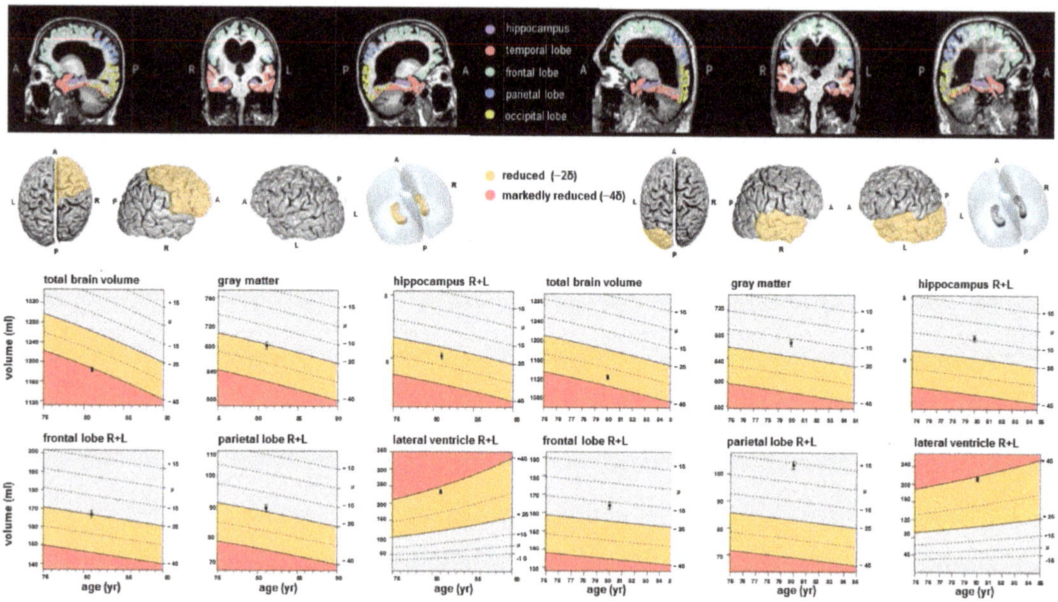

Figure 2. Sample excerpt of the software output (headings modified by the authors for translation from German) of two exemplary patients. The black dot within the percentile image indicates the patient's individual percentile in the corresponding brain area. Abbreviations: A = anterior, P = posterior, R = right and L = left. It is noteworthy that volume measurements for symmetrical structures were conducted individually for each counterpart. The analyzed cortical and subcortical brain structures and ventricles included the following: whole brain volume, whole brain white matter, whole brain gray matter, cortical gray matter, frontal volume right and left, parietal volume right and left, precuneus volume right and left, occipital volume right and left, temporal volume right and left, hippocampus volume right and left, parahippocampal gyrus volume right and left, regio entorhinalis volume right and left, nucleus caudatus volume right and left, putamen volume right and left, pallidum volume right and left, thalamus volume right and left, brainstem volume, mesencephalon volume, pons volume, cerebellar gray matter volume, left and right ventricle, third and fourth ventricle.

The automated volumetry process includes the following steps [23,24]:

1. Segmenting the structures of interest. A custom deep learning segmentation model, based on the 3D U-Net architecture [25], is utilized for this purpose. Input to the model is a 3D T1 MRI scan (cropped to include only the head and resampled to a uniform size), and the output of the model is a 3D segmentation mask, containing segmentation of various brain structures as listed above. The model was trained using the Adam optimization algorithm, a variant of stochastic gradient descent. The training dataset consisted of 2869 gender-balanced MRI scans and corresponding segmentation annotations created through a proprietary multi-rater annotation process. Various augmentation techniques, such as contrast, resolution, rotation, and elastic deformation adjustments, were applied during training to enhance the model's generalizability. The algorithm was validated on a test dataset consisting of MRI scans from 121 subjects and the corresponding manual segmentations by human experts.
2. Calculating the volume of the segmented structures by counting the number of voxels in the segmentation mask and multiplying by the voxel volume.
3. Comparing the calculated volumes to a reference population of healthy individuals (n = 6099, balanced for gender, mean age 41 ± 23 years, age range 10–97 years,

with diverse image sources from Europe, the United States, Australia, and China) to determine percentiles, accounting for age, sex, and total intracranial volume.

As per the instructions for use of mdbrain, the AI software has the following limitations:

- Use is restricted to certain allowed MR scan parameter ranges (for example, the spacing between slices must be below 2 mm);
- Quality of results cannot be guaranteed for patients with an age outside the range of 10–99 years or those with tumor or stroke;
- Comparability of results is not guaranteed if the image data were recorded with different MR sequences.

2.5. Statistical Analysis

Statistical analyses were conducted using SPSS statistical software (v27 and above, IBM Corp., Armonk, NY, USA). All applicable demographic and imaging data are given as mean ± standard deviation, unless otherwise specified. The statistical significance level was set at $p = 0.05$. Mann–Whitney U and Kruskal–Wallis tests were used for pairwise and multiple group comparisons of independent clinical and imaging data, and one-way ANOVA with post hoc testing after Bonferroni correction was used for pairwise inter-class comparisons. Spearman's rank-order correlation was used to assess for potential associations between the measured Evans' index and corpus callosal angle with the AI-based volumetrically measured total ventricle volume. Receiver operating characteristic (ROC) curves were plotted to determine the optimal Evans' index, corpus callosal angle, and total ventricle volume thresholds in order to differentiate iNPH from healthy controls and AD patients. The sensitivity, specificity, and diagnostic accuracy of both semi-quantitative and quantitative imaging parameters were calculated using these cut-off values.

3. Results

The general patients' demographics including age and gender distribution showed no statistically significant differences among the three groups of iNPH, AD, and healthy controls (CG), as shown in Table 1.

3.1. Brain Volume Deviations in iNPH Patients as Compared to a Norm Collective

In a comparison of the individual brain (sub)volumes of iNPH patients with the embedded population-based norm collective, a significant proportion of subjects with volume deviations from the norm collective was found, especially in the lateral ventricles and third ventricle, both thalami, hippocampi, and putamina, as detailed in Figure 3 (Supplementary Table S1).

3.2. Artificial Intelligence-Based Volumetric Results

The volumetry software successfully processed all MR imaging studies, and the measurement results of brain area and ventricle volumes in mL and the corresponding relative measurement values normalized to the total intracranial volume (ICV, sum of total brain volume and total ventricle volume) for the three participant groups, iNPH, AD, and controls, are shown in Table 3 (absolute volume measurements in mL) and Table 4 (relative volumes normalized to ICV). In the comparative analysis of brain volumetric measurements in mL among groups, several significant differences were observed across various brain regions: Overall, ICV in mL was significantly higher in iNPH patients compared to both controls and AD groups, with mean ICV values of 1231.81 ± 144, 1181.17 ± 98, and 1115.96 ± 115, respectively ($p < 0.001$). These differences represented relative volume increases of 4% compared to controls and 10% compared to the AD patients. Regarding total brain volume, iNPH patients exhibited lower volumes compared to controls but higher volumes compared to AD patients, with mean mL values of 1102.46 ± 122, 1138.41 ± 93, and 1050.72 ± 110, respectively ($p = 0.003$). The relative volume reduction was 3% for iNPH compared to controls, and the relative volume increase was 5% compared to AD patients. In terms of specific brain regions, notable differences were observed in gray matter and white matter volumes in mL: iNPH patients displayed a mean gray matter volume of 638.69 ± 64 and a mean white matter volume of 462.07 ± 83. These volumes were lower compared to controls (gray matter: 641.54 ± 55; white matter: 496.87 ± 46), higher compared to AD gray matter (576.48 ± 59), and lower compared to AD white matter (474.24 ± 56). The relative volume increases in gray matter were 0% compared to controls and 10% compared to AD patients, while in white matter, volume reductions were 8% compared to controls and 3% compared to AD patients. Significant differences were also found in specific brain regions such as the frontal lobe, parietal lobe, precuneus, occipital lobe, temporal lobe, hippocampus, parahippocampus, entorhinal lobe, caudate, putamen, pallidum, thalamus, brainstem, mesencephalon, pons, and cerebellum among the three groups. For instance, in the temporal lobe, iNPH patients exhibited a mean mL volume of 116.03 ± 16, which was lower than that of controls (130.11 ± 12) and higher than that of AD patients (103.79 ± 14). This corresponded to a relative volume reduction of 12% compared to controls and to an increase of 11% compared to AD patients ($p < 0.001$). In the frontal lobe, iNPH patients exhibited a 1% reduction compared to controls and a 6% increase compared to AD patients. In the hippocampus, iNPH patients had a mean mL volume of 5.77 ± 1, whereas controls and AD patients showed volumes of 6.72 ± 1 and 5.79 ± 4, respectively. This indicated a relative volume reduction of 14% compared to controls and no difference compared to AD patients ($p < 0.001$).

Regarding ventricular volumes, iNPH patients showed substantial enlargement, with a 67% increase in total ventricular volume compared to controls. Specifically, the lateral ventricles displayed a 68% increase in volume, with percentage changes of 69% for the right lateral ventricle and 67% for the left lateral ventricle, while the third and fourth ventricles exhibited percentage changes of 38% and 31%, respectively, compared to controls. AD patients also displayed ventricular enlargement, with a 50% increase in total ventricular volume compared to controls. Specifically, the lateral ventricles exhibited a 50% increase in volume, with percentage changes of 50% for the right lateral ventricle and 51% for the left lateral ventricle, while the third and fourth ventricles demonstrated percentage changes of 28% and 29%, respectively, compared to controls.

In the relative measurements normalized to ICV, the most prominent differences between the groups are seen in the relative volumes of mesial structures such as the hippocampus and the parahippocampus. The relative volume of the hippocampus in iNPH is 0.0050, while it is 0.0057 in CG and 0.0053 in AD. Similarly, the relative volume of the parahippocampus is 0.0041 in iNPH compared to CG (0.0055) and AD (0.0049). These differences correspond to percentage changes of −17% and −8% for the hippocampus and −34% and −20% for the parahippocampus in iNPH compared to CG and AD.

Additional subgroup analysis revealed significant differences in various brain regions and ventricular volumes among patients with iNPH, AD, and controls (Tables 5 and 6). Notable differences were observed in total ventricular volume, lateral ventricular volume, and the volumes of the third and fourth ventricles, with significant mean differences found between iNPH and both other groups (controls and AD) ($p < 0.001$). The mean differences for total ventricle volume between iNPH and CG, iNPH and AD, and controls and AD were 86.60, 64.11, and 22.49, respectively, with corresponding standard errors of 7.16. Similarly, for the lateral ventricular volume, the mean differences were 85.05, 62.85, and 22.20, with standard errors of 7.02. Regarding brain volumes, while differences in total brain volume were not statistically significant between iNPH and controls, a borderline significant difference was detected between iNPH and AD groups (mean difference = 51.74, SE = 24.15, $p = 0.103$). White matter volume exhibited significant differences between iNPH and control groups (mean difference = −34.80, SE = 14.04, $p = 0.044$), although no significant differences were observed between iNPH and AD or control and AD groups. Cortical gray matter volume showed a marked discrepancy between iNPH and AD groups (mean difference = 39.21, SE = 11.85, $p = 0.004$), along with significant differences between control and AD groups (mean difference = −59.31, SE = 11.85, $p < 0.001$). Notably, significant distinctions were observed in various lobes, including the frontal, parietal, precuneus, temporal, and hippocampal regions.

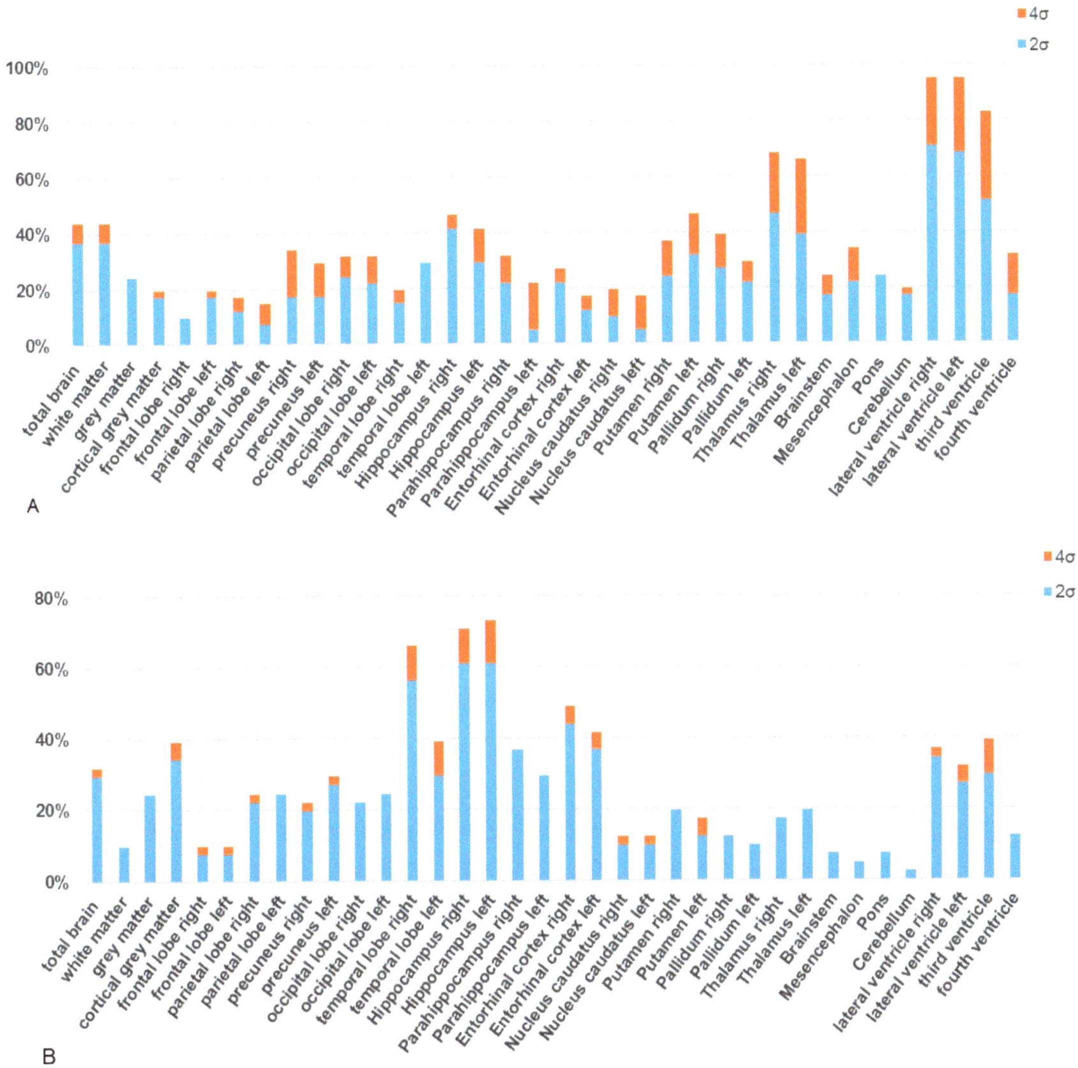

Figure 3. Proportion of subjects with volume deviations among iNPH patients (**A**) and AD patients (**B**) is presented. The incidence of these deviations, either by 2 (illustrated in blue) or 4 (in orange) standard deviations, is compared with a healthy, population-based control group embedded in the AI-based volumetric software. The deviations of the supratentorial volumes are provided separately for each side of the brain. Deviations in cortical and subcortical volumes indicate atrophy, whereas deviations in ventricular volumes suggest enlargement.

Table 3. Results of volumetric MRI analysis in iNPH and AD patients and the control group, given in absolute volumes [mL].

	iNPH (n = 41)		Control Group (CG) (n = 41)		Alzheimer's Disease (AD) (n = 41)		iNPH vs. CG	iNPH vs. AD	p-Value #
	Mean	SD	Mean	SD	Mean	SD	% Δ	% Δ	
ICV	1231.8122	144.1960	1181.1659	97.8317	1115.9610	115.0844	4	10	0.000
Total brain volume	1102.4610	122.2758	1138.4146	93.2713	1050.7220	110.5777	−3	5	0.003
White matter	462.0683	82.6898	496.8659	45.7924	474.2415	56.4358	−8	−3	0.025
Gray matter	638.6902	64.4441	641.5390	54.7634	576.4780	58.6856	0	10	0.000
Cortical gray matter	417.1683	46.5077	437.2683	39.6168	377.9537	70.0438	−5	9	0.000
Frontal lobe									
Total	155.5732	18.5901	157.3951	15.6656	145.6000	17.3142	−1	6	0.006
Right	79.5854	9.5334	79.6146	8.1767	73.5463	8.7155	0	8	0.002
Left	75.9878	9.2762	77.7805	7.5898	72.0537	8.8034	−3	5	0.018
Parietal lobe									
Total	84.8756	12.3904	86.3463	8.0252	78.1098	9.4089	−2	8	0.000
Right	41.8171	6.4713	41.9829	3.9911	38.4707	4.8411	0	8	0.002
Left	43.0585	6.3699	44.3634	4.1813	39.6390	4.9422	−3	8	0.000
Precuneus lobe									
Total	17.8976	5.8758	20.9902	2.4871	18.1268	3.0642	−16	0	0.000
Right	8.8927	3.0532	10.2854	1.2165	9.0951	1.5105	−14	0	0.001
Left	9.0049	2.9051	10.7049	1.3405	9.0317	1.6951	−18	0	0.000
Occipital lobe									
Total	60.6756	11.5576	63.3610	7.1624	57.9146	8.0108	−5	5	0.004
Right	29.4439	5.8061	30.3220	3.3806	27.9439	4.1954	−3	6	0.013
Left	31.2317	6.1485	33.0390	4.0683	29.9707	4.0727	−7	4	0.006
Temporal lobe									
Total	116.0268	15.5967	130.1073	11.7281	103.7902	14.0070	−12	11	0.000
Right	58.7976	9.0286	66.2951	6.2910	52.5171	7.4310	−13	10	0.000
Left	57.2293	7.4943	63.8122	5.7790	51.2732	7.1145	−11	11	0.000
Hippocampus									
Total	5.7683	1.0999	6.7171	0.7466	5.7927	3.8273	−14	0	0.000
Right	2.9854	0.5677	3.5683	0.4245	3.3171	3.8189	−29	−14	0.000
Left	2.7829	0.5749	3.1488	0.3565	2.4756	0.4603	−13	12	0.000
Parahippocampus									
Total	5.3683	1.4093	6.4756	0.8040	5.4220	0.8027	−21	−1	0.000
Right	2.6146	0.7562	3.1902	0.4364	2.6439	0.4411	−22	0	0.000
Left	2.7537	0.7036	3.2854	0.3997	2.7780	0.3947	−19	−1	0.000
Entorhinal lobe									
Total	3.9732	0.9260	4.7585	0.7050	3.4732	0.7934	−20	12	0.000
Right	1.9878	0.5109	2.4220	0.3883	1.7341	0.4059	−23	13	0.000
Left	1.9854	0.4503	2.3366	0.3569	1.7390	0.4254	−19	13	0.000
Caudate									
Total	6.0902	2.1805	6.0634	1.2397	5.9049	1.5079	1	3	0.662
Right	3.1878	1.1860	3.1976	0.6436	3.0805	0.8054	0	4	0.534
Left	2.9024	1.0613	2.8659	0.6315	2.8244	0.7599	2	3	0.713
Putamen									
Total	7.0854	1.6249	7.8927	0.9863	2.5976	0.3453	−12	64	0.731
Right	3.5707	0.7329	3.9049	0.5059	3.7439	0.5714	−9	−6	0.136
Left	3.5146	0.9372	3.9878	0.5154	3.7878	0.6282	−13	−7	0.069

Table 3. *Cont.*

	iNPH (n = 41)		Control Group (CG) (n = 41)		Alzheimer's Disease (AD) (n = 41)		iNPH vs. CG	iNPH vs. AD	*p*-Value #
	Mean	SD	Mean	SD	Mean	SD	% Δ	% Δ	
Pallidum									
Total	2.4902	0.4415	2.5512	0.3487	2.5463	0.3805	−2	−2	0.731
Right	1.2098	0.2300	1.2634	0.1959	1.2951	0.1788	−5	−8	0.446
Left	1.2805	0.2283	1.2878	0.1646	1.3024	0.1753	0	−3	0.807
Thalamus									
Total	12.7439	2.6397	14.6122	1.1752	15.5951	9.0255	−14	−22	0.000
Right	6.3390	1.3687	7.1244	0.6041	6.9951	0.6753	−12	−10	0.001
Left	6.4049	1.3638	7.4878	0.6133	8.6000	9.0765	−17	−34	0.000
Brainstem	25.4146	3.6199	26.2366	2.3313	27.0537	2.6888	−3	−6	0.055
Mesencephalon	7.1073	0.9429	7.3390	0.6016	8.1707	3.4995	−3	−15	0.016
Pons	13.9805	2.4003	14.9317	3.9517	14.2951	2.5724	−7	−2	0.304
Cerebellum	90.5610	11.1429	99.7610	9.8351	94.0366	17.9692	−10	−4	0.001
Ventricle volume	129.3512	50.0321	42.7512	16.5050	65.2390	19.4649	67	50	0.000
Lateral ventricle									
Total	124.8488	49.0248	39.8000	16.0337	61.9976	19.1541	68	50	0.000
Right	60.6195	23.0739	18.6829	8.0216	30.5244	10.8250	69	50	0.000
Left	64.2293	26.8961	21.1171	8.3129	31.4732	9.7894	67	51	0.000
Third ventricle	2.4902	0.7816	1.5390	0.5098	1.8049	0.4779	38	28	0.000
Fourth ventricle	2.0122	0.8051	1.4122	0.4008	1.4366	0.3352	31	29	0.000

Results of volumetric MRI analysis in iNPH and AD patients and the control group, given in absolute volumes. Mean absolute volumes [mL] of the different brain volumes for iNPH, control group, and AD patients. # *p*-value of the Kruskal–Wallis test. *p*-value < 0.005 was considered statistically significant. % Δ percentage difference of the mean of iNPH patients to the mean of healthy patients in the control group (CG) and to the mean of AD patients. Abbreviations: SD = standard deviation. ICV = intracranial volume. iNPH = idiopathic normal pressure hydrocephalus. AD = Alzheimer's disease. CG = control group.

Table 4. Results of volumetric MRI analysis in iNPH and AD patients and the control group in relative volumes, normalized to the total intracranial volume (ICV) of each subgroup.

	iNPH (n = 41)		Control Group (CG) (n = 41)		Alzheimer's Disease (AD) (n = 41)		iNPH vs. CG	iNPH vs. AD	*p*-Value #
	Mean	SD	Mean	SD	Mean	SD	% Δ	% Δ	
ICV	1.000		1.000		1.000				
Total brain volume	0.8961	0.0341	0.9640	0.0128	0.9414	0.0161	−9	−5	0.000
White matter	0.3743	0.0397	0.4207	0.0175	0.4246	0.1753	−11	−11	0.000
Gray matter	0.5214	0.0398	0.5433	0.0167	0.5168	0.1521	0	0	0.000
Cortical gray matter	0.3391	0.0257	0.3702	0.1404	0.3382	0.0509	−13	0	0.000
Frontal lobe									
Total	0.1271	0.0118	0.1333	0.0074	0.1305	0.0085	−6	−3	0.005
Right	0.0653	0.0060	0.0674	0.0039	0.0659	0.0039	0	0	0.028
Left	0.0622	0.0060	0.0659	0.0036	0.0646	0.0048	−7	0	0.001
Parietal lobe									
Total	0.0693	0.0095	0.0731	0.0037	0.0699	0.0041	−6	0	0.009
Right	0.0342	0.0050	0.0356	0.0019	0.0345	0.0026	−13	0	0.204
Left	0.0352	0.0048	0.0376	0.0020	0.0355	0.0021	0	0	0.000

Table 4. *Cont.*

	iNPH (n = 41)		Control Group (CG) (n = 41)		Alzheimer's Disease (AD) (n = 41)		iNPH vs. CG	iNPH vs. AD	*p*-Value #
	Mean	SD	Mean	SD	Mean	SD	% Δ	% Δ	
Precuneus lobe									
Total	0.0151	0.0049	0.0178	0.0015	0.0162	0.0017	0	1	0.000
Right	0.0071	0.0026	0.0087	0.0008	0.0081	0.0010	−24	−16	0.009
Left	0.0070	0.0024	0.0091	0.0008	0.0081	0.0009	−29	−16	0.000
Occipital lobe									
Total	0.0502	0.0086	0.0536	0.0038	0.0519	0.0047	−7	−4	0.063
Right	0.0243	0.0042	0.0257	0.0019	0.0250	0.0025	−7	−3	0.243
Left	0.0263	0.0047	0.0279	0.0023	0.0269	0.0026	−6	−3	0.042
Temporal lobe									
Total	0.0944	0.0072	0.1102	0.0046	0.0929	0.0067	−17	1	0.000
Right	0.0484	0.0050	0.0561	0.0027	0.0469	0.0038	−17	8	0.000
Left	0.0470	0.0035	0.0540	0.0025	0.0459	0.0038	−15	3	0.000
Hippocampus									
Total	0.0050	0.0008	0.0057	0.0005	0.0053	0.0040	−17	−8	0.000
Right	0.0021	0.0004	0.0030	0.0003	0.0031	0.0039	−43	−49	0.000
Left	0.0022	0.0004	0.0027	0.0002	0.0022	0.0004	−23	0	0.000
Parahippocampus									
Total	0.0041	0.0011	0.0055	0.0005	0.0049	0.0006	−34	−20	0.000
Right	0.0023	0.0006	0.0027	0.0003	0.0024	0.0003	−17	−4	0.000
Left	0.0024	0.0005	0.0028	0.0002	0.0025	0.0003	−17	−4	0.000
Entorhinal lobe									
Total	0.0031	0.0007	0.0040	0.0005	0.0031	0.0006	−29	0	0.000
Right	0.0022	0.0004	0.0020	0.0003	0.0015	0.0003	9	27	0.000
Left	0.0021	0.0003	0.0020	0.0003	0.0016	0.0003	5	24	0.000
Caudate									
Total	0.0051	0.0018	0.0051	0.0010	0.0053	0.0012	0	−4	0.748
Right	0.0032	0.0010	0.0027	0.0005	0.0028	0.0007	16	13	0.985
Left	0.0021	0.0009	0.0024	0.0005	0.0025	0.0006	−14	−19	0.478
Putamen									
Total	0.0061	0.0014	0.0067	0.0007	0.0068	0.0008	−9	−11	0.000
Right	0.0031	0.0007	0.0033	0.0004	0.0034	0.0004	−6	−10	0.001
Left	0.0032	0.0008	0.0034	0.0004	0.0034	0.0005	−6	−6	0.002
Pallidum									
Total	0.0022	0.0004	0.0022	0.0002	0.0023	0.0002	0	−4	0.000
Right	0.0012	0.0002	0.0011	0.0001	0.0012	0.0001	9	0	0.000
Left	0.0013	0.0002	0.0011	0.0001	0.0012	0.0001	15	8	0.002
Thalamus									
Total	0.0111	0.0023	0.0124	0.0008	0.0115	0.0094	−11	−3	0.000
Right	0.0053	0.0012	0.0060	0.0004	0.0063	0.0003	−13	−19	0.000
Left	0.0052	0.0012	0.0064	0.0004	0.0079	0.0097	−23	−52	0.000
Brainstem	0.0211	0.0025	0.0223	0.0016	0.0243	0.0019	−6	−15	0.000
Mesencephalon	0.0063	0.0008	0.0062	0.0004	0.0073	0.0031	2	−16	0.000
Pons	0.0114	0.0016	0.0127	0.0032	0.0129	0.0022	−12	−13	0.000
Cerebellum	0.0743	0.0083	0.0846	0.0068	0.0846	0.0156	−14	−14	0.000
Ventricle volume	0.1042	0.0341	0.0360	0.0128	0.0586	0.0161	65	44	0.000
Lateral ventricle									
Total	0.1001	0.0335	0.0335	0.0125	0.0556	0.0159	67	44	0.000
Right	0.0491	0.0159	0.0157	0.0063	0.0274	0.0089	68	44	0.000
Left	0.0523	0.0185	0.0178	0.0065	0.0283	0.0085	66	46	0.000

Table 4. *Cont.*

	iNPH (n = 41)		Control Group (CG) (n = 41)		Alzheimer's Disease (AD) (n = 41)		iNPH vs. CG	iNPH vs. AD	p-Value #
	Mean	SD	Mean	SD	Mean	SD	% Δ	% Δ	
Third ventricle	0.0022	0.0006	0.0013	0.0004	0.0016	0.0004	41	27	0.000
Fourth ventricle	0.0022	0.0006	0.0012	0.003	0.0013	0.0003	45	41	0.003

Mean relative volumes of the different brain volumes for iNPH, control group and AD patients. All volumes have been normalized to the intracranial volume (ICV). # p-value of the Kruskal–Wallis test. p-value < 0.005 was considered statistically significant. % Δ percentage difference of the mean of iNPH patients to the mean of healthy patients in the control group (CG) and to the mean of AD patients. Abbreviations: SD = standard deviation. ICV = intracranial volume. iNPH = idiopathic normal pressure hydrocephalus. AD = Alzheimer's disease. CG = control group.

Table 5. Subgroup comparisons of absolute volumetric measurements in iNPH and AD patients and the healthy control group (CG).

Variable	Subgroups iNPH, CG, and AD	Mean Difference	SE	p
total ventricle volume	iNPH vs. CG	86.60	7.16	0.000 *
	iNPH vs. AD	64.11	7.16	0.000 *
	CG vs. AD	22.49	7.16	0.006 *
total lateral ventricle volume	iNPH vs. CG	85.05	7.02	0.006 *
	iNPH vs. AD	62.85	7.02	0.000 *
	CG vs. AD	22.20	7.02	0.006 *
right ventricle volume	iNPH vs. CG	41.94	3.41	0.000 *
	iNPH vs. AD	30.10	3.41	0.000 *
	CG vs. AD	11.84	3.41	0.002 *
left ventricle volume	iNPH vs. CG	43.11	3.80	0.000 *
	iNPH vs. AD	32.76	3.80	0.000 *
	CG vs. AD	10.36	3.80	0.022 *
third ventricle volume	iNPH vs. CG	0.95	0.13	0.000 *
	iNPH vs. AD	0.69	0.13	0.000 *
	CG vs. AD	0.27	0.13	0.147
fourth ventricle volume	iNPH vs. CG	0.60	0.12	0.000 *
	iNPH vs. AD	0.58	0.12	0.000 *
	CG vs. AD	0.02	0.12	1.000
total brain volume	iNPH vs. CG	−35.95	24.15	0.418
	iNPH vs. AD	51.74	24.15	0.103
	CG vs. AD	−87.69	24.15	0.001 *
white matter	iNPH vs. CG	−34.80	14.04	0.044 *
	iNPH vs. AD	−12.17	14.04	1.000
	CG vs. AD	−22.62	14.04	0.329
cortical gray matter	iNPH vs. CG	−20.10	11.85	0.277
	iNPH vs. AD	39.21	11.85	0.004 *
	CG vs. AD	−59.31	11.85	0.000 *

Table 5. *Cont.*

Variable	Subgroups iNPH, CG, and AD	Mean Difference	SE	p
frontal lobe right	iNPH vs. CG	−0.03	1.95	1.000
	iNPH vs. AD	6.04	1.95	0.007 *
	CG vs. AD	−6.07	1.95	0.007 *
frontal lobe left	iNPH vs. CG	−1.79	1.90	1.000
	iNPH vs. AD	3.93	1.90	0.120
	CG vs. AD	−5.73	1.90	0.009 *
parietal lobe right	iNPH vs. CG	−0.166	1.15	1.000
	iNPH vs. AD	3.35	1.15	0.013 *
	CG vs. AD	−3.51	1.15	0.008 *
parietal lobe left	iNPH vs. CG	−1.30	1.16	0.786
	iNPH vs. AD	3.42	1.16	0.011 *
	CG vs. AD	−4.72	1.16	0.000 *
precuneus lobe right	iNPH vs. CG	−1.39	0.46	0.009 *
	iNPH vs. AD	−0.20	0.46	1.000
	CG vs. AD	−1.19	0.46	0.033 *
precuneus lobe left	iNPH vs. CG	−1.70	0.46	0.001 *
	iNPH vs. AD	−0.027	0.46	1.000
	CG vs. AD	−1.67	0.46	0.001 *
temporal lobe right	iNPH vs. CG	−7.50	1.69	0.000 *
	iNPH vs. AD	6.28	1.69	0.001 *
	CG vs. AD	−13.78	1.69	0.000 *
temporal lobe left	iNPH vs. CG	−6.58	1.51	0.000 *
	iNPH vs. AD	5.96	1.51	0.000 *
	CG vs. AD	−12.54	1.51	0.000 *

*, statistically significant at $p < 0.05$. Mean difference given in units of mL. SE, standard error. Abbreviations: iNPH = idiopathic normal pressure hydrocephalus. AD = Alzheimer's disease. CG = control group.

Table 6. Subgroup comparisons of volumetric analysis in iNPH patients (iNPH) and AD patients and the control group in relative volumes, normalized to the total intracranial volume (ICV) of each subgroup.

Variable	Subgroups iNPH, CG, and AD	Mean	SE	p
total ventricle volume	iNPH vs. CG	129.35 vs. 42.75	50.03 vs. 16.51	0.000 *
	iNPH vs. AD	129.35 vs. 65.24	50.03 vs. 19.46	0.000 *
total lateral ventricle volume	iNPH vs. CG	124.85 vs. 39.80	49.02 vs. 16.03	0.000 *
	iNPH vs. AD	124.85 vs. 62.00	49.02 vs. 19.15	0.000 *
right ventricle volume	iNPH vs. CG	60.62 vs. 18.68	23.07 vs. 8.02	0.000 *
	iNPH vs. AD	60.62 vs. 30.52	23.07 vs. 10.83	0.000 *
left ventricle volume	iNPH vs. CG	64.23 vs. 21.12	26.90 vs. 8.31	0.000 *
	iNPH vs. AD	64.23 vs. 31.47	26.90 vs. 9.79	0.000 *
third ventricle volume	iNPH vs. CG	2.49 vs. 1.54	0.78 vs. 0.51	0.000 *
	iNPH vs. AD	2.49 vs. 1.80	0.78 vs. 0.48	0.000 *

Table 6. *Cont.*

Variable	Subgroups iNPH, CG, and AD	Mean	SE	*p*
fourth ventricle volume	iNPH vs. CG	2.01 vs. 1.41	0.81 vs. 0.40	0.000 *
	iNPH vs. AD	2.01 vs. 1.44	0.81 vs. 0.34	0.000 *
total brain volume	iNPH vs. CG	1102.46 vs. 1138.41	122.28 vs. 93.27	0.206
	iNPH vs. AD	1102.46 vs. 1050.72	122.28 vs. 110.58	0.046 *
white matter	iNPH vs. CG	462.07 vs. 496.87	82.69 vs. 45.79	0.017 *
	iNPH vs. AD	462.07 vs. 474.24	82.69 vs. 56.44	0.319
gray matter	iNPH vs. CG	638.69 vs. 641.54	64.44 vs. 54.76	0.952
	iNPH vs. AD	638.69 vs. 576.48	64.44 vs. 58.69	0.000 *
cortical gray matter	iNPH vs. CG	417.17 vs. 437.27	46.51 vs. 39.62	0.070 *
	iNPH vs. AD	417.17 vs. 377.95	46.51 vs. 70.04	0.003 *
frontal lobe right	iNPH vs. CG	79.59 vs. 79.61	9.53 vs. 8.18	0.792
	iNPH vs. AD	79.59 vs. 73.55	9.53 vs. 8.72	0.003 *
frontal lobe left	iNPH vs. CG	75.99 vs. 77.78	9.28 vs. 7.59	0.603
	iNPH vs. AD	75.99 vs. 72.05	9.28 vs. 8.80	0.067
total frontal lobe	iNPH vs. CG	155.57 vs. 157.40	18.59 vs. 15.67	0.853
	iNPH vs. AD	155.57 vs. 145.60	18.59 vs. 17.31	0.012 *
parietal lobe right	iNPH vs. CG	41.82 vs. 41.98	6.47 vs. 3.99	0.842
	iNPH vs. AD	41.82 vs. 38.47	6.47 vs. 4.84	0.005 *
parietal lobe left	iNPH vs. CG	43.06 vs. 44.36	6.37 vs. 4.18	0.412
	iNPH vs. AD	43.06 vs. 39.64	6.37 vs. 4.94	0.004 *
total parietal lobe	iNPH vs. CG	84.88 vs. 86.35	12.39 vs. 8.03	0.767
	iNPH vs. AD	84.88 vs. 78.11	12.39 vs. 9.41	0.003 *
precuneus lobe right	iNPH vs. CG	8.89 vs. 10.29	3.05 vs. 1.22	0.073 *
	iNPH vs. AD	8.89 vs. 9.10	3.05 vs. 1.51	0.200
precuneus lobe left	iNPH vs. CG	9.00 vs. 10.70	2.91 vs. 1.34	0.003 *
	iNPH vs. AD	9.00 vs. 9.03	2.91 vs. 1.70	0.177
total precuneus lobe	iNPH vs. CG	17.90 vs. 20.99	5.88 vs. 2.49	0.013 *
	iNPH vs. AD	17.90 vs. 18.13	5.88 vs. 3.06	0.196
occipital lobe right	iNPH vs. CG	29.44 vs. 30.32	5.81 vs. 3.38	0.693
	iNPH vs. AD	29.44 vs. 27.94	5.81 vs. 4.20	0.053 *
occipital lobe left	iNPH vs. CG	31.23 vs. 33.04	6.15 vs. 4.07	0.235
	iNPH vs. AD	31.23 vs. 29.97	6.15 vs. 4.07	0.072
total occipital lobe	iNPH vs. CG	60.68 vs. 63.36	11.56 vs. 7.16	0.386
	iNPH vs. AD	60.68 vs. 57.91	11.56 vs. 8.01	0.068
temporal lobe right	iNPH vs. CG	58.80 vs. 66.30	9.03 vs. 6.29	0.000 *
	iNPH vs. AD	58.80 vs. 52.52	9.03 vs. 7.43	0.001 *
temporal lobe left	iNPH vs. CG	57.23 vs. 63.81	7.49 vs. 5.78	0.000 *
	iNPH vs. AD	57.23 vs. 51.27	7.49 vs. 7.11	0.001 *
total temporal lobe	iNPH vs. CG	116.03 vs. 130.11	15.60 vs. 11.73	0.000 *
	iNPH vs. AD	116.03 vs. 103.79	15.60 vs. 14.01	0.000 *

Table 6. *Cont.*

Variable	Subgroups iNPH, CG, and AD	Mean	SE	*p*
hippocampus right	iNPH vs. CG	2.99 vs. 3.57	0.57 vs. 0.42	0.000 *
	iNPH vs. AD	2.99 vs. 3.32	0.57 vs. 3.82	0.030 *
hippocampus lobe left	iNPH vs. CG	2.78 vs. 3.15	0.57 vs. 0.36	0.005 *
	iNPH vs. AD	2.78 vs. 2.48	0.57 vs. 0.46	0.005 *
total volume hippocampus	iNPH vs. CG	5.77 vs. 6.72	1.10 vs. 0.75	0.000 *
	iNPH vs. AD	5.77 vs. 5.79	1.10 vs. 3.83	0.015 *
gyrus parahippocampalis right	iNPH vs. CG	2.61 vs. 3.19	0.76 vs. 0.44	0.000 *
	iNPH vs. AD	2.61 vs. 2.64	0.76 vs. 0.44	0.372
gyrus parahippocampalis left	iNPH vs. CG	2.75 vs. 3.29	0.70 vs. 0.40	0.000 *
	iNPH vs. AD	2.75 vs. 2.78	0.70 vs. 0.39	0.424
total volume parahippocampal	iNPH vs. CG	5.37 vs. 6.48	1.41 vs. 0.80	0.000 *
	iNPH vs. AD	5.37 vs. 5.42	1.41 vs. 0.80	0.380
regio entorhinalis right	iNPH vs. CG	1.99 vs. 2.42	0.51 vs. 0.39	0.000 *
	iNPH vs. AD	1.99 vs. 1.73	0.51 vs. 0.41	0.008 *
regio entorhinalis left	iNPH vs. CG	1.99 vs. 2.34	0.45 vs. 0.36	0.001 *
	iNPH vs. AD	1.99 vs. 1.74	0.45 vs. 0.43	0.005 *
total volume entorhinal	iNPH vs. CG	3.97 vs. 4.76	0.93 vs. 0.70	0.000 *
	iNPH vs. AD	3.97 vs. 3.47	0.93 vs. 0.79	0.004 *
nucleus caudatus right	iNPH vs. CG	3.19 vs. 3.20	1.19 vs. 0.64	0.662
	iNPH vs. AD	3.19 vs. 3.08	1.19 vs. 0.81	0.330
nucleus caudatus left	iNPH vs. CG	2.90 vs. 2.87	1.06 vs. 0.63	0.411
	iNPH vs. AD	2.90 vs. 2.82	1.06 vs. 0.76	0.568
total volume nucleus caudatus	iNPH vs. CG	6.09 vs. 6.06	2.18 vs. 1.24	0.516
	iNPH vs. AD	6.09 vs. 5.90	2.18 vs. 1.51	0.441
putamen right	iNPH vs. CG	3.57 vs. 3.90	0.73 vs. 0.51	0.055 *
	iNPH vs. AD	3.57 vs. 3.74	0.73 vs. 0.57	0.521
putamen left	iNPH vs. CG	3.51 vs. 3.99	0.94 vs. 0.52	0.024 *
	iNPH vs. AD	3.51 vs. 3.79	0.94 vs. 0.63	0.318
total volume putamen	iNPH vs. CG	2.49 vs. 2.55	0.44 vs. 0.35	0.929
	iNPH vs. AD	2.49 vs. 2.60	0.44 vs. 0.35	0.496
pallidum right	iNPH vs. CG	1.21 vs. 1.26	0.23 vs. 0.20	0.756
	iNPH vs. AD	1.21 vs. 1.30	0.23 vs. 0.18	0.218
pallidum left	iNPH vs. CG	1.28 vs. 1.29	0.23 vs. 0.16	0.536
	iNPH vs. AD	1.28 vs. 1.30	0.23 vs. 0.18	0.996
total volume pallidum	iNPH vs. CG	2.49 vs. 2.55	0.44 vs. 0.35	0.929
	iNPH vs. AD	2.49 vs. 2.60	0.44 vs. 0.35	0.496
thalamus right	iNPH vs. CG	6.34 vs. 7.12	1.37 vs. 0.60	0.000 *
	iNPH vs. AD	6.34 vs. 7.00	1.37 vs. 0.68	0.007 *
thalamus left	iNPH vs. CG	6.40 vs. 7.49	1.36 vs. 0.61	0.000 *
	iNPH vs. AD	6.40 vs. 8.60	1.36 vs. 9.08	0.003 *

Table 6. *Cont.*

Variable	Subgroups iNPH, CG, and AD	Mean	SE	p
total volume thalamus	iNPH vs. CG	12.74 vs. 14.61	2.64 vs. 1.18	0.000 *
	iNPH vs. AD	12.74 vs. 15.60	2.64 vs. 9.03	0.002 *
brainstem volume	iNPH vs. CG	25.41 vs. 26.24	3.62 vs. 2.33	0.292
	iNPH vs. AD	25.41 vs. 27.05	3.62 vs. 2.69	0.026 *
mesencephalon volume	iNPH vs. CG	7.11 vs. 7.34	0.94 vs. 0.60	0.336
	iNPH vs. AD	7.11 vs. 8.17	0.94 vs. 3.50	0.009 *
pons volume	iNPH vs. CG	13.98 vs. 14.93	2.40 vs. 3.95	0.260
	iNPH vs. AD	13.98 vs. 14.30	2.40 vs. 2.57	0.165
cerebellum volume	iNPH vs. CG	90.56 vs. 99.76	11.14 vs. 9.84	0.000 *
	iNPH vs. AD	90.56 vs. 94.04	11.14 vs. 17.97	0.032 *

*, statistically significant at $p < 0.05$. Mean difference given in units of mL. SE, standard error.

3.3. Semi-Quantitative Markers for the Diagnosis of iNPH

The mean Evans' index and corpus callosal angle were statistically significantly different between the iNPH, AD, and control groups (Evans' index: iNPH 0.34 $\pm$ 0.04 vs. AD 0.28 $\pm$ 0.03 vs. control group 0.23 $\pm$ 0.01, $p \leq 0.001$; corpus callosal angle: iNPH 82.22 $\pm$ 22.12 vs. AD 117.60 $\pm$ 15.23 vs. control group 111.59 $\pm$ 9.59, $p \leq 0.001$). As graphically illustrated in Figure 4, the total ventricle volume of the iNPH patients and the measured Evans' index and corpus callosal angle were found to correlate moderately (Evans' index: right lateral ventricle r = 0.788, $p = 0.000$, and left lateral ventricle r = 0.781, $p \leq 0.001$; corpus callosal angle: right lateral ventricle r = −0.650, $p \leq 0.001$, and left lateral ventricle r = −0.636, $p \leq 0.001$).

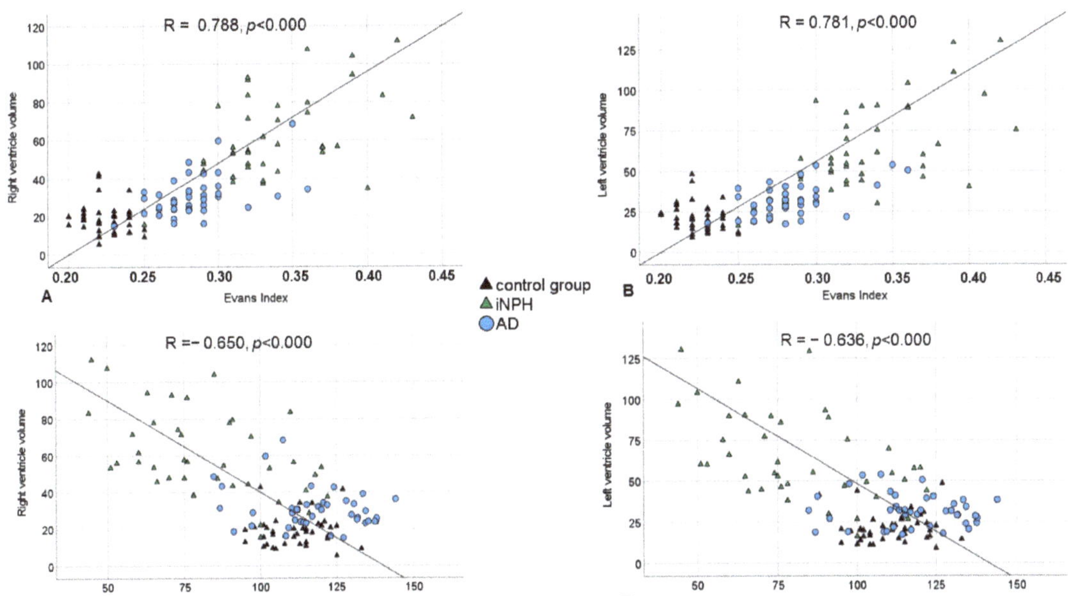

Figure 4. Scatterplot of the right (**A**,**C**) and left (**B**,**D**) ventricle volumes and the values of the Evans' index (**A**,**B**) and corpus callosum angle (**C**,**D**).

Both semi-quantitative measurement ratios, the Evans' index and the corpus callosal angle, and the automatically determined total ventricular volume allowed for a differentiation between the iNPH and healthy control groups with a moderate to high diagnostic accuracy, as shown in the corresponding ROC curves in Figure 5 and further specified in Table 7.

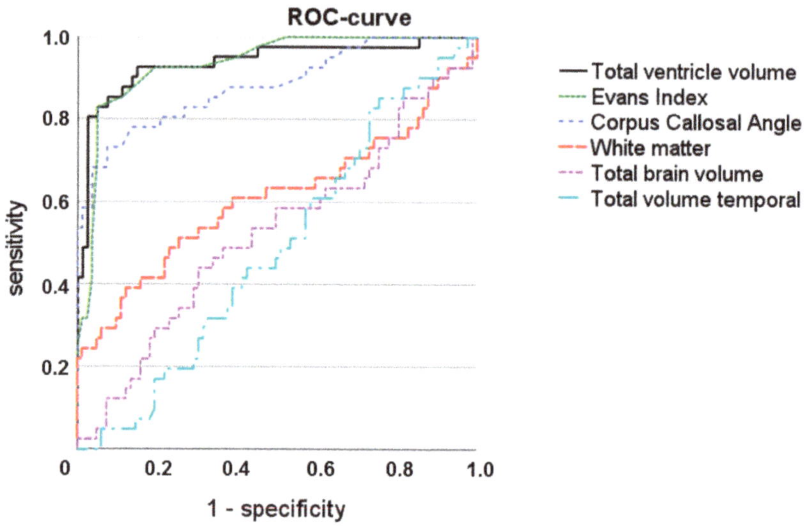

Figure 5. ROC curve with total ventricle volume, Evans' index, corpus callosum angle, white matter, total brain volume, and total temporal volume.

Table 7. Diagnostic performance of Evans' index, corpus callosum angle, and total ventricle volume in mL for diagnosing idiopathic normal pressure hydrocephalus.

Parameter	AUC	SE	p	CI1	CI2	Cut-Off	Sen	Spec	Acc
Evans' index [a]	0.935	0.023	<0.001	0.890	0.979	0.3	0.829	0.951	0.9
Corpus callosum angle [a]	0.613	0.049	0.049	0.517	0.710	90	1	0.464	0.68
Total ventricle volume [a]	0.939	0.012	<0.001	0.888	0.990	76.6	0.927	0.854	0.87

The optimal cut-off point of each ROC analysis was selected according to the maximum Youden index. Cut-off points are given unitless for Evans' index, in units of degrees (°) for corpus callosum angle, and in units of mL for total ventricle volume. AUC—area under the curve, SE—standard error, p—significance level, CI—95% confidence interval, Sen—sensitivity (true positive rate), Spec—specificity (true negative rate), Acc—accuracy (rate of correctly identified cases). [a] Means that a higher test result indicates a more positive test.

4. Discussion

Idiopathic NPH stands as one of the limited number of curable sources of dementia, requiring a collaborative effort among neurologists, neurosurgeons, and neuroradiologists for accurate diagnosis. Identifying iNPH proves to be challenging due to its inconsistent symptoms and progression and the fact that it often coexists or is confused with other neurological conditions. Therefore, a significant amount of research has been dedicated to creating a reliable, noninvasive predictive test for iNPH.

In this study, we evaluated brain volumes of patients with iNPH compared to those with AD and healthy controls. Using automated, AI-based brain MRI volumetry, we aimed to identify morphometric changes across these groups. This study represents the largest MRI investigation of volumetric analysis in iNPH patients to date, as far as we are aware. Our volumetric analyses unveiled substantial differences in ventricular size and small to moderate, yet statistically significant, volumetric disparities in cortical and subcortical brain regions among iNPH patients compared to those with AD and healthy controls.

These differences primarily manifested as global ventriculomegaly alongside concurrent atrophy of the entire white matter, thalamus, brainstem, and midbrain, with notable reductions in brain volume observed particularly in temporomesial and midline structures. These patterns of ventricular enlargement and focal brain atrophy were corroborated by corresponding alterations in normalized percentiles derived from a normal population database integrated into our analysis software, thus confirming the existence of structural brain changes commonly reported in the literature. Notably, iNPH patients exhibited significant enlargement of all ventricles including the third and fourth ventricle alongside reductions in whole brain white matter, midbrain, and mesial structures, while largely preserving the neocortical gray matter in comparison to healthy controls and AD. This contrasts with the significant reduction observed in both the gray and white matter of the cerebrum in AD. These findings align with previous studies and provide quantitative evidence supporting the notion that larger global ventricular size with preservation of neocortical structures is indicative of iNPH.

In our iNPH patients, the volumes of the third and fourth ventricles were significantly enlarged compared to AD patients, while there was no difference in the lateral ventricles among these two entities. Moreover, the whole and supratentorial grey matter volumes were significantly lower in AD than in iNPH patients, with a notable volume reduction of AD patients in the frontal and parietal regions compared to iNPH. In contrast, both groups demonstrated substantial volume reductions in the temporal region. The findings are broadly consistent with the current literature, highlighting a decrease in volume surrounding the expanded ventricles in iNPH patients as CSF studies in patients with iNPH have shown subsequent neuronal degeneration [26]. The pathophysiology underlying these volume changes remains an area of ongoing investigation, with some studies suggesting that the compression exerted by the enlarged ventricles may contribute to brain tissue reduction [27]. Additionally, it points out that AD patients typically exhibit pronounced cortical atrophy, especially in the parietotemporal and frontal regions [28–30]. Our findings additionally demonstrated focal brain atrophy of the temporal, precuneus, and frontal lobes in iNPH patients as compared to healthy controls, potentially indicating the cerebrospinal fluid pressure exerted on these brain areas that are thought to probably cause the Hakim triad.

In clinical practice, the assessment of ventricular size and potential structural changes in iNPH typically relies on subjective visual inspection or semi-quantitative measurements, such as the Evans' index or the corpus callosum angle, in computed tomography (CT) and/or MRI scans [16–19]. With regard to traditional semi-quantitative parameters, our measurements of the Evans' index and the corpus callosum angle in iNPH showed moderate correlations with the lateral ventricular width, with the correlation of the corpus callosum angle being slightly weaker than the correlation with the Evans' index. These observations may suggest that the two aforementioned parameters represent a practical measure for estimating ventricular enlargement. However, visual assessment and semi-quantitative estimation of ventricular volume are prone to error due to sensitivity to the choice and angle of multiplanar image reconstructions, particularly with indices like the Evans' index and corpus callosal angle. These limitations, combined with individual anatomical variations, can significantly affect the correlation between these semi-quantitative radiological markers and ventricular volume. While recent studies have explored the utility of the Evans' index in distinguishing individuals with iNPH from healthy subjects [31–34], its application necessitates strict adherence to specific image reconstruction planes, and its efficacy in post-surgical patient monitoring remains uncertain. Similarly, morphological imaging indicators associated with iNPH, such as a decreased corpus callosum angle and disproportionately enlarged subarachnoid space, are reliant on operator expertise. Although these indicators demonstrated some correlation with ventricular volume and the ventricular volume to intracranial volume (ICV) ratio, the correlation was shown to be not robust: variability in brain CT and MRI protocols across different imaging facilities, including differences in slice thickness and angulations, undermines the reliability of comparing the Evans' index across scans, even for the same patient [35]. This variability complicates the comparison of pre- and postoperative scans, as well as the comparison of ventricular volumes between iNPH patients and those without iNPH. While this may not greatly affect patients with significantly enlarged ventricles, it holds critical implications for patients with borderline ventricle sizes, both in research on iNPH and in clinical management [35]. The Evans' index previously demonstrated a high combined sensitivity of 96%, but its specificity was relatively low at 83% in diagnosing iNPH [36]. Similarly, the corpus callosal angle showed a comparable combined sensitivity and specificity of 89% and 92%, respectively [36]. Consequently, while the Evans' index and the corpus callosum angle are useful in most instances for substantiating a morphologically suspected diagnosis with semi-quantitative parameters, they ultimately do not allow one to reliably diagnose and differentiate iNPH from other neurodegenerative diseases, such as Alzheimer's dementia, in terms of their combined sensitivity and specificity.

Advancements in artificial intelligence (AI) have facilitated the development of automated volumetry tools, which have shown promise in providing a more nuanced understanding of iNPH. These AI-driven tools are thought to assess brain volume with high precision and consistency, potentially identifying subtle changes that may be overlooked in visual assessments. By integrating machine learning algorithms with traditional imaging data, it becomes possible to generate more accurate and reliable diagnostic criteria. The advent of AI and machine learning in neuroimaging has added a layer of precision to the assessment of brain volume loss and its relationship with radiological signs in iNPH and may explain the significant brain volume differences observed in our cohort involving gray and white matter, brainstem and midline volume, and ventricular size. The investigator-independent objectivity that can be achieved by an automated volumetry of MR images may hold the potential to substantially improve both diagnosis and follow-up of iNPH patients while adding just a few minutes of automated postprocessing time to the whole examination. To date, some semi-automated volumetric techniques have been effectively employed in research focusing on changes in ventricular volume post-surgery in iNPH [37,38]. It can be expected that fully automated segmentation methods such as the CE-marked AI-based software used in this study will further improve the accuracy and time efficiency of cerebral and ventricular volume measurements in patients with iNPH, including postoperative assessments. According to our volumetric data, even small but statistically significant decreases in cerebral white matter volume in the presence of global ventriculomegaly and largely preserved age-normalized cerebral gray matter are strongly suggestive of the presence of iNPH—findings that, due to their often subtle nature, may elude visual assessment even by an experienced examiner. Knowledge of this focal pattern of structural cerebral alteration, as opposed to the global and particularly frontoparietal accentuated cortical atrophy in AD patients, should potentially enhance diagnostic accuracy. However, larger trials should be conducted to identify specific volumetric cut-off values to eventually substantiate the value of AI volumetrization for diagnosing iNPH.

We acknowledge several limitations of this study. The retrospective nature generally limits the conclusions to be drawn since the reproducibility of the obtained clinical and imaging data cannot be determined. Also, we are aware of the relatively small sample size. However, we included only those patients who had a standardized, in-house MRI scan before ventriculoperitoneal shunt placement and a positive cerebrospinal fluid opening pressure. Furthermore, all of our AD patients were classified at a lower level (Levels 3 (mild cognitive impairment) and 4 (mild dementia)) according to the criteria of the Diagnostic and Statistical Manual of Mental Disorders, fourth edition (DSM-IV), and of the National Institute of Neurologic and Communicative Disorders and Stroke and the Alzheimer's Disease and Related Disorders Association [39]. It is, therefore, uncertain whether AD patients in a more advanced stage of the disease show more or less distinct brain structural differences from iNPH. The limiting factor of AI is the "black box" nature of many AI models, particularly deep neural networks, which complicates their integration into clinical decision-making. The lack of transparency in AI's decision-making processes can be a significant hurdle, as medical professionals rely on clear, understandable reasons behind diagnostic decisions. AI models can also suffer from overfitting, where they excel on training data but perform poorly on new, unseen data, or underfitting, where the model is too simplistic to capture the data's complexities. However, our software's algorithm and its encompassed normative database were refined based on a broad dataset of multi-center imaging data, surpassing the confines of a single institution. Lastly, the commercial AI-based mdbrain package used in this study is a proprietary software solution, so reproducibility with other software solutions is not readily possible. In order to determine what the underlying network structure actually looks like and how specific hyperparameters were defined—information that is not available to us in detail—one would have to contact the manufacturer, Mediaire. However, it will be easy to reproduce our results using the commercial software.

5. Conclusions

In conclusion, iNPH patients exhibited distinct brain volume alterations with patterns of global ventriculomegaly and white matter loss in addition to thalamic, brainstem, and temporomesial volume loss, while neocortical gray matter was largely preserved. Our volumetric results are supported by a concomitant decrease in population-normalized percentiles in the corresponding brain areas. These findings could potentially facilitate the radiological differentiation of patients with iNPH from healthy controls and AD, requiring only a few minutes of additional software-based post-processing. While traditional radiologic signs probably remain central to the understanding and diagnosis of iNPH in clinical practice, the inclusion of AI-driven measurements may provide an additional level of precision. Future research should focus on refining and validating volumetric markers and developing a comprehensive approach that aligns clinical and radiologic findings. This would not only improve diagnostic accuracy but also pave the way for more effective therapeutic interventions in iNPH.

Supplementary Materials: The following supporting information can be downloaded at https://www.mdpi.com/article/10.3390/diagnostics14131422/s1, Table S1: Subgroup comparisons in absolute volumes.

Author Contributions: Conceptualization, Z.B. and F.C.S.; methodology, Z.B., R.H. and F.C.S.; software, R.H. and F.C.S.; validation, Z.B. and F.C.S.; formal analysis, Z.B. and F.C.S.; investigation, Z.B., V.P., R.H., S.Z., C.K., V.B., M.B. and F.C.S.; resources, F.D., U.W., A.R. and F.C.S.; data curation, Z.B. and F.C.S.; writing—original draft preparation, Z.B. and F.C.S.; writing—review and editing, all authors; visualization, Z.B. and F.C.S.; supervision, A.R. and F.C.S.; project administration, F.C.S.; funding acquisition, A.R. and F.C.S. All authors have read and agreed to the published version of the manuscript.

Funding: This research was supported by the Open Access publication support fund of the University of Bonn (approval No. 2024063053000495).

Institutional Review Board Statement: The research received approval from the Ethics Committee for Clinical Trials on Humans and Epidemiological Research with Personal Data at the Faculty of Medicine of the Rheinische Friedrich-Wilhelms-Universität Bonn (IRB Number 118/22).

Informed Consent Statement: Patient consent was waived due to the retrospective nature of this study; written informed consent was not required.

Data Availability Statement: All relevant scientific data are included in the manuscript or the Supplementary Materials.

Acknowledgments: We gratefully acknowledge the help and technical expertise of all radiographers at the Neuroradiology Department of Bonn University Hospital.

Conflicts of Interest: The authors declare no conflicts of interest regarding the subject matter of this article.

References

1. Adams, R.D.; Fisher, C.M.; Hakim, S.; Ojemann, R.G.; Sweet, W.H. Symptomatic occult hydrocephalus with "normal" cerebrospinal-fluid pressure: A treatable syndrome. *N. Engl. J. Med.* **1965**, *273*, 117–126. [CrossRef] [PubMed]
2. Relkin, N.; Marmarou, A.; Klinge, P.; Bergsneider, M.; Black, P.M. Diagnosing idiopathic normal-pressure hydrocephalus. *Neurosurgery* **2005**, *57*, S4–S16. [CrossRef] [PubMed]
3. Hebb, A.O.; Cusimano, M.D. Idiopathic normal pressure hydrocephalus: A systematic review of diagnosis and outcome. *Neurosurgery* **2001**, *49*, 1166–1186. [CrossRef] [PubMed]
4. Silverberg, G.D. Normal pressure hydrocephalus (NPH): Ischaemia, CSF stagnation or both. *Brain* **2004**, *127*, 947–948. [CrossRef]
5. Ries, M.L.; Carlsson, C.M.; Rowley, H.A.; Sager, M.A.; Gleason, C.E.; Asthana, S.; Johnson, S.C. Magnetic resonance imaging characterization of brain structure and function in mild cognitive impairment: A review. *J. Am. Geriatr. Soc.* **2008**, *56*, 920–934. [CrossRef] [PubMed]
6. Gallia, G.L.; Rigamonti, D.; Williams, M.A. The diagnosis and treatment of idiopathic normal pressure hydrocephalus. *Nat. Clin. Pract. Neurol.* **2006**, *2*, 375–381. [CrossRef] [PubMed]
7. Kuroda, T.; Honma, M.; Mori, Y.; Futamura, A.; Sugimoto, A.; Kasai, H.; Yano, S.; Hieda, S.; Kasuga, K.; Ikeuchi, T.; et al. White Matter Lesions May Aid in Differentiating Idiopathic Normal Pressure Hydrocephalus and Alzheimer's Disease. *J. Alzheimer's Dis.* **2022**, *85*, 851–862. [CrossRef] [PubMed]
8. Owler, B.K.; Pickard, J.D. Normal pressure hydrocephalus and cerebral blood flow: A review. *Acta Neurol. Scand.* **2001**, *104*, 325–342. [CrossRef]
9. Evans, W.A. An encephalographic ratio for estimating the size of the cerebral ventricles: Further experience with serial observations. *Am. J. Dis. Child.* **1942**, *64*, 820–830. [CrossRef]
10. Hashimoto, M.; Ishikawa, M.; Mori, E.; Kuwana, N. Study of INPH on neurological improvement (SINPHONI). Diagnosis of idiopathic normal pressure hydrocephalus is supported by MRI-based scheme: A prospective cohort study. *Cerebrospinal Fluid. Res.* **2010**, *7*, 18. [CrossRef]
11. Kitagaki, H.; Mori, E.; Ishii, K.; Hirono, N.; Imamura, T. CSF spaces in idiopathic normal pressure hydrocephalus: Morphology and volumetry. *AJNR Am. J. Neuroradiol.* **1998**, *19*, 1277–1284. [PubMed]
12. Ishii, K.; Kanda, T.; Harada, A.; Miyamoto, N.; Kawaguchi, T.; Shimada, K.; Ohkawa, S.; Uemura, T.; Yoshikawa, T.; Mori, E. Clinical impact of the callosal angle in the diagnosis of idiopathic normal pressure hydrocephalus. *Eur. Radiol.* **2008**, *18*, 2678–2683. [CrossRef] [PubMed]
13. Wikkelso, C.; Andersson, H.; Blomstrand, C.; Lindqvist, G.; Svendsen, P. Normal pressure hydrocephalus: Predictive value of the cerebrospinal fluid tap-test. *Acta Neurol. Scand.* **1986**, *73*, 566–573. [CrossRef]
14. Damasceno, B.P. Normal pressure hydrocephalus: Diagnostic and predictive evaluation. *Dement. Neuropsychol.* **2009**, *3*, 8–15. [CrossRef] [PubMed]
15. Damasceno, B.P. Neuroimaging in normal pressure hydrocephalus. *Dement. Neuropsychol.* **2015**, *9*, 350–355. [CrossRef]
16. Nakajima, M.; Yamada, S.; Miyajima, M.; Ishii, K.; Kuriyama, N.; Kazui, H.; Kanemoto, H.; Suehiro, T.; Yoshiyama, K.; Kameda, M.; et al. Guidelines for management of idiopathic normal pressure hydrocephalus (third edition): Endorsed by the Japanese society of normal pressure hydrocephalus. *Neurol. Med. Chir.* **2021**, *61*, 63–97. [CrossRef] [PubMed]
17. Evans, W. An Encephalographic Ratio for Estimating Ventricular Enlargement and Cerebral Atrophy. *Arch. Neurol. Psychiatry* **1942**, *47*, 931. [CrossRef]
18. Brix, M.; Westman, E.; Simmons, A.; Ringstad, G.A.; Eide, P.K.; Wagner-Larsen, K.; Page, C.M.; Vitelli, V.; Beyer, M.K. The Evans' Index Revisited: New Cut-Off Levels for Use in Radiological Assessment of Ventricular Enlargement in the Elderly. *Eur. J. Radiol.* **2017**, *95*, 28–32. [CrossRef] [PubMed]
19. Virhammar, J.; Laurell, K.; Cesarini, K.G.; Larsson, E.M. The callosal angle measured on MRI as a predictor of outcome in idiopathic normal-pressure hydrocephalus. *J. Neurosurg.* **2014**, *120*, 178–184. [CrossRef]
20. *ISO 13485:2016*; Medical Devices—Quality Management Systems—Requirements for Regulatory Purposes. ISO: Geneva, Switzerland, 2016. Available online: https://www.iso.org/obp/ui/en/#iso:std:59752:en (accessed on 1 July 2024).
21. Dieckmeyer, M.; Roy, A.G.; Senapati, J.; Wachinger, C.; Grundl, L.; Dopfert, J.; Bertran, P.F.; Lemke, A.; Zimmer, C.; Kirschke, J.S.; et al. Effect of MRI acquisition acceleration via compressed sensing and parallel imaging on brain volumetry. *Magn. Reson. Mater. Phys. Biol. Med.* **2021**, *34*, 487–497. [CrossRef]
22. Haase, R.; Lehnen, N.C.; Schmeel, F.C.; Deike, K.; Rüber, T.; Radbruch, A.; Paech, D. External evaluation of a deep learning-based approach for automated brain volumetry in patients with huntington's disease. *Sci. Rep.* **2024**, *14*, 9243. [CrossRef] [PubMed]
23. Ronneberger, O.; Fischer, P.; Brox, T. U-net: Convolutional networks for biomedical image segmentation. In *Medical Image Computing and Computer-Assisted Intervention—Proceedings of the 2015 18th International Conference, Munich, Germany, 5–9 October 2015*; Proceedings, Part III 18; Springer International Publishing: Cham, Switzerland, 2015; Volume 9351, pp. 234–241.
24. Kingma, D.P.; Ba, J. Adam: A Method for Stochastic Optimization. In Proceedings of the International Conference on Learning Representations (ICLR), San Diego, CA, USA, 7–9 May 2015.
25. Çiçek, Ö.; Abdulkadir, A.; Lienkamp, S.S.; Brox, T.; Ronneberger, O. 3D U-Net: Learning dense volumetric segmentation from sparse annotation. In *Medical Image Computing and Computer-Assisted Intervention—Proceedings of the MICCAI 2016: 19th International Conference, Athens, Greece, 17–21 October 2016*; Proceedings, Part II 19; Springer International Publishing: Cham, Switzerland, 2016; pp. 424–432.

26. Tullberg, M.; Rosengren, L.; Blomsterwall, E.; Karlsson, J.E.; Wikkelso, C. CSF neurofilament and glial fibrillary acidic protein in normal pressure hydrocephalus. *Neurology* **1998**, *50*, 1122–1127. [CrossRef] [PubMed]
27. Zhao, Z.; He, J.; Chen, Y.; Wang, Y.; Wang, C.; Tan, C.; Liao, J.; Xiao, G. The pathogenesis of idiopathic normal pressure hydrocephalus based on the understanding of AQP1 and AQP4. *Front. Mol. Neurosci.* **2022**, *15*, 952036. [CrossRef] [PubMed]
28. Traini, E.; Carotenuto, A.; Fasanaro, A.M.; Amenta, F. Volume analysis of brain cognitive areas in Alzheimer's disease: Interim 3-year results from the ASCOMALVA trial. *J. Alzheimer's Dis.* **2020**, *76*, 317–329. [CrossRef] [PubMed]
29. Harper, L.; Bouwman, F.; Burton, E.J.; Barkhof, F.; Scheltens, P.; O'Brien, J.T.; Fox, N.C.; Ridgway, G.R.; Schott, J.M. Patterns of atrophy in pathologically confirmed dementias: A voxelwise analysis. *J. Neurol. Neurosurg. Psychiatry* **2017**, *88*, 908–916. [CrossRef] [PubMed]
30. Ishii, K.; Soma, T.; Kono, A.K.; Sofue, K.; Miyamoto, N.; Yoshikawa, T.; Mori, E.; Murase, K. Comparison of regional brain volume and glucose metabolism between patients with mild dementia with lewy bodies and those with mild Alzheimer's disease. *J. Nucl. Med.* **2007**, *48*, 704–711. [CrossRef] [PubMed]
31. Yun, S.Y.; Choi, K.S.; Suh, C.H.; Kim, S.C.; Heo, H.; Shim, W.H.; Jo, S.; Chung, S.J.; Lim, J.S.; Lee, J.H.; et al. Risk estimation for idiopathic normal-pressure hydrocephalus: Development and validation of a brain morphometry-based nomogram. *Eur. Radiol.* **2023**, *33*, 6145–6156. [CrossRef]
32. Anderson, R.C.; Grant, J.J.; de la Paz, R.; Frucht, S.; Goodman, R.R. Volumetric measurements in the detection of reduced ventricular volume in patients with normalpressure hydrocephalus whose clinical condition improved after ventriculoperitoneal shunt placement. *J. Neurosurg.* **2002**, *97*, 73–79. [CrossRef] [PubMed]
33. Toma, A.K.; Holl, E.; Kitchen, N.D.; Watkins, L.D. Evans' index revisited: The need for an alternative in normal pressure hydrocephalus. *Neurosurgery* **2011**, *68*, 939–944. [CrossRef]
34. Ambarki, K.; Israelsson, H.; Wahlin, A.; Birgander, R.; Eklund, A.; Malm, J. Brain ventricular size in healthy elderly: Comparison between Evans' index and volume measurement. *Neurosurgery* **2010**, *67*, 94–99; discussion 99. [CrossRef]
35. Williams, M.A.; Relkin, N.R. Diagnosis and management of idiopathic normal-pressure hydrocephalus. *Neurol. Clin. Pract.* **2013**, *3*, 375–385. [CrossRef] [PubMed]
36. Park, H.Y.; Kim, M.; Suh, C.H.; Lee, D.H.; Shim, W.H.; Kim, S.J. Diagnostic performance and interobserver agreement of the callosal angle and Evans' index in idiopathic normal pressure hydrocephalus: A systematic review and meta-analysis. *Eur. Radiol.* **2021**, *31*, 5300–5311. [CrossRef] [PubMed]
37. Takagi, K.; Watahiki, R.; Machida, T.; Onouchi, K.; Kato, K.; Oshima, M. Reliability and interobserver variability of Evans" index and disproportionately enlarged subarachnoid space hydrocephalus as diagnostic criteria for idiopathic normal pressure hydrocephalus. *Asian J. Neurosurg.* **2020**, *15*, 107–112. [CrossRef] [PubMed]
38. Tsunoda, A.; Mitsuoka, H.; Bandai, H.; Arai, H.; Sato, K.; Makita, J. Intracranial cerebrospinal fluid distribution and its postoperative changes in normal pressure hydrocephalus. *Acta Neurochir.* **2001**, *143*, 493–499. [CrossRef]
39. Reisberg, B.; Ferris, S.H.; de Leon, M.J.; Crook, T. The Global Deterioration Scale for assessment of primary degenerative dementia. *Am. J. Psychiatry* **1982**, *139*, 1136–1139. [CrossRef]

diagnostics

Article

Automatic Segmentation of Type A Aortic Dissection on Computed Tomography Images Using Deep Learning Approach

Xiaoya Guo [1,*], Tianshu Liu [1], Yi Yang [1], Jianxin Dai [1], Liang Wang [2], Dalin Tang [2,3] and Haoliang Sun [4,*]

1 School of Science, Nanjing University of Posts and Telecommunications, Nanjing 210023, China; tianshu271@163.com (T.L.); yang61yi@163.com (Y.Y.); daijx_js@126.com (J.D.)
2 School of Biological Science and Medical Engineering, Southeast University, Nanjing 210096, China; liangwang@seu.edu.cn (L.W.); dtang@wpi.edu (D.T.)
3 Mathematical Sciences Department, Worcester Polytechnic Institute, Worcester, MA 01609, USA
4 Department of Cardiovascular Surgery, First Affiliated Hospital of Nanjing Medical University, Nanjing 210029, China
* Correspondence: guoxiaoya1990@163.com (X.G.); shlsky@126.com (H.S.)

Abstract: Purpose: Type A aortic dissection (TAAD) is a life-threatening aortic disease. The tear involves the ascending aorta and progresses into the separation of the layers of the aortic wall and the occurrence of a false lumen. Accurate segmentation of TAAD could provide assistance for disease assessment and guidance for clinical treatment. Methods: This study applied nnU-Net, a state-of-the-art biomedical segmentation network architecture, to segment contrast-enhanced CT images and quantify the morphological features for TAAD. CT datasets were acquired from 24 patients with TAAD. Manual segmentation and annotation of the CT images was used as the ground-truth. Two-dimensional (2D) nnU-Net and three-dimensional (3D) nnU-Net architectures with Dice- and cross entropy-based loss functions were utilized to segment the true lumen (TL), false lumen (FL), and intimal flap on the images. Four-fold cross validation was performed to evaluate the performance of the two nnU-Net architectures. Six metrics, including accuracy, precision, recall, Intersection of Union, Dice similarity coefficient (DSC), and Hausdorff distance, were calculated to evaluate the performance of the 2D and 3D nnU-Net algorithms in TAAD datasets. Aortic morphological features from both 2D and 3D nnU-Net algorithms were quantified based on the segmented results and compared. Results: Overall, 3D nnU-Net architectures had better performance in TAAD CT datasets, with TL and FL segmentation accuracy up to 99.9%. The DSCs of TLs and FLs based on the 3D nnU-Net were 88.42% and 87.10%. For the aortic TL and FL diameters, the FL area calculated from the segmentation results of the 3D nnU-Net architecture had smaller relative errors (3.89–6.80%), compared to the 2D nnU-Net architecture (relative errors: 4.35–9.48%). Conclusions: The nnU-Net architectures may serve as a basis for automatic segmentation and quantification of TAAD, which could aid in rapid diagnosis, surgical planning, and subsequent biomechanical simulation of the aorta.

Keywords: type A aortic dissection; deep learning; computed tomography; image segmentation; nnU-Net

1. Introduction

Aortic dissection is a severe cardiovascular disease caused by a tear in the intimal layer of the aorta, which results in the separation of the layers of the aortic wall and congestion in the tunica media, leading to blood flow deficiency and even mortality. When the dissection involves the ascending aorta, it is defined as type A aortic dissection (TAAD), which is usually managed with an emergency open-chest surgery [1]. Computed tomography (CT) is a common imaging modality used for screening and evaluating patients with aortic dissection in clinical settings [2,3], as it can provide important morphological information on the dissected aorta to guide the clinical diagnosis and treatment decision-making of acute aortic dissection.

The diameter of the ascending aorta is one of the key morphological features for aortic dissection management. The guidelines of the European Society of Cardiology on the diagnosis and treatment of aortic diseases recommend prophylactic replacement of the ascending aorta to prevent acute TAAD when the diameter of the ascending aorta is larger than 55 mm [4,5]. Furthermore, the diameter of the dissected ascending aorta serves as a basis for selecting artificial aortic grafts of the right size in the aortic replacement operation [6]. Other morphological information, including the true lumen (TL), false lumen (FL), and intimal flap, also have important prognostic significance [7]. The intimal flap fenestration is commonly implemented to depressurize the FL by puncture of the intimal flap from the TL into the FL for patients with malperfusion syndrome [4]. Hence, accurately segmenting these morphological features of TAAD from CT images is vital for assessment of aortic dissection.

A large number of cross-sectional images of the dissected aorta are generated from CT scans. The manual segmentation of these images is too time-consuming to meet the clinical need. Recently, deep learning approaches were used to complete the task [8]. Li et al. applied a 2D cascaded convolutional network to extract the contours of the aorta and TL from contrast-enhanced CT datasets of 45 patients with aortic dissection. Their proposed method showed good performance, with Dice similarity coefficients (DSCs) of 0.989 for the aorta and 0.925 for the true lumen [9]. This was the first study to target aortic dissection segmentation using deep learning. However, few studies on the segmentation of TAAD exist in the current literature, compared to type B aortic dissection (TBAD, dissection that does not involve the ascending aorta). Many excellent deep learning methods were proposed to segment TBAD [10–13]. Hahn et al. proposed a stepwise convolutional neural network (CNN)-based approach to segment the TL and FL using 153 CT datasets from 45 patients with TBAD, with the DSC ranging from 0.87 to 0.90 for segmenting the TL and FL [11]. Considering the significant prognostic value of false lumen thrombosis for predicting late adverse events, Wobben et al. utilized 3D residual U-Net-based models for segmenting the TL, FL, and false lumen thrombosis [12]. Their best results for TL and FL segmentation achieved median (interquartiles) DSCs of 0.85 (0.77 0.88) and 0.84 (0.82 0.87), respectively. Xiang et al. proposed a flap-attention-based deep learning approach for segmenting the TL and FL (flap simplified to the boundary between the TL and FL) of TBAD, with DSCs of 91.1% and 88.4% [14]. In addition to the segmentation of the TL or TL + FL, some studies focused on the segmentation of the intimal flap [15,16]. Lyu et al. constructed a CNN-based two-step algorithm to capture the aortic structure and intimal flaps without distinguishing the TL and FL, and they achieved an average DSC of over 92% [16]. U-Net, as the deep learning architecture of a pixel-level classification, is well suited for segmentation tasks and is widely used in image segmentation [17]. Cheng et al. introduced a semantic segmentation algorithm based on the U-Net framework to segment the aortic TL using contrast-enhanced CT images obtained from 20 patients with TAAD, with 85.00% accuracy (90.00% sensitivity, 80.00% specificity) [18].

nnU-Net (no new U-Net) is a state-of-the-art network architecture for biomedical image segmentation and is extensively applied to the segmentation of many different organs and tissues, including the lung and heart; it can automatically adapt its pre-processing strategies and network architecture to a given image dataset without additional manual tuning [19]. However, similar designs based on nnU-Net have not been employed to segment TAAD on CT images. For TAAD, segmentation and morphological quantification of the dissected ascending aorta, including the TL, FL, and intimal flap, are essential for surgical interventions such as aortic valve replacement or ascending aortic replacement [20]. In this study, nnU-Net was adopted to segment contrast-enhanced CT data from 24 patients with TAAD to characterize the TL, FL, and intimal flap, and quantify the aortic morphology to examine its performance. Aortic morphological factors, including the diameter of the dissected aorta, the thickness of the intimal flaps, and the areas of the TL and FL, were measured. These are crucial indicators for surgical planning and follow-up studies, such as computational simulation and modeling.

2. Materials and Methods

2.1. Data Acquisition and Processing

A total of 24 patients with TAAD (20 males and 4 females, mean age ± standard deviation: 56.3 ± 9.4) who underwent surgical repair of the aorta following Sun's procedure were enrolled at Jiangsu Province Hospital, with informed consent obtained [20]. The diagnosis of aortic dissection was evaluated by pre-operative CT examinations using a CT system with 64-row detectors (SOMATOM Definition AS, Siemens AG). All patients were scanned through CT after injection of intravenous contrast and with a 1.25 mm slice width. One CT sequence was selected from each patient to perform subsequent image segmentation. The CT data were collected following the protocol approved by the Medical Ethics Committee of Jiangsu Province Hospital (approval number: 2022-SR-730). The original images in DICOM format with a slice size of 512 × 512 pixels were converted to the NIfTI format as the standard data format for the segmentation task. Manual annotation was performed to label the TL and FL regions of the aorta on the CT slices (see Figure 1). The unlabeled region on the CT slice was perceived as the background. Of note, the intimal flap region was not labeled at this stage, and would be extracted after image segmentation using deep learning. For the consistency and reliability of manual annotation, the labeling was accomplished by three experts following the standard procedure described in Isensee et al. and Yao et al. [13,19]. The 24 CT datasets were randomly divided into two groups. The segmentation task of two groups was assigned to two different experts, and then checked by each other, slice-by-slice, after segmentation and annotation, with the disagreement and inconsistency between the two experts resolved by the third expert. Manual annotation took approximately 2.5 h per dataset. The conversion of medical image format and manual labeling were executed with the open source ITK-SNAP software (version 4.0, http://www.itksnap.org/ (accessed on 1 June 2023)) [21].

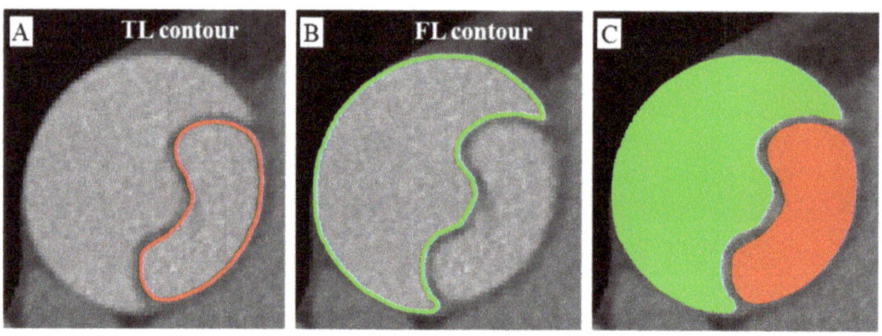

Figure 1. Manual annotation procedure. (**A**) The manual TL contour; (**B**) the manual FL contour; (**C**) the labeled areas of TL and FL in aorta. Red color, TL; green color, FL.

2.2. Segmentation Network Architecture

Both two-dimensional (2D) and three-dimensional (3D) nnU-Net frameworks with Dice and cross entropy-based loss functions were utilized to segment the aortic region of contrast-enhanced CT images from 24 patients with TAAD, and the performances of these two networks were compared. The nnU-Net architecture integrates with pre- and post-processing. For pre-processing, data augmentation contains rotations, scaling, Gaussian noise and blur, brightness, contrast and simulation of low resolution, gamma correction, and mirroring. Intensity normalization was performed on all the CT images by clipping them to an intensity range between 0.5 and 99.5 percentiles, and then the images were normalized using a global foreground mean and a standard deviation. All images were resampled to 1.25 × 0.8242 × 0.8242 mm. The stochastic gradient method with Nesterov's momentum was used to optimize the learning rate, in which the initial values of the learning rate and Nesterov's momentum were set at 0.01 and 0.99, respectively. Then, the learning rate was decayed throughout the training process, following the 'poly' learning rate policy: $(1 - \text{epoch}/\text{epoch}_{max})^{0.9}$. More details can be found in [19]. The number of epochs was set at 600 according to the performance of the model on the validation set. All experiments were conducted with Python 3.8.0, Pytorch 2.2.1 and NVIDIA CUDA 11.8 on NVIDIA RTX 4090 GPU with 24 GB of video memory. The batch size, patch size, and epoch time were 12, 512 × 512, and 34 s (per epoch) for the 2D nnU-Net network, and were 2, 56 × 224 × 192, and 42 s (per epoch) for the 3D nnU-Net network, respectively. The training times per fold of the 2D and 3D nnU-Net networks were 5.8 h and 7.3 h. The basic flowchart of TAAD segmentation using the 2D and 3D nnU-Net is shown in Figure 2.

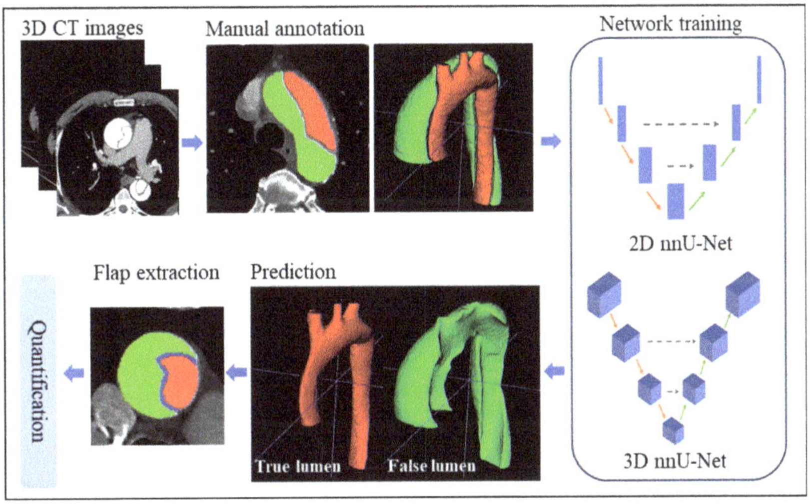

Figure 2. Basic flowchart of TAAD segmentation based on nnU-Net architectures.

2.3. Validation and Performance Evaluation

Twenty-four CT datasets were randomly divided into four groups. Four-fold cross validation was employed, with any three out of four groups used as training data, while the remaining one group was utilized as the testing data. Six metrics, including accuracy, precision, recall, Intersection of Union (IoU), Dice similarity coefficient (DSC), and Hausdorff distance (HD), were calculated to evaluate the performance of 2D and 3D nnU-Net algorithms on the TAAD dataset. The six metrics were calculated according to the following formulas:

$$\text{accuracy} = (TP + TN)/(TP + TN + FP + FN), \quad (1)$$

$$\text{precision} = TP/(TP + FP), \quad (2)$$

$$\text{recall} = TP/(TP + FN), \quad (3)$$

$$IoU = TP/(TP + FP + FN), \quad (4)$$

$$DSC = 2 \times TP/(2 \times TP + FP + FN), \quad (5)$$

$$HD = \max(h(GT, UN), h(UN, GT)), \quad (6)$$

where TP is the number of true-positive outcomes, FP is the number of false-positive outcomes, TN is the number of true-negative outcomes, FN is the number of false-negative outcomes, GT is ground-truth, UN is the nnU-Net method result, and $h(GT, UN) = \max_{g \in GT}$

$\min_{u \in UN} \|g - u\|$, $\|\cdot\|$ is the L2 norm. The six metrics were calculated using the TP, TN, FP, FN, GT, and UN of the CT images on a per-voxel basis.

2.4. Post-Processing and Quantification of Aortic Morphological Information

Each CT slice was segmented by both 2D and 3D nnU-Net architectures to output the segmented TL and FL regions. Then, the contours of the TL and FL were extracted in segmented CT images with TL and FL output labels, respectively, using the active contour models [22]. To obtain the contours of the whole aorta, active contour models were applied again to the segmented image by treating TL + FL as the same labeling region. The intimal flap region was determined by the remaining region after removing the TL and FL in the whole aortic area. After that, the cross-sectional whole aorta was categorized into three regions, including TL, FL, and intimal flap. The maximum diameter of the aorta, the areas of the TL and FL, and the average thickness of the flap were quantified for clinical assessment. Relative errors from both nnU-Net architectures were calculated and compared to determine which one had better performance in quantifying aortic morphological information.

2.5. Statistical Analysis

All output segmentation results of the CT slices were gathered to present patient-level results. The Shapiro–Wilk test was used to test the normality of the patient-level results. The differences of six metrics between 2D and 3D nnU-Net networks all obey normal distribution. Therefore, the differences between the results of the 2D and 3D nnU-Net networks were compared using the paired *t*-test. A *p*-value of <0.05 is considered as statistically significant. Statistical analysis was performed with MATLAB (MathWorks Inc., Natick, MA, USA).

3. Results

3.1. Performance of nnU-Net Architectures

The TL and FL regions of the 2D and 3D nnU-Net architectures were used to calculate the six metrics at the patient level: accuracy, precision, recall, IoU, DSC, and HD. The statistics of the six metrics at the patient level were listed in Table 1. For TL prediction, the 3D nnU-Net was significantly higher in accuracy, precision, IoU, DSC, and HD, compared to the 2D nnU-Net (all *p*-values < 0.05). For FL prediction, the 3D nnU-Net was significantly higher only in accuracy, with *p*-value 0.0380. All the prediction slices from the 24 CT datasets were stacked into a 3D vessel structure for aortic geometry visualization (see Figure 3D–L). The prediction results of the 3D nnU-Net architecture had better smoothness. Overall, the 3D nnU-Net architecture had better performance in TAAD CT datasets, and the segmentation of the TLs was better than the segmentation of the FLs for all metrics. Figure 3 shows the output segmentation of the 2D and 3D nnU-Net architectures from one sample patient, including segmentation of two CT slices, and the 3D structures of the TLs and FLs.

Table 1. TL and FL segmentations using 2D and 3D nnU-Net architectures: mean and standard deviation values of six metrics for 24 TAAD patients.

Net	Label	Accuracy (%)	Precision (%)	Recall (%)	IoU (%)	DSC (%)	HD
2D	TL	99.90 ± 0.07	86.28 ± 11.42	84.44 ± 11.94	74.51 ± 13.68	84.68 ± 9.56	24.10 ± 10.92
	FL	99.85 ± 0.09	86.34 ± 9.66	83.79 ± 13.30	73.56 ± 12.25	84.17 ± 8.84	30.40 ± 19.51
3D	TL	99.92 ± 0.06	89.71 ± 11.18	87.94 ± 9.11	80.20 ± 12.71	88.41 ± 8.97	19.40 ± 10.10
	FL	99.89 ± 0.07	88.57 ± 10.03	86.94 ± 12.51	78.19 ± 13.13	87.10 ± 9.39	28.68 ± 20.49
p-value	TL	0.0225	0.0483	0.0845	0.0223	0.0266	0.0280
	FL	0.0380	0.1853	0.2107	0.1001	0.1592	0.4594

p-value from paired *t*-test.

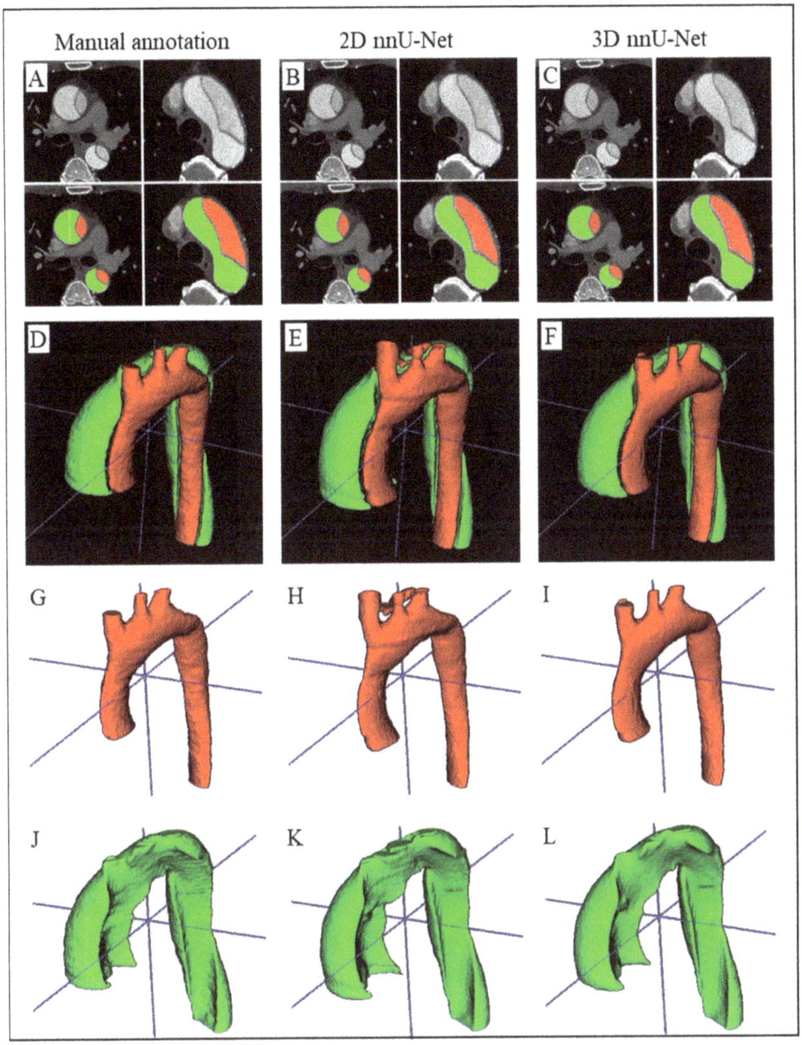

Figure 3. The 3D visualization of ascending aorta from manual annotation, 2D nnU-Net, and 3D nnU-Net. The four subgraphs in the first row (**A–C**) are two CT images of ascending aorta and arch and corresponding segmentation. (**D–F**) are the 3D visualization of ascending aorta from manual annotation, and prediction of 2D and 3D nnU-Net. (**G–I**) are the 3D visualization of TL. (**J–L**) are the 3D visualization of FL. Red region, TL; green region, FL.

3.2. *Quantification of Aortic Morphological Information*

Figure 4 shows the cross-section of the ascending aorta after extracting the intimal flap from the output segmentation. The region between the TL and FL was characterized as the intimal flap. CT slices containing the TL, FL, and intimal flap in the section of the ascending aorta were chosen to calculate the diameter, area, etc. The outputs from twenty-one patients were adopted to quantify the morphological features of dissected aortas, while the other three patients were not processed due to the tiny size of the tear in the ascending aorta region compared to the large dissection size in the aortic arch region. The cross-section images of the aortic arch (Figure 3A) could not provide accurate aortic morphology, such as aortic diameter and the TL and FL diameters of the ascending aorta, and thus the three patients were not included in the analysis of the quantifying aortic morphological features. The average values of the aortic morphological factors were used to quantify the aortic morphology of the ascending aorta for patients with TAAD. Table 2 shows the quantifications of TL, FL, and intimal flap according to manual annotation (ground-truth), and the output segmentation of the 2D nnU-Net and 3D nnU-Net, respectively. Relative error was adopted to compare the quantification of the TL, FL, and intimal flap from the two architectures, with manual annotation as the reference. The TL area and intimal flap thickness calculated from the output segmentation of the 2D nnU-Net architecture had smaller relative errors: 6.42% and 12.55%. The aortic, TL, and FL diameters, and the FL area calculated from the output segmentation of the 3D nnU-Net architecture all had smaller relative errors (3.89–6.80%), compared to the 2D nnU-Net architecture (relative errors: 4.35–9.48%).

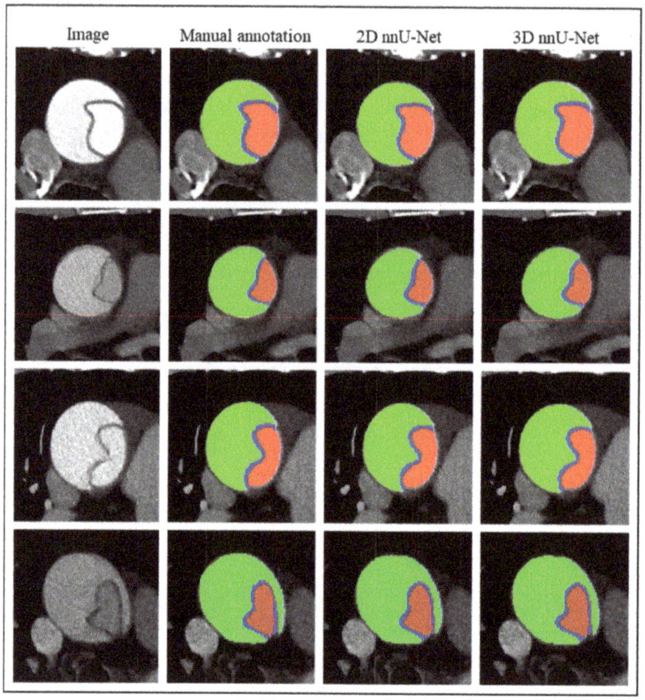

Figure 4. Three regions in the cross-section of ascending aorta segmented by ground-truth, 2D nnU-Net, and 3D nnU-Net from four sample patients. Red region, TL; green region, FL; blue region, intimal flap.

Table 2. TL, FL, and intimal flap quantifications according to manual annotation (ground-truth), output segmentation of 2D nnU-Net, and output segmentation of 3D nnU-Net, respectively: the diameters of whole aorta, TL, and FL; areas of TL and FL; and intimal flap thickness.

		Aortic Diameter (mm)	TL Diameter (mm)	FL Diameter (mm)	TL Area (mm^2)	FL Area (mm^2)	Flap Thickness (mm)
GT	Ave.	49.71	36.07	47.32	566.83	927.47	2.35
2D	Ave.	48.91	36.02	45.60	566.73	869.15	2.46
	RE Ave.	4.58%	5.55%	4.35%	6.42%	9.48%	12.55%
3D	Ave.	49.61	36.47	46.74	577.93	892.84	2.41
	RE Ave.	4.42%	4.89%	3.89%	6.94%	6.80%	14.08%

Legend: GT, ground-truth; Ave., average; RE, relative error; TL, true lumen; FL, false lumen; Flap, intimal flap.

4. Discussion

4.1. Segmentation Based on Artificial Intelligence

Medical image segmentation plays an increasingly important role in disease diagnosis, assessment, and treatment. Contrast-enhanced CT imaging has significant clinical diagnostic value and advantages for TAAD. The large number of CT images not only bring a wealth of aortic information, but also pose a huge burden on image segmentation processing. TAAD is a dangerous and urgent aortic disease, and it is unable to withstand the time-consuming manual segmentation in a clinical setting. With the rapid development of artificial intelligence, deep learning has been applied to a variety of medical fields, such as image segmentation, disease screening, identification, and prediction [8,23]. The nnU-Net proposed by Isensee et al. is a self-configuring method for deep learning-based segmentation, and is the clear winner of the Medical Segmentation Decathlon challenge [19,24]. The CT datasets of TAAD and the task segmentation of the TL and FL are applicable to nnU-Net due to the versatility, ease of use, and robustness of this network. In this study, the 2D and 3D nnU-Nets were employed to segment the aortic TLs and FLs, using contrast-enhanced CT images obtained from 24 patients with TAAD, and they achieved over 99% accuracy. Particularly, precision and recall based on the 3D nnU-Net exceeded 88.57% and 86.94%, respectively. Cheng et al. compared the performance of seven deep learning algorithms in a TL segmentation task of the TAAD dataset. The results showed that the accuracy and DSCs of the seven algorithms ranged from 93.88% to 99.42%, and 80.08% to 90.58%, respectively [18]. The performance of the 3D nnU-Net in the TL and FL segmentation in this study was good, with accuracy up to 99.9%. Additionally, the DSCs of the TLs and FLs based on the 3D nnU-Net in our TAAD dataset are 88.42% and 87.10%. Compared to DSC values ranging from 84% to 90% when segmenting the TLs and FLs based on the CNN algorithm and the 3D residual U-Net using the TBAD dataset [10,12], the performance of the 3D nnU-Net in TL and FL segmentation is still excellent in the TAAD dataset. These TAAD datasets included a variety of tear configurations, as shown by the different false lumen shapes on cross-sectional CT images, which should be a good representation of the patient population of TAAD, but there still needs to be more validation for these results when generalizing to a larger dataset.

4.2. Significance of Quantification of Aortic Morphological Information

The quantification of dissected aorta morphology is an integral part of the diagnostic workflow in clinical routines. It provides fundamental information for the selection of clinical treatment options and important aortic structures for subsequent biomechanical simulation. The diameter of the ascending aorta is an important morphological factor in diagnosis and aortic replacement for TAAD. Among aortic, TL, and FL diameters, the 2D and 3D nnU-Net architectures performed best in quantification of the FL diameter, with the average of relative error less than 4.35%, and worst in quantification of the TL diameter, with the average of relative error above 4.89%. The FL area and intimal flap thickness had worse performances from the two nnU-Net architectures. This may be affected by the false lumen thrombus, which is regarded as the false lumen region in manual annotation because of the small number of thrombi in this study. Accurate quantification could contribute to diagnosis and treatment, and it also provides essential information for the computational simulation of aortic biomechanics. Since the pathogenesis of TAAD is still obscure, many studies focused on the biomechanical modeling of the aorta to investigate the biomechanical mechanism of TAAD [25,26]. The TL, FL, and intimal flap are the structures of the dissected aorta. The segmentation of the three regions in the aorta is necessary for more accurate biomechanical simulation. From the aspect of clinical utility, streamlining the whole segmentation process is necessary. Our work has completed the FL/TL segmentation, flap extraction, quantification, and visualization of TAAD. If these discrete efforts are integrated into clinical software, it would be beneficial to the diagnosis of aortic dissection.

4.3. Limitations

One major limitation of this study is that the segmentation of the false lumen thrombus was not performed. For simplification, the false lumen thrombus was treated as the false lumen region, and not regarded as an additional region different from the TL, FL, and intimal flap regions. The false lumen thrombus should be treated as a different label from the existing FL and TL labels, and its segmentation should be taken into consideration in follow-up studies. Another limitation is the small sample size of CT datasets studied. Large-scale studies with more CT datasets are needed to validate and improve the significance of the prediction method.

5. Conclusions

This study demonstrates that deep learning is a valuable, broadly applicable tool for segmentation and quantification of the aorta with TAAD in a clinical setting. The nnU-Net architecture may serve as a basis for automatic segmentation and quantification of TAAD, which could aid in rapid diagnosis, surgical planning, and subsequent biomechanical simulation of the aorta.

Author Contributions: Conceptualization, X.G.; methodology, T.L., Y.Y. and X.G.; data curation, H.S. and L.W.; writing—original draft preparation, X.G.; writing—review and editing, J.D., D.T. and L.W.; funding acquisition, H.S. and X.G. All authors have read and agreed to the published version of the manuscript.

Funding: This research was funded by the National Natural Sciences Foundation of China, grant number 82100254, and the Foundation of Jiangsu Provincial Double Innovation Doctor Program, grant number CZ007SC20006.

Institutional Review Board Statement: This study was conducted in accordance with the Declaration of Helsinki, and approved by the Medical Ethics Committee of Jiangsu Province Hospital (protocol code 2022-SR-730 and 30, December 2022).

Informed Consent Statement: Informed consent was obtained from all subjects involved in the study.

Data Availability Statement: The datasets presented in this article are not readily available because the data are part of an ongoing study. Requests to access the datasets should be directed to the corresponding author.

Conflicts of Interest: The authors declare no conflicts of interest.

References

1. El Hussein, M.T.; Green, T. Acute Aortic Dissection: Pathophysiology and Antihypertensive Therapy. *J. Nurse Pract.* **2022**, *18*, 383–387. [CrossRef]
2. McMahon, M.A.; Squirrell, C.A. Multidetector CT of aortic dissection: A pictorial review. *Radiographics* **2010**, *30*, 445–460. [CrossRef] [PubMed]
3. Fleischmann, D.; Afifi, R.O.; Casanegra, A.I.; Elefteriades, J.A.; Gleason, T.G.; Hanneman, K.; Roselli, E.E.; Willemink, M.J.; Fischbein, M.P.; American Heart Association Council on Cardiovascular Radiology and Intervention; et al. Imaging and Surveillance of Chronic Aortic Dissection: A Scientific Statement from the American Heart Association. *Circ Cardiovasc. Imaging* **2022**, *15*, e000075. [CrossRef]
4. Erbel, R.; Aboyans, V.; Boileau, C. The Task Force for the Diagnosis and Treatment of Aortic Diseases of the European Society of Cardiology (ESC). Document covering acute and chronic aortic diseases of the thoracic and abdominal aorta of the adult. 2014 ESC Guidelines on the diagnosis and treatment of aortic diseases. *Eur. Heart J.* **2014**, *35*, 2873–2926.
5. Tozzi, P.; Gunga, Z.; Niclauss, L.; Delay, D.; Roumy, A.; Pfister, R.; Colombier, S.; Patella, F.; Qanadli, S.D.; Kirsch, M. Type A aortic dissection in aneurysms having modelled pre-dissection maximum diameter below 45 mm: Should we implement current guidelines to improve the survival benefit of prophylactic surgery? *Eur. J. Cardiothorac. Surg.* **2021**, *59*, 473–478. [CrossRef] [PubMed]
6. Huang, F.; Li, X.; Zhang, Z.; Li, C.; Ren, F. Comparison of two surgical approaches for acute type A aortic dissection: Hybrid debranching versus total arch replacement. *J. Cardiothorac. Surg.* **2022**, *17*, 166. [CrossRef]
7. Patterson, B.O.; Cobb, R.J.; Karthikesalingam, A.; Holt, P.J.; Hinchliffe, R.J.; Loftus, I.M.; Thompson, M.M. A systematic review of aortic remodeling after endovascular repair of type B aortic dissection: Methods and outcomes. *Ann. Thorac. Surg.* **2014**, *97*, 588–595. [CrossRef]

8. Pepe, A.; Li, J.; Rolf-Pissarczyk, M.; Gsaxner, C.; Chen, X.; Holzapfel, G.A.; Egger, J. Detection, segmentation, simulation and visualization of aortic dissections: A review. *Med. Image Anal.* **2020**, *65*, 101773. [CrossRef]
9. Li, Z.; Feng, J.; Feng, Z.; An, Y.; Gao, Y.; Lu, B.; Zhou, J. *Lumen Segmentation of Aortic Dissection with Cascaded Convolutional Network*; Springer International Publishing: Cham, Switzerland, 2019; pp. 122–130.
10. Cao, L.; Shi, R.; Ge, Y.; Xing, L.; Zuo, P.; Jia, Y.; Liu, J.; He, Y.; Wang, X.; Luan, S.; et al. Fully automatic segmentation of type B aortic dissection from CTA images enabled by deep learning. *Eur. J. Radiol.* **2019**, *121*, 108713. [CrossRef]
11. Hahn, L.D.; Mistelbauer, G.; Higashigaito, K.; Koci, M.; Willemink, M.J.; Sailer, A.M.; Fischbein, M.; Fleischmann, D. CT-based True- and False-Lumen Segmentation in Type B Aortic Dissection Using Machine Learning. *Radiol. Cardiothorac. Imaging* **2020**, *2*, e190179. [CrossRef]
12. Wobben, L.D.; Codari, M.; Mistelbauer, G.; Pepe, A.; Higashigaito, K.; Hahn, L.D.; Mastrodicasa, D.; Turner, V.L.; Hinostroza, V.; Baumler, K.; et al. Deep Learning-Based 3D Segmentation of True Lumen, False Lumen, and False Lumen Thrombosis in Type-B Aortic Dissection. In Proceedings of the 43rd Annual International Conference of the IEEE-Engineering-in-Medicine-and-Biology-Society (IEEE EMBC), Virtual, 1–5 November 2021; pp. 3912–3915.
13. Yao, Z.; Xie, W.; Zhang, J.; Dong, Y.; Qiu, H.; Yuan, H.; Jia, Q.; Wang, T.; Shi, Y.; Zhuang, J.; et al. ImageTBAD: A 3D Computed Tomography Angiography Image Dataset for Automatic Segmentation of Type-B Aortic Dissection. *Front. Physiol.* **2021**, *12*, 732711. [CrossRef]
14. Xiang, D.; Qi, J.; Wen, Y.; Zhao, H.; Zhang, X.; Qin, J.; Ma, X.; Ren, Y.; Hu, H.; Liu, W.; et al. ADSeg: A flap-attention-based deep learning approach for aortic dissection segmentation. *Patterns* **2023**, *4*, 100727. [CrossRef] [PubMed]
15. Morariu, C.A.; Terheiden, T.; Dohle, D.S.; Tsagakis, K.; Pauli, J. Increasing the feasibility of minimally invasive procedures in type A aortic dissections: A framework for segmentation and quantification. *Int. J. Comput. Assist. Radiol. Surg.* **2015**, *11*, 243–252. [CrossRef] [PubMed]
16. Lyu, T.; Yang, G.; Zhao, X.; Shu, H.; Luo, L.; Chen, D.; Xiong, J.; Yang, J.; Li, S.; Coatrieux, J.L.; et al. Dissected aorta segmentation using convolutional neural networks. *Comput. Methods Programs Biomed.* **2021**, *211*, 106417. [CrossRef] [PubMed]
17. Siddique, N.; Paheding, S.; Elkin, C.P.; Devabhaktuni, V. U-Net and Its Variants for Medical Image Segmentation: A Review of Theory and Applications. *IEEE Access* **2021**, *9*, 82031–82057. [CrossRef]
18. Cheng, J.; Tian, S.; Yu, L.; Ma, X.; Xing, Y. A deep learning algorithm using contrast-enhanced computed tomography (CT) images for segmentation and rapid automatic detection of aortic dissection. *Biomed. Signal Process. Control.* **2020**, *62*, 102145. [CrossRef]
19. Isensee, F.; Jaeger, P.F.; Kohl, S.A.; Petersen, J.; Maier-Hein, K.H. nnU-Net: A self-configuring method for deep learning-based biomedical image segmentation. *Nat. Methods* **2021**, *18*, 203–211. [CrossRef] [PubMed]
20. Ma, W.G.; Zhu, J.M.; Zheng, J.; Liu, Y.M.; Ziganshin, B.A.; Elefteriades, J.A.; Sun, L.Z. Sun's procedure for complex aortic arch repair: Total arch replacement using a tetrafurcate graft with stented elephant trunk implantation. *Ann. Cardiothorac. Surg.* **2013**, *2*, 642–648. [PubMed]
21. Yushkevich, P.A.; Piven, J.; Hazlett, H.C.; Smith, R.G.; Ho, S.; Gee, J.C.; Gerig, G. User-guided 3D active contour segmentation of anatomical structures: Significantly improved efficiency and reliability. *Neuroimage* **2006**, *31*, 1116–1128. [CrossRef]
22. Kass, M.; Witkin, A.; Terzopoulos, D. Snakes: Active contour models. *Int. J. Comput. Vis.* **1988**, *1*, 321–331. [CrossRef]
23. Mastrodicasa, D.; Codari, M.; Baumler, K.; Sandfort, V.; Shen, J.; Mistelbauer, G.; Hahn, L.D.; Turner, V.L.; Desjardins, B.; Willemink, M.J.; et al. Artificial Intelligence Applications in Aortic Dissection Imaging. *Semin. Roentgenol.* **2022**, *57*, 357–363. [CrossRef]
24. Antonelli, M.; Reinke, A.; Bakas, S.; Farahani, K.; Kopp-Schneider, A.; Landman, B.A.; Litjens, G.; Menze, B.; Ronneberger, O.; Summers, R.M. The medical segmentation decathlon. *Nat. Commun.* **2022**, *13*, 4128. [CrossRef]
25. Guo, X.; Gong, C.; Zhai, Y.; Yu, H.; Li, J.; Sun, H.; Wang, L.; Tang, D. Biomechanical characterization of normal and pathological human ascending aortic tissues via biaxial testing Experiment, constitutive modeling and finite element analysis. *Comput. Biol. Med.* **2023**, *166*, 107561. [CrossRef]
26. Zhu, Y.; Mirsadraee, S.; Rosendahl, U.; Pepper, J.; Xu, X.Y. Fluid-structure interaction simulations of repaired type A aortic dissection: A comprehensive comparison with rigid wall models. *Front. Physiol.* **2022**, *13*, 913457. [CrossRef]

diagnostics MDPI

Article

Critical Risk Assessment, Diagnosis, and Survival Analysis of Breast Cancer

Shamiha Binta Manir * and Priya Deshpande

Department of EECE, Marquette University, Milwaukee, WI 53233, USA; priya.deshpande@marquette.edu
* Correspondence: shamihabinta.manir@marquette.edu

Abstract: Breast cancer is the most prevalent type of cancer in women. Risk factor assessment can aid in directing counseling regarding risk reduction and breast cancer surveillance. This research aims to (1) investigate the relationship between various risk factors and breast cancer incidence using the BCSC (Breast Cancer Surveillance Consortium) Risk Factor Dataset and create a prediction model for assessing the risk of developing breast cancer; (2) diagnose breast cancer using the Breast Cancer Wisconsin diagnostic dataset; and (3) analyze breast cancer survivability using the SEER (Surveillance, Epidemiology, and End Results) Breast Cancer Dataset. Applying resampling techniques on the training dataset before using various machine learning techniques can affect the performance of the classifiers. The three breast cancer datasets were examined using a variety of pre-processing approaches and classification models to assess their performance in terms of accuracy, precision, F-1 scores, etc. The PCA (principal component analysis) and resampling strategies produced remarkable results. For the BCSC Dataset, the Random Forest algorithm exhibited the best performance out of the applied classifiers, with an accuracy of 87.53%. Out of the different resampling techniques applied to the training dataset for training the Random Forest classifier, the Tomek Link exhibited the best test accuracy, at 87.47%. We compared all the models used with previously used techniques. After applying the resampling techniques, the accuracy scores of the test data decreased even if the training data accuracy increased. For the Breast Cancer Wisconsin diagnostic dataset, the K-Nearest Neighbor algorithm had the best accuracy with the original dataset test set, at 94.71%, and the PCA dataset test set exhibited 95.29% accuracy for detecting breast cancer. Using the SEER Dataset, this study also explores survival analysis, employing supervised and unsupervised learning approaches to offer insights into the variables affecting breast cancer survivability. This study emphasizes the significance of individualized approaches in the management and treatment of breast cancer by incorporating phenotypic variations and recognizing the heterogeneity of the disease. Through data-driven insights and advanced machine learning, this study contributes significantly to the ongoing efforts in breast cancer research, diagnostics, and personalized medicine.

Keywords: resampling; principal component analysis; random forest; K-nearest neighbors; logistic regression

1. Introduction

Breast cancer is a significant public health concern among women. According to the American Cancer Society, high body weight, physical inactivity, and alcohol consumption are risk factors that contribute to 30% of breast cancer cases [1]. Even while these variables may be changed to potentially lower risk, determining a woman's unique risk profile is crucial for directing counseling regarding risk reduction, genetic testing, and breast cancer surveillance.

In recent years, substantial advancements have been made in the risk assessment, diagnosis, and treatment of breast cancer. Early detection significantly improves survival rates, with a 99% 5-year relative survival rate for localized cases [2]. Consequently, accurate risk assessment is crucial to determining whether a woman has an average or elevated chance of developing breast cancer. Women at different risk levels can benefit from customized management techniques based on this categorization [3].

This study's main goal is to improve knowledge about breast cancer and how it is managed by employing a thorough and multifaceted approach. Through the utilization of large-scale datasets and machine learning skills, our goal is to investigate multiple aspects of breast cancer research, including risk evaluation, improved diagnosis, and a detailed examination of survivability variables. This study also focuses on breast cancer risk prediction, identifying individuals at varying risk levels by integrating multiple risk factors into our models to facilitate early diagnosis [4].

It is worth noting that artificial intelligence (AI) systems have shown the ability to improve breast cancer diagnosis by outperforming traditional radiological methods by 11.5% [5]. Consequently, there is an urgent need for enhanced automated screening and diagnosis procedures. Using the Breast Cancer Wisconsin (diagnostic) dataset, this study demonstrates the application of diverse machine learning algorithms to the task of breast cancer diagnosis.

Furthermore, the impact of breast cancer is not uniform across all demographic groups, leading to clusters with varying incidence and mortality rates [6]. Effective screening and prognostic identification are pivotal in addressing this heterogeneity. The objective is to determine and depict relevant demographic and prognostic characteristics that affect breast cancer survivability by using machine learning techniques. Breast cancer survivability rates highlight the value of routine screening [7]. The SEER Breast Cancer Dataset enables a thorough analysis of survivability from both supervised and unsupervised perspectives.

Several studies have contributed significantly to breast cancer diagnosis, risk assessment, and survivability analysis. The literature review that follows focuses on several significant studies that provide insight into risk factors, predictive models, and diagnostic precision in breast cancer research.

Kabir et al. conducted experiments on a breast cancer dataset with imbalanced data, using various resampling techniques to adjust the training data. They used resampling techniques like random undersampling (RUS) of the majority class, random oversampling (ROS) of the minority class, synthetic minority oversampling technique (SMOTE), edited nearest neighbors (ENN), SMOTE combined with ENN, and SMOTE combined with Tomek Link. They also used Decision Trees (DTs), Random Forests (RFs), and extreme gradient boosting (XGBoost) classifiers. The results showed that using resampling techniques improved performance, particularly for the minority class. The highest accuracy achieved for DT was 90.69% with ENN, for RF it was 88.55% with SMOTE + ENN, and for XGBoost it was 91.49% with ENN. However, the overall performance was better without applying any resampling method for the minority class [8].

Using information gathered during screening, Louro et al. created a risk prediction model to estimate the short- and long-term risk of breast cancer in women undergoing mammography. They used partially conditional Cox proportional hazard regression including covariates like age, family history, past benign breast disease, and mammographic characteristics. Over a period of two to twenty years, the E/O ratio (the ratio of expected to observed cases in the target group) varied from 0.99 to 1.02. The 4-year risk estimate had the lowest AUC (area under the curve) (58.7%, 95% CI: 55.9–61.5%) and the 18-year risk estimate had the highest AUC (64.7%, 95% CI: 62.5–66.9%) [9].

Behravan et al. proposed a method that used the XGBoost machine learning model with an adaptive iterative search algorithm to identify the interactions between genetic and demographic risk factors that provide the best accuracy in predicting breast cancer (BC). The first module provides a list of potential BC-risk-predictive traits, while the second module captures the interacting characteristics that produce the best BC risk prediction accuracy on validation data. By combining interacting genetic features and family history features, the proposed approach achieved a mean average precision (mAP) of 77.78 on the Kuopio Breast Cancer Project (KBCP) dataset, which was better than the mAPs of 74.19 and 73.65 obtained by using only Group 1 features and interacting SNPs, respectively. When using only estrogen metabolism (Group 2) features, the mAP was 72.57, but by combining interacting genetic and Group 2 features, the mAP increased to 78.00, outperforming the former [10].

In 2021, Gupta et al. investigated the effects of several class balancing methods on models created from an unbalanced mammographic dataset. In 2022, Sood et al. created a breast cancer prognostic modeling method using the minimum necessary data from mammography screening [11].

In 2022, Sood et al. created a breast cancer prognostic modeling method using the minimum necessary data from mammography screening. The paper discussed the use of machine learning techniques for predicting breast cancer in women. The researchers experimented with seven different machine learning algorithms, including Naive Bayes, SVM, K-Nearest Neighbors, Decision Trees, Random Forests, Adaboost, and deep learning. They evaluated the performance of these algorithms based on their ability to correctly identify cancerous and noncancerous cases. The researchers noted a significant improvement in the accuracy of the models after addressing the imbalance in the data through resampling techniques. The best accuracy achieved was 92.46% on the actual datasets [12].

Lavanya et al. conducted a study on multiple breast cancer datasets to evaluate the performance of the CART (Classification and Regression Tree) Decision Tree classifier with and without feature selection. The outcomes demonstrated that a certain feature selection made using the CART had improved the classification precision of a specific dataset [13]. The study by Naji et al. utilized the Breast Cancer Wisconsin diagnostic dataset and five machine learning algorithms, namely, Support Vector Machines, Random Forests, Logistic Regression, Decision Trees, and K-Nearest Neighbors, to accurately predict and diagnose breast cancer [14]. The research by Salama et al. uses accuracy in classification and confusion matrices to compare the performance of various classifiers (Decision Trees, Multi-Layer Perception, Naive Bayes, Sequential Minimal Optimization, and Instance-Based for K-Nearest Neighbors) on three separate sets of data of breast cancer. In order to obtain the best multi-classifier strategy for each dataset, the study also introduces a combination at the classification level among these classifiers [15]. The accuracy of various deep learning algorithms for predicting breast cancer patients' post-operative survival is compared in the research by Gupta et al. Artificial neural networks, restricted Boltzmann machines, deep autoencoders, and convolutional neural networks (CNNs) are some of the techniques that were investigated [16].

Our paper on breast cancer research is informed by a range of methodologies and perspectives, as outlined in this literature review. This comprehensive approach supports our efforts to thoroughly address the topic.

The paper adheres to a conventional structure; Sections 2 and 3 describe the methodology and the analysis of the results obtained, respectively. Finally, Section 4 discusses future directions and concludes the research.

2. Methodology

2.1. Risk Assessment Using BCSC Risk Factor Dataset

The BCSC (Breast Cancer Surveillance Consortium) Risk Factor Dataset includes risk factor data from mammograms performed at the BCSC between January 2005 and December 2017. To build this dataset, one exam per woman per calendar year and per age was chosen.

In the data from the 6,788,436 mammograms, the 13 attributes are age, race/ethnicity, family history of breast cancer, age at menarche, age at first birth, breast density, use of hormone replacement therapy, menopausal status, body mass index, history of biopsy, and history of breast cancer. There is also an attribute named count, which shows the frequency count of that combination of features [17]. For data pre-processing, after applying several approaches like removing the year column, rows with unknown values, and data scaling the dataset, 527 k records were left (Figure 1).

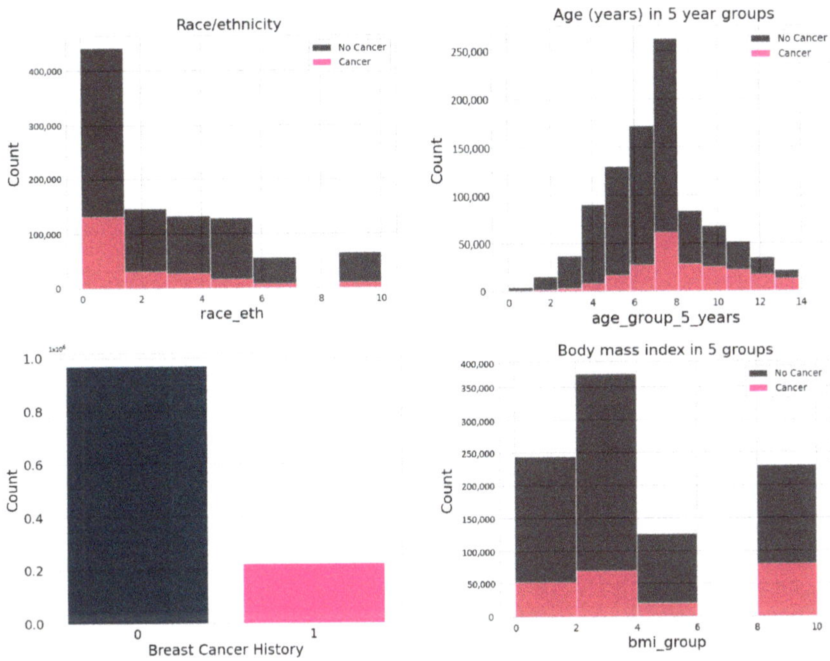

Figure 1. Data Visualization of the BCSC risk factor dataset.

We addressed the dataset imbalance through resampling techniques—random undersampling of the majority class and oversampling of the minority class. The goal was to generate a balanced training dataset and mitigate classifier bias toward the dominant class. We then trained various classifier models using both the imbalanced and resampled training data.

The prediction performance of our models is directly affected by the data quality and relevance, which were ensured by this thorough pre-processing stage. Machine learning models that followed were able to zero in on real patterns, not data collecting artifacts or representation biases, because we scaled the dataset and corrected imbalances. Our study's reliability is enhanced and it aligns with best practices in data science for health-related research because of this careful methodology.

The following sections provide detailed information on the selected machine learning algorithms and their performance indicators.

Step 1: First, we obtained breast cancer risk factor data for the classification. Training data contain 70% and test data contain the remaining 30% of the dataset.

Step 2: In this step, the training data are resampled because they are of unequal distribution in the dataset. The test data are not resampled and kept the same.

Step 3: In this step, different classifier models are trained first using the imbalanced training data. After that, we use resampled training data which had been modified by resampling methods for the classifiers.

Step 4: Apply test data on the models to predict the risk of developing breast cancer.

During Step 2, we used various resampling strategies to balance the BCSC Risk Factor Dataset. To balance the majority class, we randomly undersampled, and to enhance the minority class, we used the synthetic minority oversampling method (SMOTE). For Step 3's classification problem, we employed machine learning techniques such as Decision Trees (DTs), Random Forests (RFs), and extreme gradient boosting (XGBoost), and evaluated their performance using key measures like accuracy, sensitivity, specificity, and precision (Figure 2).

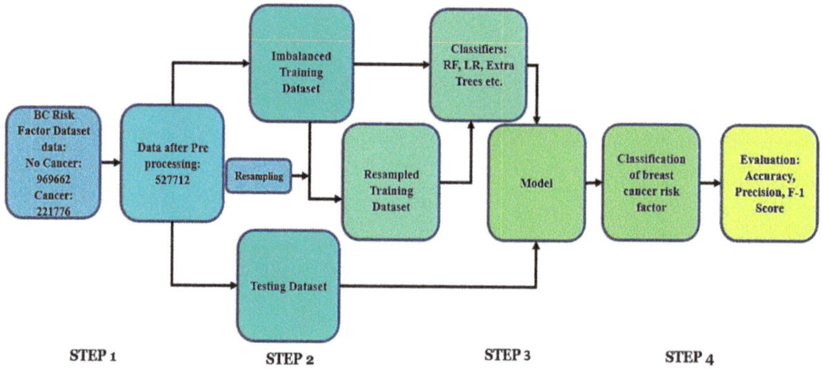

Figure 2. Flowchart of the proposed approach for dataset 1.

2.2. Diagnosis Using Breast Cancer Wisconsin (Diagnostic) Dataset

In the Breast Cancer Wisconsin (diagnostic) dataset, features were extracted to describe the characteristics of the cell nuclei from the digitized images of FNA (fine needle aspirate) biopsy. The dataset contains 569 data points, 357 benign and 212 malignant. Cell nucleus radius, texture, perimeter, area, smoothness, compactness, concavity, concave points, symmetry, and fractal dimension were the 10 features calculated in the dataset (Figure 3). Each attribute contains three data elements: the mean, the standard deviation, and the greatest or "worst" value (the mean of the three highest values), resulting in a dataset with a total of 30 features [18].

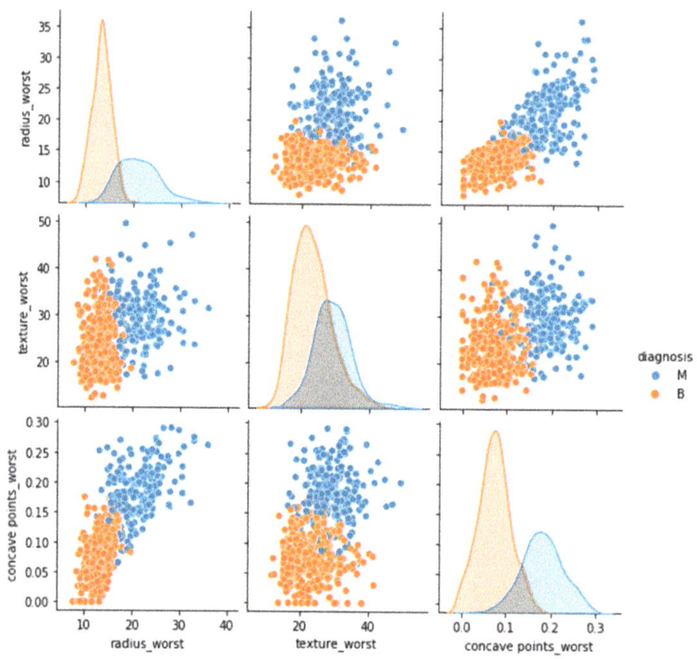

Figure 3. Pairwise relationship of malignant and benign tumors based on three important features from the Breast Cancer Wisconsin diagnostic dataset.

Before running the Decision Tree (DT) and K-Nearest Neighbor (KNN) algorithms, a principal component analysis was run on the dataset, and PC1 to PC7 was able to maintain more than 91% of the total variance. So, dimensionality was reduced from 30 features to 7 principal components. The efficiency of principal component analysis (PCA) in simplifying complex data without significantly reducing information loss encouraged our decision to use it. In addition to lowering computing requirements, this method lowered the risk of overfitting by zeroing in on the dataset's most important features. More accurate and generalizable results were obtained after applying machine learning models to this optimized dataset, demonstrating the need for careful data preparation in diagnostic investigations.

Then, both the original dataset and the new principal component dataset were split into a 30% validation set and a 70% training set and 10-fold cross-validation was performed. Then, both datasets underwent classification using Decision Trees and KNN. We applied the KNN model, which uses K = 10, on the validation dataset of the original dataset and applied the KNN model, which uses K = 6, on the validation dataset of the PC dataset. The Decision Trees were pruned to minimize their size by removing the branches that could not classify cases. Best pruned and minimum error trees were found using xerror and xstd.

These models' performance was assessed using the criteria of accuracy, sensitivity, specificity, and precision. A receiver operating characteristic (ROC) curve was used to graphically display performance.

2.3. Survival Analysis Using SEER Breast Cancer Dataset

The SEER Breast Cancer Dataset of breast cancer patients was made available in the NCI's Surveillance, Epidemiology and End Results (SEER) program's November 2017 update. SEER provides information on population-based cancer statistics. Age, race, marital status, tumor size, estrogen status, and progesterone status are some of the 15 features in this dataset. The 4024 data points in the dataset have a survivability status of 616 dead and 3408 living [19] (Figure 4).

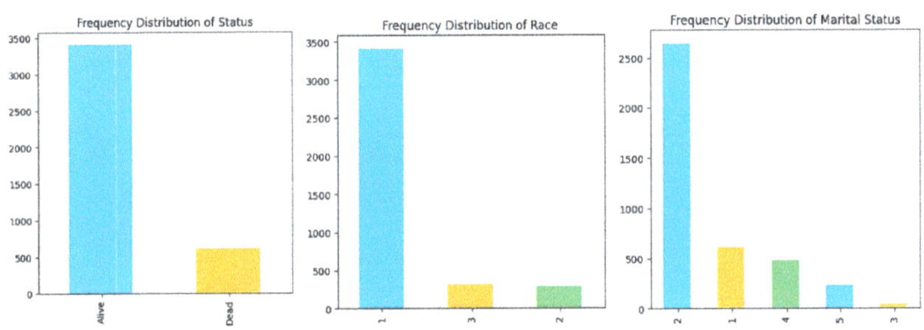

Figure 4. Data visualization of the SEER Breast Cancer Dataset.

The various categorical feature types were converted into numeric values for data pre-processing. For instance, the race categories White, Black, Other (American Indian/AK Native, Asian/Pacific Islander), and Unknown were denoted by 1, 2, 3, and 4, respectively. Similarly, positive estrogen progesterone status was represented as 1, and negative estrogen progesterone status was represented as 2. This was performed for all the feature columns. The survivability status, either alive or dead, was transformed to 1 s and 0 s. After that, dealing with missing values, scaling the data, and eliminating duplicate data and unnecessary columns from the dataset were all steps of preparing the data.

In order to prepare the dataset for advanced machine learning analysis, it was necessary to transform categorical features into numerical values and fix any missing data. The accuracy of our models in representing the complicated facts of breast cancer survival outcomes is ensured by these processes, which involve interpreting and analyzing the data. Our rigorous methodology lays the groundwork for investigating the interplay of survivability-related variables, demonstrating the promise of machine learning as a tool for improving patient care.

Unsupervised data mining techniques such as cluster analysis can be used to divide the patients into groups to determine which factors are more strongly associated with breast cancer survival [20]. In this project, different clustering methods, including agglomerative hierarchical clustering and K-means clustering, were used to find similarities between the patient features for breast cancer survivability. Data points are organized using K-means into discrete, non-overlapping groupings. So, K-means clustering was one of the machine learning algorithms used in this project. Tree structures can be created from related datasets using hierarchical clustering techniques such as AGNES. The relationships between various sub-clusters and the distances between data points are visible. For both clustering techniques in R, random sample sets from the data were taken, and then clustering was applied to sample data. As the sample sets were selected randomly, they can be considered a fairly accurate representation of the whole dataset.

Clustering, an unsupervised method, is frequently combined with other kinds of analysis. Two supervised learning techniques were used: Logistic Regression and the Decision Tree algorithm. Logistic Regression was used as the outcome of survivability status was binary: either dead or alive. For the Logistic Regression algorithm, data were divided into a test set and a training set, with 70% of the dataset going into the training set and 30% going into the test set. The Decision Trees were pruned to minimize their size by removing the branches that could not classify cases. Best pruned and minimum error trees were found using xerror and xstd. Time series cross-validation was employed to forecast the survivability of breast cancer patients. The time series was based on survival months and was created using complete dates, including days. There were 107 months of survival data.

3. Results

3.1. Results of Risk Assessment

To evaluate the performance of the different classification techniques applied to the BCSC Dataset, such as Random Forests, Logistic Regression, Extra Trees, etc., the accuracy, precision, and F-1 scores were derived. The accuracy scores are given in Table 1 and Figure 5.

Table 1. Overall performance of the classifiers on BCSC Risk Factor Dataset.

Classifier	Hyperparameters	Training Data Accuracy	Test Data Accuracy
Decision Tree	default	0.971250087	0.723443296
Random Forest	default	0.971236011	0.766493249
XG-Boost	default	0.734065414	0.733957811
Decision Tree	default	0.860784303	0.822131966
Random Forest	default	0.860759939	0.820369645
Logistic Regression	default	0.83735429	0.836679005
Extra Trees	criterion = 'gini', n_estimators = 128	0.915882057	0.756212338
Extra Trees	criterion = 'gini', n_estimators = 128	0.93007542	0.875298457
Extra Trees	criterion = 'entropy', n_estimators = 128	0.93007542	0.875285824
Extra Trees	criterion = 'gini', n_estimators = 96	0.93007542	0.875386889
Logistic Regression	C = 100, max_iter = 500, solver = 'newton-cg'	0.839243851	0.838536074
Logistic Regression	C = 10, max_iter = 500, solver = 'newton-cg'	0.839243851	0.838536074
Logistic Regression	C = 1.0, max_iter = 500, solver = 'lbfgs'	0.839243851	0.838536074
Logistic Regression	C = 0.01, max_iter = 500, solver = 'newton-cg'	0.83928175	0.83839711
Logistic Regression	C = 100, max_iter = 500, solver = 'lbfgs'	0.839243851	0.838536074
Logistic Regression	C = 10, max_iter = 500, solver = 'lbfgs'	0.839251972	0.838523441
Random Forest	criterion = 'entropy', n_estimators = 150	0.93007542	0.874824715
Random Forest	criterion = 'gini', n_estimators = 120	0.93007542	0.87433834
Random Forest	criterion = 'entropy', n_estimators = 80	0.930067299	0.873713001
Random Forest	criterion = 'entropy', n_estimators = 200	0.93007542	0.875260558

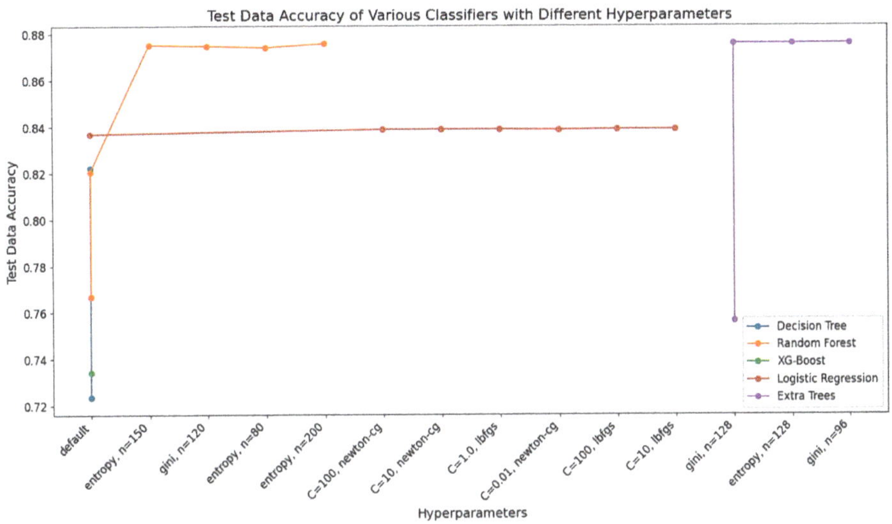

Figure 5. Comparison of classifier performance on the BCSC Dataset.

With respect to their individual hyperparameter settings, the test data accuracy of several classifiers used on the Breast Cancer Surveillance Consortium (BCSC) Dataset is shown in this figure. Different hyperparameter configurations, including g'ini, n = 128', e'ntropy, n = 150', and others, were tested for each classifier, which includes Decision Trees, Random Forests, XG-Boost, Logistic Regression, and Extra Trees. In addition to demonstrating the influence of hyperparameter modification on model performance, the graph attempts to visually evaluate how well these classifiers predict the risk of breast cancer. Finding the most accurate model configurations to improve the prediction accuracy in the assessment of breast cancer risk is dependent on this comparison.

Table 2 shows the accuracy of the Random Forest classifier after applying the resampling techniques SMOTE, RUS, ROS, and Tomek Link. The Random Forest algorithm had the best performance out of all the classifiers with an accuracy of 87.53%. Out of the different resampling techniques applied to the training dataset for using it to train the Random Forest classifier, the Tomek Link had the best test accuracy, at 87.47%.

Table 2. Performance of Random Forest classifier after various resamplings.

Pre-Processing	Training Data Accuracy	Test Data Accuracy
SMOTE	0.9364	0.8526
RUS	0.9404	0.8064
ROS	0.9362	0.8494
Tomek Link	0.9300	0.8747

The BCSC Dataset was used to evaluate various classification techniques, including Decision Trees, Random Forests, XG-Boost, Logistic Regression, and Extra Trees. The accuracy of each classifier was measured using both training and test data to determine how well the models learned from the training data and generalized to new data. The Decision Tree and Random Forest models had a high training data accuracy, but a notable overfitting issue was observed. The XG-Boost model had comparatively lower accuracy scores, while the Logistic Regression model performed well on both the training and test data. The Extra Trees classifier showed promising results after fine-tuning with specific hyperparameters.

As an example, while the Decision Tree classifier achieved a training data accuracy of 97.1%, its test data accuracy of 72.3% indicates that it may struggle to generalize to new data points. On the other hand, the Random Forest classifier achieved a similar training data accuracy but demonstrated a better generalization ability, with a higher test data accuracy of 76.6%. These discrepancies in accuracy can be attributed to the inherent complexity of the models and their susceptibility to overfitting or underfitting.

The Tomek Link approach improved the Random Forest classifier's accuracy by removing noisy and ambiguous examples from the training data, resulting in better class distinction and more accurate predictions.

3.2. Results of Diagnosis

The overall performance of the classifiers on the Breast Cancer Wisconsin diagnostic datasets is given in Table 3. From this, we can see that the KNN classifier exhibited the best accuracy out of the three. And in the case of the dataset, the classifiers had a better accuracy, sensitivity, specificity, and precision for the validation set by being trained with the seven-principal-component dataset training set rather than the original dataset training set.

Table 3. Overall performance of the classifiers on Breast Cancer Wisconsin diagnostic dataset.

Classification Methods	Dataset	Accuracy	Sensitivity	Specificity	Precision
LR	Original dataset	0.9415	0.9626	0.9062	0.9450
KNN	Original dataset	0.9471	0.9365	0.9533	0.9219
	Principal components dataset	0.9529	0.9365	0.9626	0.9365
DT	Original dataset	0.9294	0.8730	0.9626	0.9322
	Principal components dataset	0.9471	0.9048	0.9720	0.9500

From the figures in Table 4, we can see lift charts and ROC curves for KNN and DT. This leads us to better understand the performance of the models. For example, in the decile-wise lift chart of the Decision Tree model for the principal component dataset, the lift value of the leftmost bar is 2.5, meaning that for the top 35% of the validation cases with the highest predicted probability of belonging to the target class, the model would identify 2.5 times as many target-class cases than if the cases were randomly selected.

An evaluation method for binary classification issues is the Receiver Operator Characteristic (ROC) curve. The values for the area under the curve (AUC) are as follows: Logistic Regression AUC = 0.9344, KNN AUC = 0.9868, KNN with PCA AUC = 0.9796, DT AUC = 0.8826, DT with PCA AUC = 0.9302. Having a higher AUC means the model is more accurate at classifying zero classes as 0 and classifying one class as 1.

Using the PCA dataset reduced the dimensionality of the dataset to less than one-third (from 30 features to 7) and retained more than 91% of the information of the original dataset. Although using PCA reduces the interpretability of a dataset, it reduces the collinearity of the dataset by taking highly correlated variables and turning them into uncorrelated variables. Other types of ensemble trees (such as Random Forests) could have been used for the classification model, but as they are a combination of several Decision Trees, they need rigorous training and a longer processing time. The use of Logistic Regression can be justified because the classification problem is binary, that is, it diagnoses whether a sample tumor cell from the FNA is benign (noncancerous) or malignant (cancerous).

Table 4. Cumulative lift charts, decile-wise lift charts, and ROC curve for KNN and DT using the Wisconsin Breast Cancer diagnostic dataset.

Classifier	Figure	Original Dataset	PC1–PC7 Dataset
KNN	Cumulative lift chart	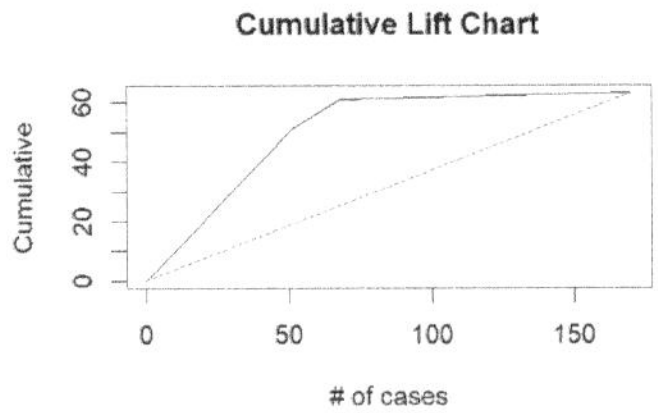	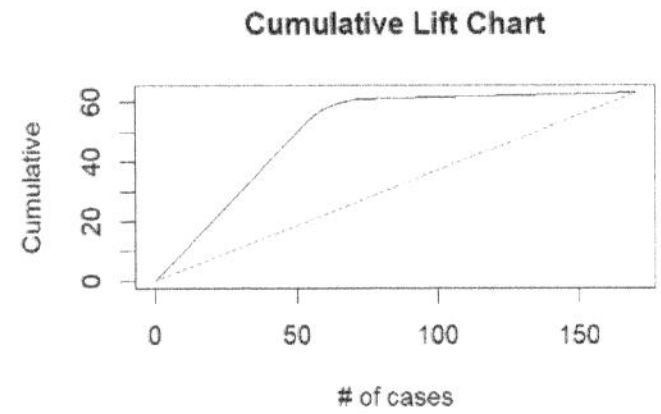
	Decile-wise lift chart	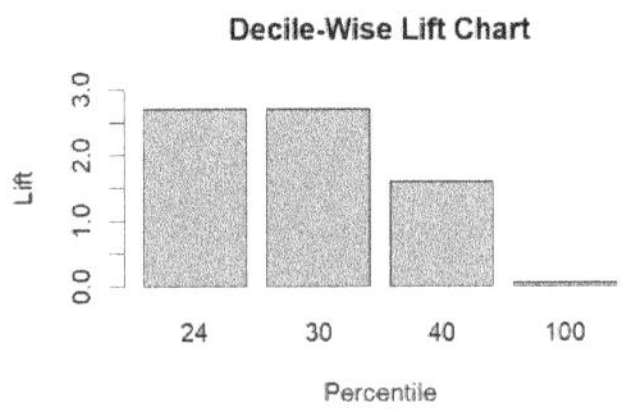	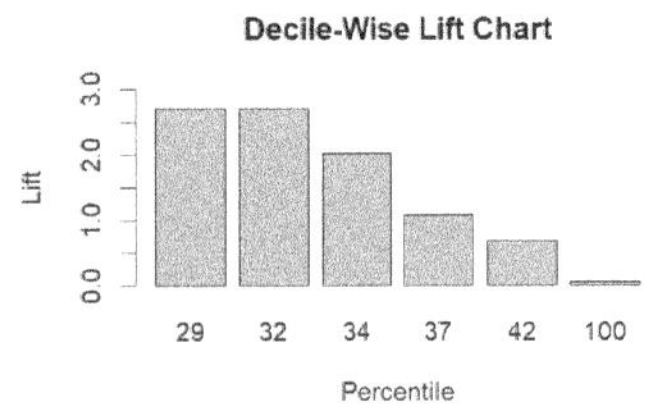
	ROC curve	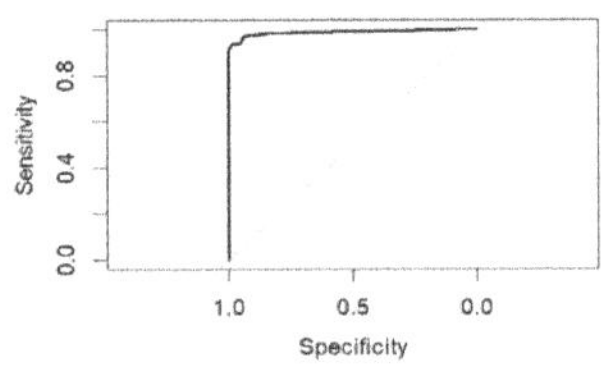	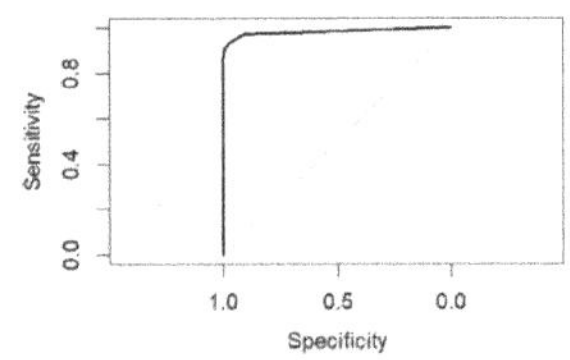
DT	Cumulative lift chart	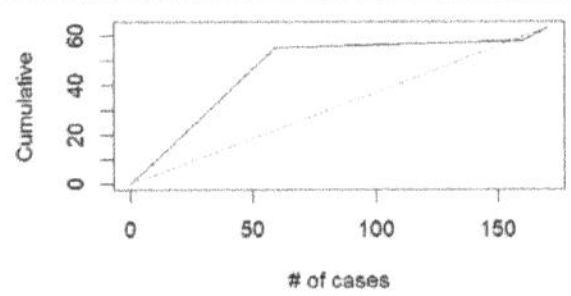	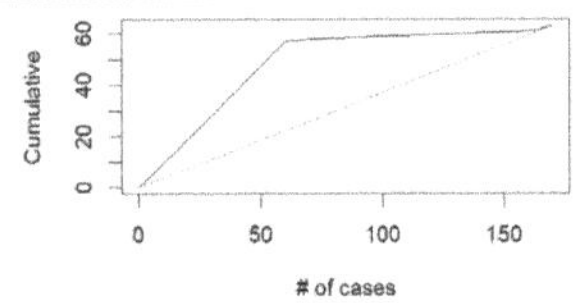
	Decile-wise lift chart	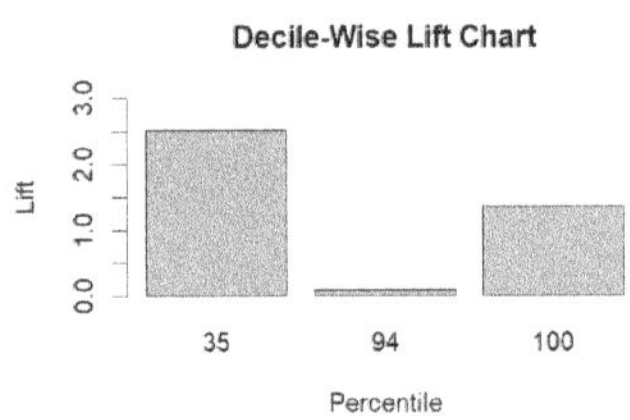	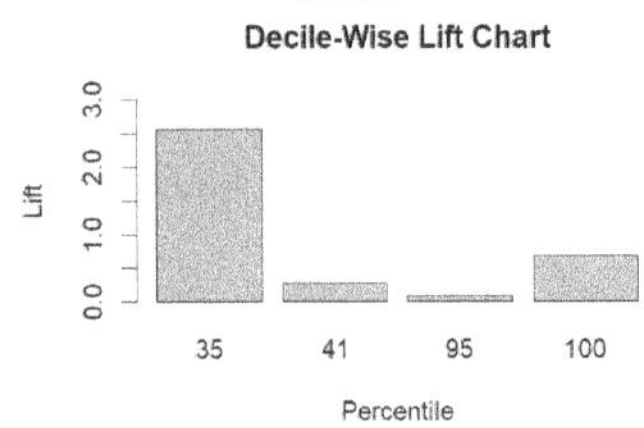
DT	ROC curve	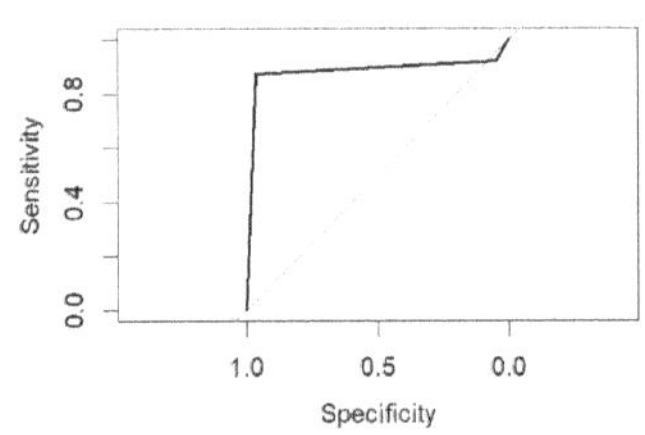	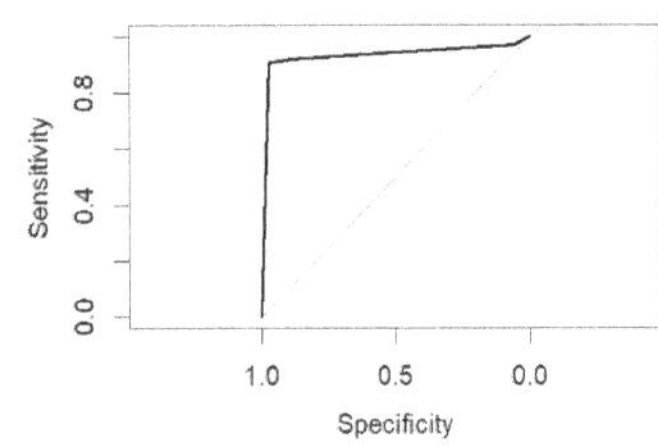

3.3. Results of Survival Analysis

The agglomerative hierarchical clustering Ward method was used on the SEER Breast Cancer Dataset, and the agglomerative coefficient was 0.981667. Clustering was performed on a sample dataset of 50.

A graphic representation known as a "banner plot" was created during the AGNES clustering procedure. In this diagram, the observations are displayed as red bars, with spaces between them denoting possible clusters. The banner plot sheds light on cluster formation by highlighting areas of separation, similar to the idea of cutting a dendrogram to obtain the desired amount of clusters (Figures 6 and 7).

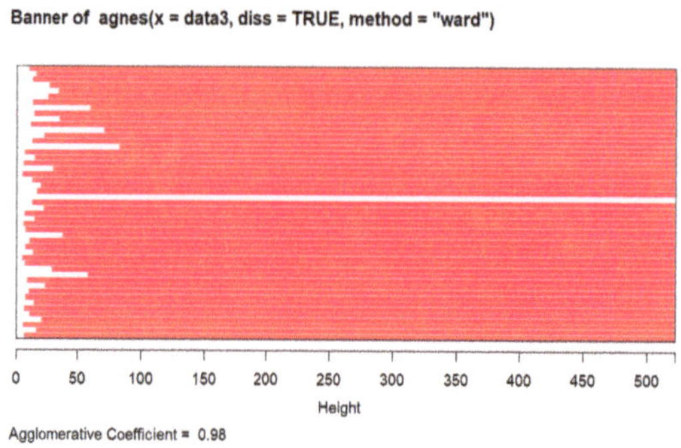

Figure 6. Banner plot made using AGNES.

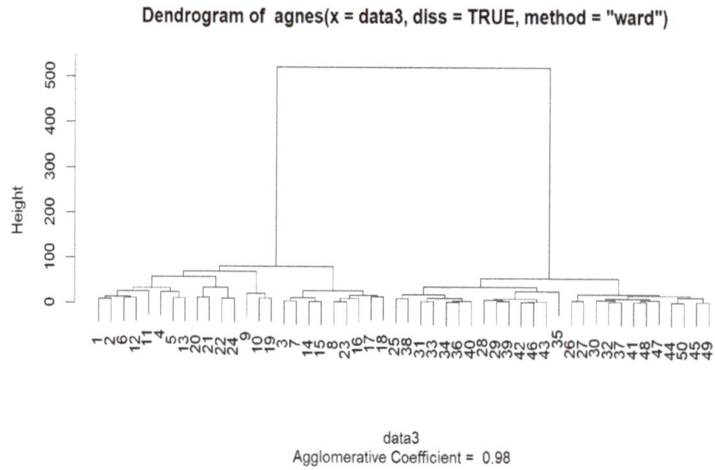

Figure 7. Dendrogram made using AGNES.

From observing the banner plot, K was set to 2. The summary of the cluster analysis is as follows (Figure 8):

```
> summary(as.factor(aClusters))
 1  2
24 26
```

Figure 8. Summary of K-means cluster analysis.

K-means clustering was performed for comparison. Here, again, K was set to 2. The following are the cluster plot and silhouette plot showing the results of the K-means clustering (Figures 9 and 10).

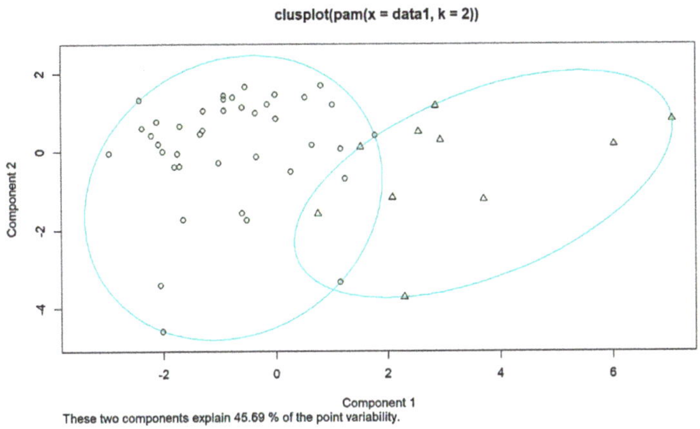

Figure 9. Cluster plot of K-means.

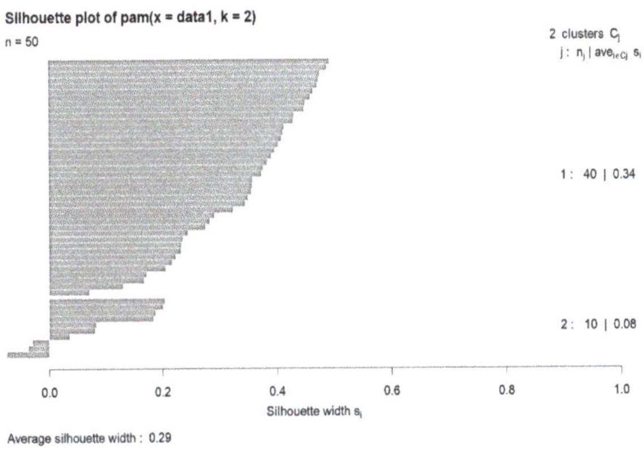

Figure 10. Silhouette plot of K-means.

Two types of supervised learning techniques, Logistic Regression and Decision Trees, were used for predicting the survivability of breast cancer. The whole dataset was used for this. The overall performance of the classifiers on the datasets is given in Table 5.

Table 5. Overall performance of the classifiers on SEER Breast Cancer Dataset.

Classification Method	Accuracy	Sensitivity	Specificity	Precision	AUC
LR	0.8998	0.4703	0.9775	0.7909	0.7239
DT	0.9080	0.9797	0.4706	0.9186	0.7412

The table shows that the Decision Tree classifier has a better accuracy and a larger area under the curve than Logistic Regression. The ROC curve for both LR and DT are given in the following Figures 11 and 12:

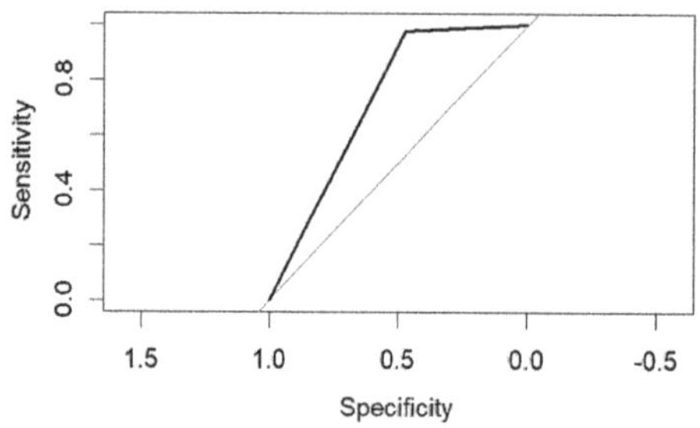

Figure 11. ROC of LR.

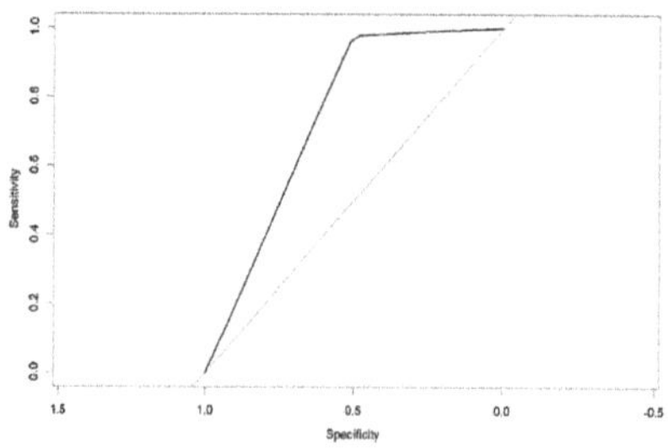

Figure 12. ROC of DT.

The pruned Decision Tree is given below (Figure 13):

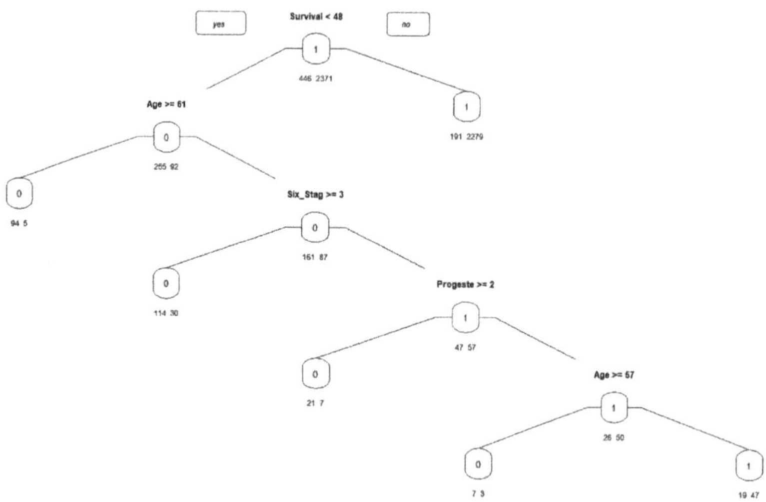

Figure 13. Decision tree.

Finally, a forecasting method, time series cross-validation, was used. The time series was survivability months. The performance measures are given in Table 6.

Table 6. Performance measures of time series cross-validation.

Time Series Cross-Validation	ME	RMSE	MAE	MASE
Training set	2.141162×10^{-17}	0.4152110	0.3448004	1.120601
Test set	2.002756×10^{-1}	0.5176724	0.3775180	1.226933

A data pre-processing technique such as PCA was not used as the dataset only had 15 columns. For applying the forecasting methods, it would be best to use a dataset related to breast cancer survivability over time. Although we used the SEER BC Dataset for time series cross-validation, it is not entirely appropriate for forecasting. Other ensemble tree types, like Random Forest, may have been used for the classification model. Still, because they combine multiple Decision Trees, they require extensive training and take longer to process.

4. Discussion

In this study, we addressed the challenge of unbalanced data in breast cancer risk factor assessment by employing various classifiers, including Decision Trees (DTs), Random Forests (RFs), Extreme Gradient Boosting (XGBoost), Support Vector Machines, and more. Although the training data accuracy was considerable, there is room for much improvement regarding the test data accuracy. Our findings, based on a thorough evaluation of the current literature, contribute to an improved understanding of machine learning applications in breast cancer risk assessment and diagnosis. Previous research by Kabir et al. [8] and Gupta et al. [11] has highlighted the impact of resampling techniques on classifier performance, but our study delves deeper into the specific effectiveness of techniques such as Tomek Link in improving Random Forest classifier outcomes for breast cancer datasets.

After applying the resampling techniques, the accuracy scores of the test data decreased even if the training data accuracy increased. This paradox highlights the nuanced challenge of balancing model generalizability and overfitting, a key area where our study contributes to the conversation beyond the findings of Louro et al. [9] and Behravan et al. [10], who emphasized predictive accuracy without a strong focus on the balancing act required for generalizable model performance.

The goal is to build better models that can predict more accurately and provide a better performance. In the future, we want to find out how different biases in the dataset toward age, race, and BMI index can affect the classification techniques and find out which of the variables has the most impact on predicting the risk of developing breast cancer.

According to the World Health Organization (WHO), 2.3 million women were diagnosed with breast cancer in 2020 globally. The use of machine learning algorithms for diagnosing breast cancer can improve the efficiency of the diagnosis and help with early detection. By addressing a gap in the use of computer-aided diagnosis systems as described by McKinney et al. [5], our comparative examination of machine learning algorithms indicates the potential for dramatically reducing the high percentage of errors and inconsistencies in radiology practice.

For this project, Logistic Regression, KNN, and Decision Tree algorithms were used with several data pre-processing techniques. The aim was for the models to classify the incidence of breast cancer from the dataset of FNA biopsy image features. This multidisciplinary approach not only improves the diagnostic process, but it also provides a road map for incorporating machine learning into clinical processes, as proposed by the scientific community.

Early precautions can be taken if demographic factors indicating high risk of breast cancer mortality are identified. It can be used to direct management for women with all breast cancer stages. Our study into the impact of demographic parameters on model accuracy and survival analysis adds to a more tailored, data-driven approach to breast cancer management, which represents a significant advancement in the field.

Defining a similarity measure between patient data is a crucial first step to stratifying patients into clinically significant subgroups and enabling individualized therapy. By utilizing advanced machine learning techniques, our research lays the groundwork for future studies to investigate individualized treatment strategies based on breast cancer's phenotypic differences, addressing the need for innovation in the personalized medicine environment.

The results must be thoroughly scrutinized for accuracy when discussing a subject like cancer, so trying different types of prediction and classification techniques would help in the diagnostic process. This study, therefore, not only presents a comprehensive assessment of machine learning's role in breast cancer research but also sets the stage for future innovations that can further refine predictive models, making them more adaptable and accurate.

Lastly, we highlight the novel characteristics of our machine learning approach to breast cancer research by placing our results in the context of the larger literature and emphasizing the particular contributions of our study. By laying the groundwork for future research that may fully utilize machine learning in oncology, our study advances the continuous development of more efficient, individualized cancer care solutions.

5. Conclusions

To summarize, this study highlights the significant capabilities of machine learning in assessing and diagnosing breast cancer risk. Additionally, it tackles important challenges such as imbalanced data and the ability of the model to be applied to different scenarios. Our comparative investigation has demonstrated that the utilization of unique resampling strategies, particularly the Tomek Link method, leads to a substantial enhancement in the performance of machine learning classifiers, such as Random Forests, when applied to breast cancer datasets.

Our research combines theoretical developments in machine learning with actual clinical applications. We argue for the integration of AI technologies in healthcare to improve diagnostic accuracy and detect diseases at an early stage. Moreover, we suggest establishing a basis for future investigations into how demographic characteristics affect the precision of models, highlighting the significance of individualized medication that takes into consideration the diverse nature of breast cancer.

This contribution seeks to expand the utilization of machine learning in the field of oncology, advocating for a patient-focused, data-oriented approach to cancer treatment. Going forward, prioritizing the improvement of these models to accurately represent the diverse patient groups they cater to is crucial, as it signifies progress in fully utilizing the possibilities of machine learning in the field of oncology.

Author Contributions: Both authors contributed to this research paper. S.B.M. worked on the methodology, software, formal analysis, and writing while P.D. supervised and directed the project. All authors have read and agreed to the published version of the manuscript.

Funding: This research received no external funding.

Institutional Review Board Statement: Not applicable.

Informed Consent Statement: Not applicable.

Data Availability Statement: Data used in this research are publicly available. Please use the following links to access them: 1. https://www.bcsc-research.org/data/rf (accessed on 1 June 2023). 2. https://archive.ics.uci.edu/ml/datasets/breast+cancer+wisconsin+(diagnostic) (accessed on 1 June 2023). 3. https://ieee-dataport.org/open-access/seer-breast-cancer-data (accessed on 1 June 2023).

Conflicts of Interest: The authors declare no conflicts of interest.

References

1. Giaquinto, A.N.; Sung, H.; Miller, K.D.; Kramer, J.L.; Newman, L.A.; Minihan, A.; Jemal, A.; Siegel, R.L. Breast Cancer Statistics. *CA A Cancer J. Clin.* **2022**, *76*, 524–541. [CrossRef] [PubMed]
2. American College of Obstetricians and Gynecologists. Breast cancer risk assessment and screening in average-risk women. *Pract. Bull.* **2017**, *179*, 2019-1.
3. Domchek, S.M.; Eisen, A.; Calzone, K.; Stopfer, J.; Blackwood, A.; Weber, B.L. Application of breast cancer risk prediction models in clinical practice. *J. Clin. Oncol.* **2003**, *21*, 593–601. [CrossRef] [PubMed]
4. Guan, Z.; Huang, T.; McCarthy, A.M.; Hughes, K.; Semine, A.; Uno, H.; Trippa, L.; Parmigiani, G.; Braun, D. Combining breast cancer risk prediction models. *Cancers* **2023**, *15*, 1090. [CrossRef] [PubMed]
5. McKinney, S.M.; Sieniek, M.; Godbole, V.; Godwin, J.; Antropova, N.; Ashrafian, H.; Back, T.; Chesus, M.; Corrado, G.S.; Darzi, A.; et al. International evaluation of an AI system for breast cancer screening. *Nature* **2020**, *577*, 89–94. [CrossRef] [PubMed]
6. Amin, R.W.; Fritsch, B.A.; Retzloff, J.E. Spatial clusters of breast cancer mortality and incidence in the contiguous USA: 2000–2014. *J. Gen. Intern. Med.* **2019**, *34*, 412–419. [CrossRef] [PubMed]
7. Breast Cancer Survival Rates Underscore Importance of Regular Screening—MN Dept. of Health. Available online: https://www.health.state.mn.us/news/pressrel/2020/breastcancer102220.html (accessed on 1 June 2023).
8. Kabir, M.F.; Ludwig, S. Classification of breast cancer risk factors using several resampling approaches. In Proceedings of the 2018 17th IEEE International Conference on Machine Learning and Applications (ICMLA), Orlando, FL, USA, 17–20 December 2018; IEEE: Piscataway, NJ, USA, 2018; pp. 1243–1248.
9. Louro, J.; Román, M.; Posso, M.; Vázquez, I.; Saladié, F.; Rodriguez-Arana, A.; Quintana, M.J.; Domingo, L.; Baré, M.; Marcos-Gragera, R.; et al. Developing and validating an individualized breast cancer risk prediction model for women attending breast cancer screening. *PLoS ONE* **2021**, *16*, e0248930. [CrossRef]
10. Behravan, H.; Hartikainen, J.M.; Tengström, M.; Kosma, V.M.; Mannermaa, A. Predicting breast cancer risk using interacting genetic and demographic factors and machine learning. *Sci. Rep.* **2020**, *10*, 11044. [CrossRef] [PubMed]
11. Gupta, R.; Bhargava, R.; Jayabalan, M. Diagnosis of Breast Cancer on Imbalanced Dataset Using Various Sampling Techniques and Machine Learning Models. In Proceedings of the 2021 14th International Conference on Developments in eSystems Engineering (DeSE), Sharjah, United Arab Emirates, 7–10 December 2021; IEEE: Piscataway, NJ, USA, 2021.
12. Sood, K.; Gundlapally, P. Risk-Based Breast Cancer Prognosis Using Minimal Patient Characteristics. In Proceedings of the 2022 IEEE 10th International Conference on Healthcare Informatics (ICHI), Rochester, MN, USA, 11–14 June 2022; IEEE: Piscataway, NJ, USA, 2022.
13. Lavanya, D.; Rani, D.K.U. Analysis of feature selection with classification: Breast cancer datasets. *Indian J. Comput. Sci. Eng. IJCSE* **2011**, *2*, 756–763.
14. Naji, M.A.; El Filali, S.; Aarika, K.; Benlahmar, E.H.; Abdelouhahid, R.A.; Debauche, O. Machine learning algorithms for breast cancer prediction and diagnosis. *Procedia Comput. Sci.* **2021**, *191*, 487–492. [CrossRef]
15. Salama, G.I.; Abdelhalim, M.; Zeid, M.A.E. Breast cancer diagnosis on three different datasets using multi-classifiers. *Breast Cancer (WDBC)* **2012**, *32*, 2.
16. Gupta, S.; Gupta, M.K. A comparative analysis of deep learning approaches for predicting breast cancer survivability. *Arch. Comput. Methods Eng.* **2021**, *29*, 2959–2975. [CrossRef]
17. Risk Factors Dataset: BCSC. Available online: https://www.bcsc-research.org/datasets/rf (accessed on 1 June 2023).
18. UCI Machine Learning Repository. Available online: https://archive.ics.uci.edu/dataset/17/breast+cancer+wisconsin+diagnostic (accessed on 1 June 2023).
19. Teng, J. SEER Breast Cancer Data. Available online: https://ieee-dataport.org/open-access/seer-breast-cancer-data (accessed on 1 June 2023).
20. Roberto Cesar, M.O.; German, L.B.; Paola Patricia, A.C.; Eugenia, A.R.; Elisa Clementina, O.M.; Jose, C.O.; Marlon Alberto, P.M.; Fabio Enrique, M.P.; Margarita, R.V. Method based on data mining techniques for breast cancer recurrence analysis. *Adv. Swarm Intell.* **2020**, *12145*, 584.

diagnostics

MDPI

Article

Conceptually Funded Usability Evaluation of an Application for Leveraging Descriptive Data Analysis Models for Cardiovascular Research

Oliver Lohaj [1,*], Ján Paralič [1], Zuzana Pella [2], Dominik Pella [3] and Adam Pavlíček [1]

[1] Department of Cybernetics and Artificial Intelligence, Faculty of Electrical Engineering and Informatics, Technical University of Košice, Letná 9, 040 01 Košice, Slovakia; jan.paralic@tuke.sk (J.P.); adam.pavlicek@student.tuke.sk (A.P.)

[2] Center of Simulator and Virtual Medicine, Department of Medical Informatics and Simulator Medicine, Pavol Jozef Šafárik University in Košice, 040 11 Košice, Slovakia; zuzana.pella@upjs.sk

[3] 1st Cardiology Clinic, Pavol Jozef Šafárik University in Košice, VÚSCH, Ondavská 8, 040 11 Košice, Slovakia; dominik.pella@gmail.com

* Correspondence: oliver.lohaj@tuke.sk

Abstract: The focus of this study, and the subject of this article, resides in the conceptually funded usability evaluation of an application of descriptive models to a specific dataset obtained from the East Slovak Institute of Heart and Vascular Diseases targeting cardiovascular patients. Delving into the current state-of-the-art practices, we examine the extent of cardiovascular diseases, descriptive data analysis models, and their practical applications. Most importantly, our inquiry focuses on exploration of usability, encompassing its application and evaluation methodologies, including Van Welie's layered model of usability and its inherent advantages and limitations. The primary objective of our research was to conceptualize, develop, and validate the usability of an application tailored to supporting cardiologists' research through descriptive modeling. Using the R programming language, we engineered a Shiny dashboard application named DESSFOCA (Decision Support System For Cardiologists) that is structured around three core functionalities: discovering association rules, applying clustering methods, and identifying association rules within predefined clusters. To assess the usability of DESSFOCA, we employed the System Usability Scale (SUS) and conducted a comprehensive evaluation. Additionally, we proposed an extension to Van Welie's layered model of usability, incorporating several crucial aspects deemed essential. Subsequently, we rigorously evaluated the proposed extension within the DESSFOCA application with respect to the extended usability model, drawing insightful conclusions from our findings.

Keywords: association rules; clustering; cardiovascular diseases; descriptive modeling; system usability scale; usability; Van Welie layered model of usability

1. Introduction

Cardiovascular disease is a term affiliated with many health problems related to the heart and the vascular system. These medical conditions comprise an abnormal medical status that directly affects the heart and all its parts. From the available information, it can be concluded that cardiovascular diseases are the most common cause of death worldwide. Heart disease is the most significant health problem today. According to the World Health Organization (WHO) [1,2], solutions to cardiovascular diseases (CVDs) should be prioritized, as they have a global primacy in regard to mortality. According to the WHO, 17.9 million deaths from CVDs have been recorded, amounting to 32% of the total number of deaths. Within CVDs, 85% of heart attack and stroke-related deaths can be attributed to CVD-related heart conditions [1]. Based on statistics from the World Health Organization, up to 31% of global deaths are attributable to diseases and problems related to the heart and vascular system [1]. According to data from the European Heart Network [3], cardiovascular diseases are responsible for nearly 4 million deaths annually in Europe, with nearly two million of these occurring within the European Union alone. This means that CVDs account for 45% of all deaths in Europe and 37% within the European Union. Across most European nations, CVDs are the primary cause of death among men, and they are also the leading cause among women in nearly all countries. Geographically, Central and Eastern Europe experience higher mortality rates from coronary heart disease and stroke compared to Northern, Southern, and Western Europe. Despite this, there has been a gradual improvement in cardiovascular health across Europe, including in less economically developed regions, although this improvement contrasts with the alarming rise in CVD-related mortality at the turn of the 21st century. Annually, approximately 11.3 million new cases of cardiovascular disease are reported in Europe.

Based on data from 2023 State of Health in the EU—Slovakia [4], it is clear that, in Slovakia, the mortality from cardiovascular diseases is almost 100% higher than the OECD (Organization for Economic Co-operation and Development) average and 300% higher than in France, the Netherlands, and Italy. Notably, in Slovakia more than 36% of all deaths, meaning more than 26 thousand people, die annually from cardiovascular diseases, i.e., ischemic heart disease and sudden stroke, which represents approximately 40% of deaths in men and 50% of deaths in women. Slovakia has one of the highest mortality rates linked to preventable and treatable causes. There is still considerable room for improvement in terms of effective public health policies to reduce premature deaths. Although prevention and health promotion are part of the political agenda, funding is still insufficient.

When examining global data, it is important to focus on two distinct geographical regions: China and the United States of America. According to statistics from the World Health Organization, non-communicable diseases (those not transmitted by infection) account for 87% of adult deaths worldwide [5], with cardiovascular disease alone contributing to nearly half, or 45%, of these fatalities. Globally, China is the most affected country, with an estimated one in five adults in the country suffering from cardiovascular disease [6]. China stands out as a concerning leader in mortality due to these diseases. In the United States, a person dies from cardiovascular disease every 37 s [7], resulting in 647,000 deaths annually, representing one in four deaths. Unlike China, coronary artery disease ranks as the primary form of heart disease in the United States [8].

Viewed from a different vantage point, the gravity of the situation and the widespread prevalence of chronic diseases are readily apparent in the continuous influx of medical records and data each day. Leveraging data analytics, particularly through the application of descriptive models, presents a pathway to deciphering the factors influencing disease severity. Furthermore, it enables the identification of distinct patient cohorts based on their medical examination results. In this landscape, the development of an intuitive and user-friendly application tailored to supporting cardiologists' research through descriptive modeling becomes paramount. The usability of the app is crucial in ensuring its effectiveness and widespread adoption. Additionally, integrating feedback mechanisms and continuous improvement processes can further refine the app's usability over time, ensuring its relevance and effectiveness in the ever-evolving healthcare landscape.

2. Related Work

The use of data mining techniques in specific application areas, including medical environments, is largely varied. In some cases, predictive models dominate, while descriptive models dominate in others.

2.1. *Use of Different Data Mining Techniques in Medicie*

A relevant perspective on this issue is provided by the study [9] conducted by Md Saiful Islam et al., who shed light on the diverse applications of data mining in the medical and healthcare domain. By systematically analyzing published studies and literature spanning over a decade, the study reveals the prevalence of various approaches to data mining tasks during the period from 2005 to 2016. Their findings underscore the significant role of predictive modeling algorithms, which were employed in 43% of the scientific publications reviewed. These algorithms enable healthcare professionals to anticipate and predict outcomes, thereby aiding in early intervention and personalized treatment strategies.

Furthermore, the study highlights the emergence of prescriptive approaches, albeit to a smaller proportion (9%), suggesting a growing emphasis on not only predicting but also recommending optimal courses of action based on data insights. Descriptive modeling algorithms, which constituted 48% of the cases analyzed, play a crucial role in unraveling patterns, trends, and associations within healthcare datasets [10]. Among these algorithms, association rules and clustering algorithms emerged as being the most frequently used, underscoring their efficacy in identifying meaningful relationships and grouping similar patient profiles.

Delving deeper into the application areas, the study reveals a notable concentration of descriptive models in domains aimed at ensuring equitable access to healthcare across diverse population groups. These models not only facilitate the analysis and interpretation of disease patterns and risks but also provide decision support to healthcare professionals during critical junctures.

A particularly compelling example (cited in the study [9]) is the application area focused on analyzing and monitoring adverse events and reactions to medicines post-marketing. Here, descriptive models play an important role in swiftly identifying and responding to potential safety concerns, thereby safeguarding public health and enhancing pharmacovigilance efforts. Overall, the study underscores the multifaceted impact of data mining in healthcare, ranging from predictive modeling for prognoses to descriptive modeling for insight generation and decision support.

2.2. *Use of Association Rules Concept in Medicine*

This can be followed up with studies that directly confirm the claims made. The research team of Towfek et al. [11] implemented a methodology that incorporated medical ontologies and improved semantic measures to boost the accuracy and interpretability of the obtained association rules. Their method delivers clinically pertinent information into patient perspectives on mental health, the accessibility of mental health resources, and the influence of physical health on mental wellbeing by mapping medical notions to ontological items and examining semantic relationships. On the other hand, the research team of Patil et al. [12] focused on finding association rules, using the Apriori algorithm to identify significant factors that point towards diabetes indications over a five-year time horizon. It is also important to note that these were exclusively the records of patients no younger than 21 years of age. On the basis of the results, it can be argued that the factors largely influencing the indication of this disease include, in particular, high plasma glucose concentration, obesity, and the age of patients between 40 and 59 years.

The Apriori algorithm can also be used in conjunction with other chronic diseases, such as COVID-19. A wide range of symptoms have been identified in COVID-19, from mild colds to severe health complications that can lead to death. This issue was addressed by the research team of Meera Tandan et al. [13]. Thus, the main aim of the mentioned study was to identify patterns in symptoms in patients with COVID-19 disease divided into groups according to age, sex, presence of chronic diseases, and whether the patient had overcome or succumbed to the disease. The most significant rules formed by Apriori in terms of age groups and the symptoms of COVID-19 disease were defined as follows:

- Patients under 20 years of age = [conjunctivitis, severe rhinitis];
- Patients aged 20 to 45 years = [dry mouth, sore throat];
- Patients aged 45 to 65 years = [nausea, weakness];
- Patients aged 65 years and older = [anorexia, fever].

In patients over 45 years of age, heart-associated symptoms, i.e., heart failure, cardiac arrhythmia, myocardial infarction, and respiratory problems such as pneumonia or sore throat, accounted for the majority of the association rules. As we can see, many identified association rules include the most significant COVID-19 risk factors, which were also proven by practice during the pandemic. We can assume, that using association rules and Apriori algorithm is likely to yield an important result within our cardiovascular research as well.

2.3. Use of Clustering Concept in Medicine

Expanding further on the significance of healthcare research, authors of the article [14] tried to improve the performance of logistic regression and YOLOv4 algorithms, proposing the use of advanced parallel k-means pre-processing, which is a clustering technique that identified patterns and structures in the data. Their results demonstrated that the combination of advanced parallel k-means pre-processing and the neural engine processor resulted in a significant improvement in the performance of logistic regression and YOLOv4, making them more reliable for use in medical applications. Another study [15], conducted by Guo Q. et al., adds valuable insights into the nuanced understanding of hypertension and its management. By employing the k-means clustering algorithm, the research team aimed to delineate distinct groups of hypertensive patients based on a comprehensive dataset encompassing not only medical records but also lifestyle and behavioral attributes. The findings of their analysis underscore the heterogeneous nature of hypertension, shedding light on the diverse profiles and characteristics exhibited by patients suffering from this condition. Through the identification of unique clusters, the study illuminates the variability in disease presentation and risk factors among hypertensive individuals.

One noteworthy revelation from the clustering analysis is the delineation of specific patient cohorts with distinct demographic and clinical profiles. For instance, the largest cluster, consisting of 172 patients comprising predominantly young male smokers, exhibits concerning cardiovascular risk factors, including elevated blood pressure levels and low HDL cholesterol, despite the absence of diagnosed coronary artery disease. This subgroup warrants targeted interventions aimed at smoking cessation and blood pressure management to mitigate their heightened cardiovascular risk. Conversely, the smallest cluster, composed of 70 mainly older women, presents a contrasting clinical profile characterized by lower diastolic blood pressure and elevated glucose and cholesterol levels. This subgroup may necessitate interventions focusing on glycemic control and lipid management to mitigate their cardiovascular risk and improve long-term outcomes.

By elucidating the heterogeneity within hypertensive populations, the study underscores the importance of personalized medicine in the management of chronic cardiovascular diseases like hypertension. With the aid of knowledge of distinct patient clusters, healthcare providers can tailor treatment strategies and interventions to address the specific needs and risk profiles of individual patients, thereby optimizing clinical outcomes and enhancing patient satisfaction. Overall, based on the research conducted by Guo Q. et al. [15], we can conclude that using descriptive models in cardiovascular research and management can yield important information and results that can be applied in practice to help save many lives.

2.4. Defining Usability

Usability was first used as a term more than 40 years ago when it replaced the term user-friendly [16]. In another study [17], authors talk about the dominant perspective that defines usability as quality of use. The quality of use perspective hypothesizes that the usability of a product varies depending on who is using the product, how they are using it, and for what purpose it is being used. The view of interaction with a product that needs to be considered when specifying usability is known as the context of use [18]. According to the International Organization for Standardization (ISO) and the norm ISO 9241-11:2018 [19], usability is defined as *"the extent to which a product can be used by specified users to achieve specified goals with effectiveness, efficiency and satisfaction in a specified context of use"*.

If we look deeper into usability, it encompasses the effectiveness, efficiency, and satisfaction with which users can interact with a product, system, app, or service. It extends beyond mere functionality to include the overall user experience, focusing on factors such as accessibility, learnability, and user satisfaction. A usable design prioritizes the needs and preferences of its users, striving to minimize cognitive load, streamline interactions, and facilitate intuitive navigation [20]. Usability testing and evaluation methodologies are crucial to identifying and addressing potential usability issues, as well as ensuring that the product or system meets the functional requirements and enhances user productivity and satisfaction. Ultimately, usability plays a vital role in determining the success and adoption of a product or service, as it directly impacts user engagement, trust, and loyalty. In fields such as healthcare, where usability can significantly affect patient safety and outcomes, a strong emphasis on usability is essential to ensuring effective and efficient delivery of care. Thus, usability serves as a main point of user-centered design principles.

To sum up the definition of usability and how we can use and implement it, in the realm of user-centered design, usability stands as a cornerstone principle. It encompasses the entire spectrum of a user's interaction with a product, system, or service. A product with high usability empowers users to accomplish their desired tasks effectively, efficiently, and with satisfaction. This translates to clear interfaces, intuitive navigation, and a design that minimizes errors and learning curves. Ultimately, strong usability fosters positive user experiences, which are critical for product adoption and long-term success. Therefore, if we want to create a high-usability product, we need to implement all important aspects of usability during the development of an app, test it rigorously, and then implement changes based on feedback received during the testing phase. This will ensure that the app we develop is going to be used and that end-users will be satisfied with it.

2.5. Defining Usability of Decision Support Systems

The usability of decision support systems (DSSs) is crucial for ensuring their effectiveness and acceptance among users in various domains, ranging from healthcare to finance and beyond. These systems are designed to assist users in making informed decisions by analyzing complex data and providing relevant insights and recommendations [21]. A key aspect of usability in DSSs involves the accessibility, comprehensibility, and interpretability [22] of the information presented. Users need to easily navigate through the system, understand the data being analyzed, and interpret the recommendations provided.

A user-friendly interface with intuitive navigation and clear visualizations can greatly enhance usability, enabling users to interact with the system seamlessly and make decisions confidently. Furthermore, the efficiency of decision-making processes is heavily influenced by the usability of DSSs [23]. Users should be able to access the information they need quickly and efficiently without being bogged down by unnecessary complexity or cumbersome interfaces. Streamlined workflows and customizable features can improve efficiency by allowing users to tailor the system to their specific needs and preferences. Additionally, usability encompasses the responsiveness and reliability of the system, ensuring that users can rely on it to deliver accurate and timely insights when needed most.

Usability testing plays a critical role in optimizing DSS usability by identifying usability issues and gathering feedback from users. Through iterative testing and refinement, designers can fine-tune the system, and, moreover, ongoing usability evaluation and monitoring are essential for ensuring that the system continues to evolve in response to changing user needs and technological advancements [24]. Usability testing plays a crucial role in the development and refinement of DSSs, ensuring that these systems effectively meet the needs of users and facilitate informed decision making. Through usability testing, designers can gather invaluable insights into how users interact with a DSS interface, identify usability issues, and iteratively improve the system's design.

One key aspect of usability testing in DSSs involves assessing the ease of use and the effectiveness of the system's features and functionalities [25]. This includes evaluating the clarity of information presentation, the intuitiveness of navigation, and the efficiency of task completion. By observing users as they perform tasks within a DSS, designers can identify areas where the interface may be confusing or cumbersome, allowing them to adjust and improve usability. Usability testing also provides an opportunity to gather feedback directly from users regarding their experience with the DSS. This feedback can offer valuable insights into user preferences, pain points, and areas for improvement. By incorporating user feedback into the design process, designers can ensure that a DSS aligns with user needs and preferences, ultimately enhancing user satisfaction and adoption. Moreover, usability testing allows designers to assess the accessibility of the DSS interface, ensuring that it is usable by individuals with diverse abilities and needs [26]. This may involve testing the system with users who have different levels of technical proficiency or who rely on assistive technologies, such as screen readers or alternative input devices. By designing for accessibility, designers can ensure that a DSS is inclusive and can be effectively used by all users.

By iteratively testing and refining the DSS interface based on user feedback [27], designers can ultimately create systems that empower users to make better decisions and achieve improved outcomes. Usability is paramount in the design and implementation of decision support systems. By prioritizing accessibility, efficiency, and reliability, designers can create DSSs that empower users (in our case, cardiologists) to make better decisions, ultimately leading to improved outcomes and organizational success.

Usability can be measured in many ways, such as with SUS, MUSiC, or TeSS. SUS, or System Usability Scale [28], is a widely used questionnaire for assessing the perceived usability of a system. It consists of 10 questions designed to evaluate the user's subjective perception of various aspects of usability, such as ease of use, learnability, and efficiency. Users rate each question on a 5-point Likert scale, and the scores are then converted and aggregated to produce an overall usability score ranging from 0 to 100. SUS provides a standardized method for quantitatively measuring usability and is commonly used in user experience research and usability testing. We define and explore SUS more in depth in chapter V.B.

The objective of the MUSiC project [29], or Metrics for Usability Standards in Computing, has been to develop a set of metrics-based methods that can be used individually and together to both specify formal requirements for the usability of a product and to assess whether a product meets those requirements. The project has defined usability in terms of the quality of use of a product, and it has developed tools and procedures for measuring usability.

In the context of usability, "TeSS" [30] refers to the "Tool for Exploring the Structure of Standards", which is used for analyzing and visualizing complex standards or ontologies. This tool aids in understanding the structure and relationships within standards, making them more accessible and usable for stakeholders.

2.6. *Van Welie Layered Model of Usability*

For the purpose of evaluating the usability of clinical decision support systems (CDSS), the Van Welie usability model according to the study [31] was used, which divides three aspects of usability into three layers, as seen in Figure 1. The first of them (counting from the top of Figure 1) includes the main three aspects of usability, e.g., effectiveness, efficiency, and satisfaction. The second layer specifies specifically Usage Indicators which are assigned to individual aspects from the first layer. These include indicators such as learnability, memorability, safety/errors, speed of performance, and satisfaction. Sheiderman [32] named this layer as *"the five measurable human factors aimed at evaluating goals"*. The third layer provides Means by which the indicators can be measured. These include, for example, consistency, availability of rollback operations, warnings, presence of feedback, and adaptability. This layer is specified by factors [31] that influence the concept to which they belong. For example, consistency positively affects learnability, and warnings can reduce mistakes/errors. It is important to study the consensus regarding usability, as this is one of the prerequisites for the success of using the DSS. This success depends on the design, functionalities, and features of the system. It addresses the needs of users and the behavior of the system that are part of the software. Improving usability is key, as poor usability is a major obstacle to successful system integration.

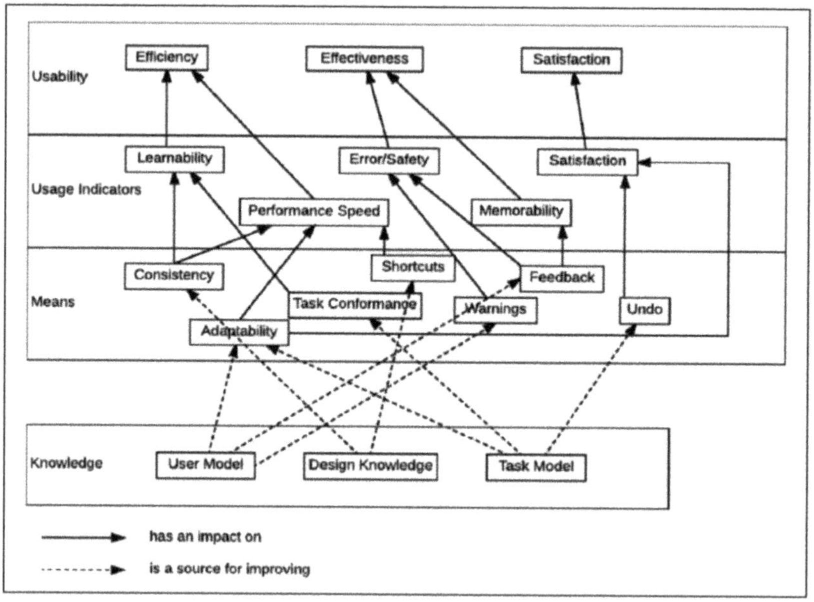

Figure 1. Van Welie layered model of usability (from [31]).

Hardenbol et al. in [33] studied 24 papers that covered at least one aspect of usability in CDSS related to medication in an outpatient setting. The findings showed positive efficacy results in 90 percent of the studies. The aim of the authors of the research was to summarize the results within three aspects of usability and to examine current usability evidence based on Van Welie indicators and means of use in CDSS evaluation research. These indicators were later analyzed to determine which usability categories in the ISO standard require further investigation.

To investigate the relationships between the three aspects of usability, Van Welie's indicators, and means of use, the authors synthesized each study according to his model in each category. The results of this research showed that some indicators of the use of the Van Welie model, such as memorability and learnability, have scarcely been studied, but also they are essential components of effectiveness and efficiency and should therefore be included and developed in future research on usability [33]. A study that looked at examining usability in software [18] points out that learning ability can be measured by comparing the quality of use for users over time or by comparing the usability of a product for experienced and inexperienced users. It follows from the article [33] that Means for measuring Usage Indicators (Figure 1) and, consequently, usability measurements vary considerably between studies, and also that most studies have focused on examining warnings rather than means of feedback.

A possible reason for this may be that providing feedback is a task that requires more knowledge of the situation and, therefore, the generation of alerts is easier to implement. It is believed that generating feedback can lead to improved CDSS because it can improve doctors' decision-making processes instead of just warning them when they perform something incorrectly. Evidence of usability of the medication-related CDSS shall be based on the study by Hardenbol et al. [33], which talks about usability aspects of medication-related decision support systems in the outpatient setting, as these could be found mainly for efficacy and showed a high percentage of positive results in reducing medication errors, which is consistent with previous studies. However, the results in the categories regarding efficiency and satisfaction diverged too much to draw conclusions, and the two categories remain insufficiently studied.

After reviewing several studies focused on usability research, its aspects, and specific metrics, we concluded that there is no universal model by which we can measure usability. As usability is a very complex concept, and we need to consider in which environment, system, and with which users we need to evaluate it, it is necessary to choose appropriate metrics and methods to measure usability. Therefore, in this part of research, we summarized current knowledge before, in the next part, proposing a new, more universal usability measurement model for a data analytics-based decision support system, according to which we can more easily determine which methods are suitable for usability measurement based on our requirements. We will start from Van Welie's model, which is still generally in use but lacks many new aspects of usability that have become important in the last few years. We will extend the model with specific indicators of usability aspects that have their own measurement methods.

3. Characteristics of Descriptive Models

In the previous section, we mentioned a few terms that should be briefly explained.

3.1. Descriptive Modeling

The descriptive approach, often referred to as descriptive modeling, constitutes a fundamental methodology in data analysis that aims to illuminate underlying patterns and structures within a dataset [34]. Unlike predictive models, which predict future outcomes based on historical data [35], descriptive models focus on elucidating the inherent characteristics and trends present in the input data themselves [34]. By employing techniques such as clustering, association rules, and summarization, descriptive analytics provides valuable insights into the nature of the data, enabling stakeholders to discern meaningful patterns and relationships.

One distinguishing feature of descriptive analytics is its emphasis on the exploration and understanding of data without necessarily predicting future outcomes. While predictive models rely on defining a specific target attribute to forecast future events [35], descriptive analytics is not constrained by such a requirement. Instead, it allows analysts to delve into the nuances of the dataset, uncovering hidden insights and informing decision-making processes. Moreover, descriptive modeling plays a pivotal role in various domains [34], including business intelligence, healthcare, and finance, where understanding past and present trends is essential for strategic planning and decision support. In essence, the descriptive approach serves as a foundational pillar in the world of data analytics, offering a comprehensive framework for exploring and interpreting data patterns, thereby facilitating informed decision making and driving organizational success. Association rules and clustering are among the most often used descriptive models.

3.2. Association Rules

Association rules serve as a powerful technique in data mining and analysis, primarily focusing on uncovering frequent and meaningful relationships between individual attributes within a dataset. By examining patterns of co-occurrence among different variables, association rules provide valuable insights into the underlying associations and dependencies present in the data [36]. This enables analysts to gain a deeper understanding of the complex interactions between variables, shedding light on hidden patterns and trends that may not be immediately apparent. One of the key advantages of association rules lies in their simplicity and ease of interpretation. Unlike some other data mining techniques that may produce complex models or require advanced statistical knowledge to interpret, association rules offer straightforward and intuitive insights into the relationships between variables, which is also important for our research.

If the insights are intuitive and easy to understand, end-users will find this app usable. This simplicity also enables stakeholders from various backgrounds to grasp the implications of the discovered associations quickly, making it a valuable tool for decision making and problem solving in diverse domains. Association rules provide a flexible framework for uncovering meaningful connections between variables. This adaptability makes association rules a valuable asset for organizations seeking to both extract actionable insights from their data and drive informed decision making.

3.3. Clustering

Clustering can be characterized as a method for finding relationships and connections between objects in a dataset before then classifying them into groups based on similar properties; these groups are also called clusters. Objects within one cluster display many more similarities to each other than to objects from different clusters [36]. Clustering is a cornerstone method in data analysis that is known for its capacity to reveal intrinsic relationships and structures within datasets. By utilizing clustering algorithms like k-means, hierarchical clustering, and density-based clustering, analysts can partition data into cohesive clusters based on similarities among data points. This enables the identification of patterns and associations, offering invaluable insights into the underlying structures of the data. From market segmentation in marketing to document clustering in natural language processing, clustering algorithms provide a flexible framework for organizing and understanding complex datasets. This adaptability makes clustering a go-to tool for exploratory data analysis, pattern recognition, and decision support across various industries. Additionally, clustering serves as an effective means of data reduction and summarization, condensing large datasets into manageable clusters. This facilitates streamlined data interpretation and visualization, empowering stakeholders to extract actionable insights and make informed decisions based on revealed patterns.

3.3.1. Hierarchical Clustering

Hierarchical clustering algorithms [36] represent a fundamental approach in data analysis, offering a hierarchical partitioning of objects within a dataset. This hierarchical structure allows for a nuanced understanding of the relationships and similarities between data points, offering valuable insights into the underlying structures of the data. These algorithms can be broadly categorized into two main groups: agglomerative and divisive.

Agglomerative hierarchical clustering [37] follows a bottom-up approach, starting with each data point as its own cluster and iteratively merging clusters based on their similarities. At each step, the algorithm identifies the two closest clusters and combines them into a single cluster, gradually forming a hierarchy of nested clusters. This process continues until all data points are grouped into a single cluster, resulting in a dendrogram that illustrates the hierarchical relationships between clusters. Conversely, divisive hierarchical clustering [38] employs a top-down approach, beginning with all data points grouped into a single cluster and recursively splitting clusters into smaller clusters based on their dissimilarity.

The choice between agglomerative and divisive hierarchical clustering depends on the nature of the dataset and the specific objectives of the analysis. Agglomerative clustering is often preferred for its simplicity and efficiency, particularly when dealing with large datasets, while divisive clustering may offer greater interpretability and control over the clustering process. Hierarchical clustering algorithms provide a powerful framework [38] for organizing and understanding complex datasets, offering insights into the hierarchical relationships between data points.

3.3.2. Clustering Based on the Selection of Cluster Representatives

Clustering algorithms that operate based on the selection of cluster representatives employ a fundamental principle where cluster centroids are initially formed randomly before individual elements are then iteratively moved within or between clusters to the nearest centroid to refine the clustering arrangement. The criterion for determining whether an element belongs to a particular cluster typically relies on its distance from the center or representative of the cluster. This distance metric is often calculated using measures, such as Euclidean distance or cosine similarity, that depend on the nature of the data and the clustering algorithm being utilized [39]. Central to this approach is the concept of the centroid, or center of gravity, which represents the average position of all data points within a cluster. The centroid serves as a reference point for assessing the similarities or dissimilarities of individual data points to the cluster they belong to.

Prominent clustering algorithms that operate based on this principle include k-means and Partitioning Around Medoids (PAM), also known as k-Medoids [40]. In k-means clustering, the centroids are iteratively updated to minimize the total intra-cluster variance, resulting in clusters with tight boundaries and well-defined centroids. PAM, on the other hand, selects representative data points, known as medoids, as the representatives of clusters, making it more robust to outliers and noise in the data. Clustering methods based on the selection of cluster representatives offer a flexible and scalable approach to partitioning data into coherent groups. By iteratively refining cluster assignments based on proximity to centroids or medoids, these algorithms enable the identification of meaningful patterns and structures within datasets, facilitating data exploration, analysis, and interpretation across various domains and applications.

4. Design, Development, and Characteristics of the Implemented System

4.1. The Idea of the Implemented System

There are many examples of applications of descriptive models. Drawing from the preceding analysis, it is evident that descriptive data mining, or descriptive models, hold significant relevance across diverse sectors within the medical and healthcare domains. Our goal and motivation were to design and develop a comprehensive system where the user would be able to search and analyze association rules while also applying clustering methods to the input dataset (either one of their choice or the one built-in) and extracting relevant information from the obtained results. The final version of the implemented system solution corresponds to this idea and can be divided into the following parts:

- Data set retrieval—although it is stated in the introduction of this article that the descriptive models implemented in the application are primarily applied to a specific group of patients from the East Slovak Institute of Heart and Vascular Diseases, we decided to add a functionality that also allows the user to work with their own dataset.
- Search for association rules—this part of the application is designed exclusively for searching and subsequent analysis of association rules, and we used the Apriori algorithm for this task.
- Applying clustering methods—within this section a user can select one of the offered clustering methods, with the choices including hierarchical clustering or clustering using PAM and k-means algorithms, and apply it to the input dataset, giving the user many outputs for obtaining valuable information.
- Search for association rules within defined clusters—the last part of the application directly follows the previous one and is based on it, whereby the user has the possibility, again using the Apriori algorithm, to search for and subsequently analyze association rules within individual clusters that have been identified by applying the chosen clustering method.

In this part of research we will focus on the presentation of key information from the design and development phases, the functionalities of the implemented software, and the characteristics of the individual parts of the application. It is also important to note that we have used an evolutionary model of the development life cycle in the design and subsequent implementation of the software solution.

4.2. Development Environment

In the implementation of the application, the choice of the R programming language proved to be highly advantageous. R is an open-source language that offers a comprehensive suite of tools for data processing, analysis, and visualization. Its extensive collection of packages and libraries makes it particularly well-suited for a wide range of tasks, from basic data manipulation to advanced statistical modeling and machine learning. One of the key strengths of R is its suitability for the development of interactive dashboards and applications. With packages such as Shiny, R users can easily create dynamic and interactive web-based interfaces that allow for real-time exploration and visualization of data. This capability proved invaluable in our application, as it enabled us to develop user-friendly interfaces that facilitated intuitive interaction with the underlying data and algorithms.

Throughout the implementation process, we leveraged a variety of R packages to streamline different aspects of the software development pipeline. For data processing and manipulation, packages such as *dplyr* and *tidyr* provided powerful tools for data wrangling and transformation. For visualization of results, ggplot2 emerged as a go-to choice, offering elegant and customizable plots that effectively communicated complex findings. Additionally, for the application of clustering algorithms, packages such as *cluster* and *factoextra* offered efficient implementations and convenient interfaces for model fitting and evaluation. Use of the R programming language, coupled with its extensive ecosystem of packages, significantly facilitated the development and deployment of our application, which we named DESSFOCA.

4.3. Application User Interface

In the implementation of our app, the utilization of the aforementioned R Shiny package proved instrumental, particularly in regard to its aptitude for crafting web applications. The user interface design was structured into three primary sections (see Figure 2), each serving distinct purposes. Firstly, the menu positioned at the top of the screen provided easy navigation and access to various core functionalities of the app. This menu served as a centralized hub for users to initiate actions and explore different features of the application. The menu also contained four main navigation buttons: *Loading data, Association rules, Clustering*, and *Association rules for clusters*. These buttons stayed on the screen throughout the duration of use and did not change, so the user could switch to a different data analytic method or decide to load a new data set and start over at any given time.

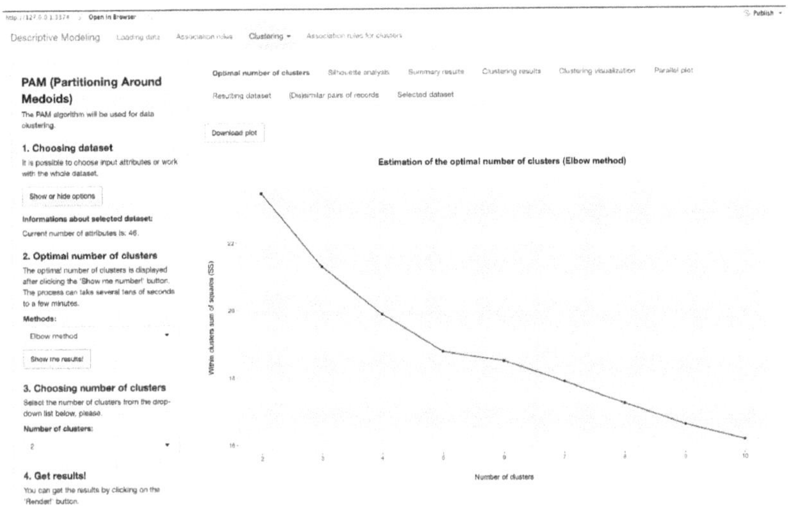

Figure 2. Screenshot from the developed app showing Clustering tab.

Adjacent to the main content area, on the left side of the screen, resided the sidebar. This sidebar served as a convenient space for housing supplementary tools, options, and controls. Its strategic placement ensured quick access to frequently used functions, enhancing user efficiency and workflow. Users could easily toggle between different sections of the application or customize settings from the sidebar, which contributed to a seamless and intuitive user experience. This sidebar on the left side of the screen was present at any given time, dynamically changing with the activity or part of the app the user was working on within that moment while always containing important settings for the given part of the app. There was also a button, *"Show or hide options"*, that we implemented so that users could view more options if they needed to try different settings (or toggle them when they were not needed anymore). We also implemented a self-explanatory numbered system in the side bar so that the users could logically proceed in settings.

The main section, occupying the rest of the screen, served as the focal point of user interaction. This expansive area accommodated the core functionalities and content of the application, providing ample space for data visualization, analysis, and exploration. Within this main section, specific components, such as the data retrieval section, were meticulously designed to facilitate user interaction and data manipulation.

For instance, the data retrieval section was the only one different from all other sections, as it featured a menu that allowed users to specify their data retrieval criteria, such as selecting datasets or defining query parameters. The main section of this data retrieval component housed various interactive elements, such as input fields, dropdown menus, and action buttons, enabling users to customize their data retrieval process according to their specific needs and preferences. This modular design approach ensured that users could efficiently navigate through the data retrieval process while making informed decisions and extracting valuable insights from the available datasets. The structured layout of the user interface (comprising the menu, sidebar, and main section, as well as specific components like the data retrieval section) played a pivotal role in enhancing user engagement, facilitating seamless navigation, and fostering an immersive user experience within our application, resulting in a high-usability application.

4.4. Characteristics of the Implemented System

The individual parts of the application can be characterized as follows:

4.4.1. Loading Data

The first step that the users must perform in the application is the aforementioned data retrieval. The users can either choose to use the data already loaded in the application (data from VÚSCH) or to load their own data. If they use the first option, they are allowed to use other functionalities of the application (which we will describe later) immediately. If the users choose to load their own data, they must perform an additional action. This involves defining how the individual numeric attributes that are part of the dataset will be loaded and transformed into categorical ones. The users have two options for such a transformation, which include direct retyping of numeric attributes to categorical attributes, which is useful if the numbers represent a different value, such as a category. The second option is discretization into intervals of equal width or depth. It is necessary to define an adjustment for each numeric attribute separately. In cases where the retrieved dataset does not contain any numeric attributes, the transformation process is not needed. After the data loading is finished, or users chose the preloaded dataset, they can proceed to the sections of the app dedicated to descriptive data analytic models.

4.4.2. Search for Association Rules

After successful data retrieval, the users can choose whichever section is required from the menu on the top of the screen. In the case of the section dedicated to finding association rules, the users have several input elements to use. These are located both in the side menu and the main panel of the user interface, and they are intended for customizing the content of the input dataset. The users are also able to define threshold values for the input model parameters in order to search for association rules such as minimum support, minimum confidence, or left-hand side rule length. They can also restrict the occurrence of selected attributes on the right-hand side of the association rules. It has several outputs that are located in the main part of the user interface.

The outputs can include the found association rules, presented in the form of a table, or interactive visualizations via which it is possible to better understand the content of the obtained results. Specifically, we mean the visualization of the relationships between the individual attribute values that occurred in the generated association rules and the association rules themselves, as seen in Figure 3. Additionally there is a dot plot, which is used in this context to show the distribution of the different association rules based on metrics such as support, confidence, and lift, which in different combinations can occur on the x and y axes. When working within the association rules tab, users can choose their own set of attributes from the dataset in the menu, located on the left part of the screen and marked with Number 1, and they also receive information about the number of currently chosen attributes. After that, pressing the button *Show data* transfers them to the main screen and to part Number 2, where users can use filters to select individual records. If they are satisfied with their choice, pressing the button *Confirm!* confirms the choice of records and transfers users to the tab marked *Selected data*. Users can then proceed to settings in the left menu (which is numbered 3), where they set metrics for association rules, e.g., minimum support, minimum confidence, and length of LHS (Left-Hand-Side of association rules). Number 4 contains restrictions of RHS (Right-Hand-Side of association rules), where users can choose which attributes will not appear in the association rules. This feature is implemented based on doctor requests, where they want to filter these rules based on the known factors of cardiovascular diseases. Number 5 in the menu contains the *Render!* button, which renders association rules based on the settings chosen in previous parts.

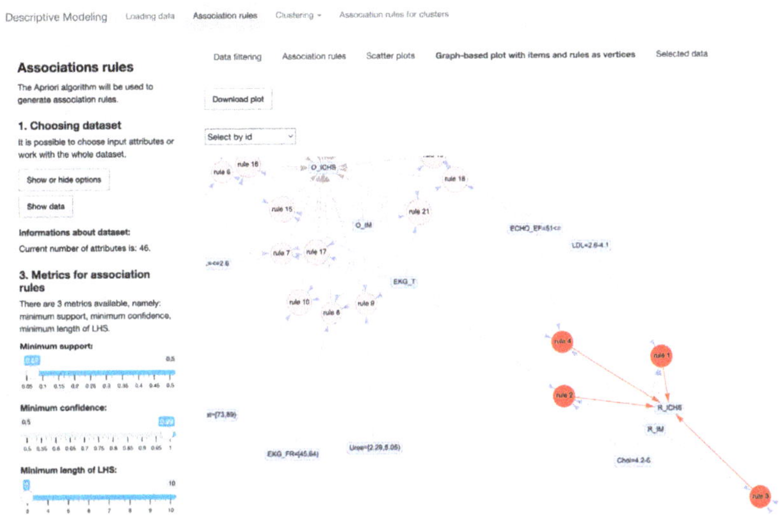

Figure 3. Graph-based plot with items and rules as vertices.

4.4.3. Application of Clustering Methods

In the implementation of our system, the second option in the top menu facilitates the application of clustering methods to the input data. This functionality is pivotal for organizing and understanding complex datasets by grouping similar records into clusters. To compute the distance between records accurately, we adopted the Gower distance metric [41], which is renowned for its efficacy in scenarios where the input dataset comprises attributes of varying data types. Users are granted the flexibility to customize the input dataset according to their specific requirements or to conduct analyses to determine the optimal number of clusters. This can be achieved through the utilization of well-established methods, such as the Silhouette method or the Elbow method, which aid in identifying the most suitable clustering configuration.

For hierarchical clustering, users are provided with additional customization options, including the selection of a specific method tailored to their analytical needs. Users can choose and filter patient records in part Number 1 in the left menu. They can proceed with Number 2 to choose the method of clustering and Number 3 to render the dendrogram. Number 4 in the left menu calculates the optimal number of clusters with the Elbow method, Silhouette method, and Silhouette analysis. In Number 5, users choose the number of clusters that they want to be generated, confirming it in Number 6 by pressing the *Render!* button. Furthermore, users have the capability to generate a dendrogram visualization, which offers a graphical representation of the hierarchical clustering process, enabling a visual understanding of cluster relationships and structures. This section of the application yields multiple outputs designed to provide comprehensive insights into the clustering results. Among the available outputs are individual validation criteria, which offer quantitative assessments of clustering quality and aid users in evaluating the effectiveness of the clustering algorithm. Additionally, users can access visualizations depicting the distribution of data points across clusters, as seen in Figure 4, facilitating the identification of cluster boundaries and patterns within the data. Furthermore, users gain access to the resulting dataset, which delineates the assignment of each record from the input dataset to specific clusters, thereby providing clarity on which record belongs to which cluster. Finally, users are presented with information regarding the records that are most representative in individual clusters, showing the characteristics of each cluster and identifying records that exhibit the lowest amount of similarities within clusters.

Figure 4. Clustering visualization.

4.4.4. Search for Association Rules within Defined Clusters

In the final segment of our app, the third part is dedicated to the discovery of association rules within the clusters defined in the previous section. Once users have reviewed and are satisfied with the clustering results, they have the option to delve deeper into each individual cluster and uncover meaningful associations among the data points. This functionality is particularly valuable for extracting actionable insights and identifying patterns specific to each cluster, thereby facilitating targeted decision making and tailored interventions.

Similar to the clustering section, users are presented with a familiar interface that allows for seamless interaction with the data and analysis results. The inputs and outputs remain consistent with those defined in the preceding paragraphs, ensuring continuity and ease of use throughout the application, which is important for high usability. However, in this part of the application, users are provided with the flexibility to focus exclusively on association rules found within the clusters of their choice. This approach enables users to refine their analysis and concentrate on the clusters that are most relevant to their specific objectives or areas of interest. As with the clustering section, users can explore a variety of outputs designed to provide comprehensive insights into the association-rule-mining process. These outputs include individual validation criteria, visualizations of rule distributions, detailed summaries of rule characteristics, and information on the most significant associations within each cluster. By offering a comprehensive suite of outputs tailored to the user's selected clusters, the application helps users to find valuable insights and derive actionable recommendations, ultimately driving informed decision making.

4.4.5. Estimation of the Optimal Number of Clusters

In the clustering application section, users are provided with the functionality to estimate the optimal number of clusters for their dataset. This critical step aids users in determining the appropriate granularity for partitioning their data and identifying meaningful patterns within the dataset. To facilitate this process, the application offers two popular methods for estimating the optimal number of clusters that we have mentioned before: the Silhouette method and the Elbow method. The Silhouette method evaluates the quality of clustering by assessing how well each data point fits within its assigned cluster relative to other clusters. This method generates silhouette scores for different numbers of clusters, with higher scores indicating better cluster cohesion and separation. By visualizing these scores in the form of bar graphs, users can gain insights into the optimal number of clusters that maximizes overall cluster quality, as seen in Figure 5.

Similarly, the Elbow method provides users with a heuristic approach to determining the optimal number of clusters based on the distortion or within-cluster sum of squares (WCSS) values. This method calculates the WCSS for different numbers of clusters and identifies the point at which adding more clusters no longer significantly reduces the WCSS. The resulting visualization, presented as bar graphs again, enables users to identify the "elbow" point, which suggests the optimal number of clusters.

The system recommends a range of clusters, from two to 10, offering users flexibility in exploring different clustering configurations. However, it is important to note that these recommendations serve as guidance rather than strict rules. Users may choose to further analyze the clustering results or adjust the number of clusters based on domain knowledge or specific project requirements. Ultimately, the goal of providing these recommendations is to assist users in making informed decisions about the optimal clustering configuration for their data, thereby facilitating more effective data analyses and interpretations.

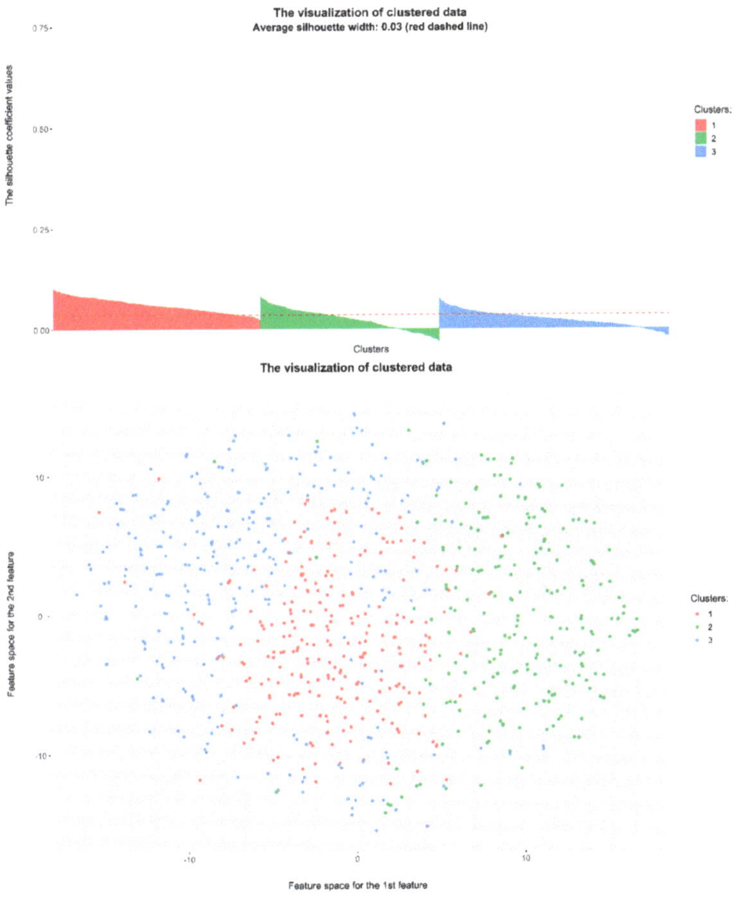

Figure 5. Optimal number of clusters according to the silhouette method.

4.4.6. Saving the Obtained Results

An important functionality of our system is the export of obtained results from the application environment, enabling users to utilize and share their findings beyond the application. This feature enhances the versatility, accessibility and usability of the analysis results, empowering users to integrate them into various workflows and presentations. For tabular data, or records displayed within the application, users have the flexibility to export them in widely used formats such as .xlsx or .csv. This capability facilitates easy integration with external tools or further analysis. Additionally, users can selectively export a subset of records, enabling focused analysis or sharing of specific data points of interest.

Graphical visualizations generated within the application can be exported in the .png format, preserving the visual representation of data patterns and insights. This facilitates the incorporation of visualizations into reports, presentations, or publications, ensuring clarity and accessibility of the analysis results. Furthermore, the application features interactive visualizations, allowing users to dynamically explore and interact with data visualizations. To preserve the interactivity of these elements outside the application environment, users can export them in the .html format. This format retains the interactive features of the visualizations, enabling users to continue exploring and analyzing the data even after exporting it. By offering a range of export options tailored to different types of analysis results, the application empowers users to effectively communicate their findings and insights to stakeholders, colleagues, or clients. This seamless integration with external tools and formats enhances the usability and impact of the analysis results, facilitating informed decision making and driving actionable outcomes.

5. Application and Usability Testing with Medical Professionals

5.1. Application Functionality Testing with Medical Professionals

According to the Nielsen Norman group [42], the best testing results come from testing with at least five users and running as many small tests as can be afforded. We decided to test the app with more than five respondents to receive more feedback. The respondent group that participated in the 1st testing of the implemented application consisted of seven members, all of whom were specifically doctors working in various departments of the Louis Pasteur University Hospital (UNLP) in Košice and in the VÚSCH (East Slovak Institute of Heart and Vascular Diseases), which is also located in Košice. The respondents were of different genders and ages, and the length of their activity in the medical sphere was also different.

The initial testing phase comprised prepared scenarios that involved tasks developed in collaboration with a specialized cardiology department physician. This testing was important, as we needed to thoroughly test all the core app functionalities with professionals. The testing scenario consisted of two tasks and involved the respondent answering by choosing one of the three options offered. The way in which the wording of the tasks was formulated corresponded to the primary target group of the implemented application, which consisted of the aforementioned physicians or medical professionals. It is also important to note that the questionnaire covered all the important parts of the application in its content and, therefore, in the way the tasks were constructed. Given that the respondents were guided, if necessary, in testing the application and thus in completing the given tasks, we can say that they were able to examine the app well and in depth. However, what was relevant was that all respondents went through the prepared tasks and had their own experience of working with all parts of the application. This fulfilled the prerequisites for qualified answers to the questions in the second part of testing the app with non-professional users.

5.2. Application Usability Testing with Medical Professionals

In the second phase of testing, our focus shifted towards evaluating the usability of the implemented application through the administration of a standardized questionnaire. To achieve this, we employed a framework consisting of ten predefined questions known as the System Usability Scale (SUS questionnaire) [28]. This approach was selected for its ability to provide valuable insight into the subjective evaluation of the application's usability from the perspective of the users. This phase of testing is closely intertwined with the previous one, as it allows testers to assess the usability of the application based on the firsthand user experience gained during the execution of the assigned tasks. By soliciting feedback through the SUS questionnaire, we aimed to gather comprehensive insights into various aspects of usability, including ease of use, efficiency, learnability, and overall user satisfaction. This structured approach enabled us to identify strengths and weaknesses in the application's usability, paving the way for targeted improvements and enhancements. Moreover, by correlating the findings from the usability evaluation with the outcomes of the task-based testing, we were able to gain a holistic understanding of the application's performance and user experience. This iterative approach towards testing and evaluation ensures that the application not only meets functional requirements but also aligns with user expectations and preferences, ultimately leading to a more user-centric and effective solution. The System Usability Scale (SUS) score is calculated, as we mentioned before, based on users' responses to a set of ten questions designed to assess the usability of a system. Each question in the SUS questionnaire is scored on a 5-point Likert scale, ranging from Strongly Disagree (1) to Strongly Agree (5). The scoring process involves the following steps:

The SUS yields a single number representing a composite measure of the overall usability of the system being studied. Note that scores for individual items are not meaningful on their own. To calculate the SUS score, first sum the score contributions from each item. Each item's score contribution will range from 0 to 4. For items 1, 3, 5, 7, and 9, the score contribution is the scale position minus 1. For items 2, 4, 6, 8 and 10, the contribution is 5 minus the scale position. Multiply the sum of the scores by 2.5 to obtain the overall value of SU. SUS scores have a range of 0 to 100.

1. Odd numbered questions: For questions 1, 3, 5, 7, and 9, responses are reverse coded, so the score contribution is the scale position (1–5) minus 1;
2. Even numbered questions: For questions 2, 4, 6, 8 and 10, the contribution is 5 minus the scale position (1–5);
3. Calculating scores: Each individual question is then assigned the score based on whether it is an odd or even numbered question;
4. Summing scores: The scores for all ten questions are summed together to obtain a total score;
5. Adjusting scores: Multiply the sum of the scores by 2.5 to obtain the overall value of usability. SUS scores have a range of 0 to 100.

Therefore, the final SUS score represents an aggregate measure of the perceived usability of the system, with higher scores indicating greater usability. A score above 68 would generally be considered above average, while a score below 68 suggests that improvements are needed to enhance usability.

With an average app rating score of 74.64 from all of seven respondents, our application falls within the range of 68 to 80.3 on the rating scale. According to the grading system, this places our application in the "B" category, indicating a high level of usability. In verbal terms, our software can be described as "good", further affirming its usability. Additionally, the positive feedback from respondents reinforces the usability of our application. Out of the seven respondents, five have reported finding the implemented system easy to use. This demonstrates a majority agreement on the user-friendly nature of our application. Furthermore, an equal proportion of users fully agree that the functions within the application are very well integrated. This cohesion and seamless integration of features contribute to a positive user experience, further validating the usability of our application. Overall, these findings affirm that our application not only meets usability standards but also provides a user-friendly and integrated experience for its users.

6. Extension of Van Welie Layered Model of Usability

After reviewing several studies focused on usability research, its aspects, and specific metrics, we concluded that there is no universal model by which we can measure usability. As usability is a very complex concept, and because it is necessary to consider in which environment, system, and for which users we need to evaluate it, it is necessary to choose appropriate metrics and methods to measure usability. Therefore, in this part of the research, we have summarized current knowledge and defined a new, more universal usability measurement model for data analytics, according to which we can more easily determine which methods are suitable for usability measurement based on our requirements. We will start from Van Welie's model, mentioned in Section 2, which is a general usability model that we will extend to specific indicators of usability aspects that have their own measurement methods (Figure 6).

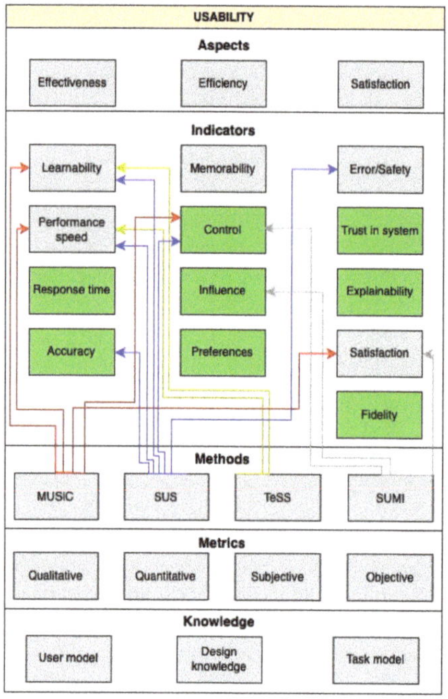

Figure 6. Proposed extension of Van Welie layered model of usability.

Based on the studies examined, we concluded that, based on the study [43], it would be appropriate to add an indicator of "explainability". The indicators of "control" and "influence" consistently appear in the studies [28,44,45]. The new indicators determining the "accuracy" of the system came from the study [18], while "fidelity" and "response time" are mentioned in the other research article [46]. In the parts highlighted with green in Figure 6, new indicators are shown with which we can better focus on usability measurement in data analytics models. Thanks to these indicators, we will focus more on the explainability of the system, its accuracy, what response it has, and whether it looks trustworthy and preferable to the user or whether its operation is easy for the user. Indicators in Figure 6. are arranged in columns under the aspects, so we know which indicators to focus on if we want to measure the usability of an aspect pertaining to our system.

After grouping the examined methods for usability measurement, we can see that we can measure some indicators by several methods. To measure objective metrics, i.e., to determine the usability of the system from a technical point of view, such as speed, simplicity, error rate or accuracy of the system, it is advisable to use the methods like SUS, MUSiC, or TeSS. Conversely, if we need to measure how the user sees the usability of a given system from a subjective point of view, it is advisable to use the SUMI method, which focuses on satisfaction, ease of operation, and user influence.

The newly developed model shows that the methods we analyze focus primarily on examining aspects such as system efficiency and effectiveness. Methods for measuring user satisfaction with the system did not focus so much on this aspect. A study [43] describing the explainability divides the methods of measuring this indicator into qualitative and quantitative categories. We can also apply these two groups of division of methods to our model, as we have divided the indicators into objective and subjective point of view. Among qualitative methods, we can, by definition, include the SUMI method, as well as the SUS method, depending on what aspect (objective/subjective) of the system the questionnaire is oriented towards. Among quantitative methods, we can, by definition, include the methods SUS, TeSS, and MUSiC.

If we were to divide the chosen measurement methods according to the evaluation of the explainability aspect, the TeSS method would belong to Human Based Evaluation, as the TeSS experiment is carried out with laymen and includes an essential element of learning in addition to the need to obtain specific information. We could include MUSiC, SUMI, and SUS methods in the application-based evaluation, as all three methods are performed only after users have experienced working in a given system. However, the SUS method can also be included in the evaluation based on the functionality of the system, as it can determine how to create a suitable, easy-to-use system based on what users would expect from a functional system.

7. Testing Usability of DESSFOCA App

The second and most important part of the testing with which respondents worked is to measure the usability of the DESSFOCA application from the non-professional users' point of view. This allows us to determine the real usability of the app, which is not biased towards domain expertise. This part is divided into subsections that measure the usability of the application in terms of working with association rules, the application of hierarchical clustering methods, and also the general usability of the application. It is also important to mention that we mapped the proposed extension of the Van Welie model on aspects, which we rigorously tested and determined, if they were really applicable and could improve usability. When evaluating usability, 17 respondents replied. Respondents were able to either choose an answer with an attributable value ("I strongly disagree"—1, "I rather disagree"—2, "I am not sure"—3, "I rather agree"—4, "I completely agree"—5) regarding which best described their view of the question, and they also had the option of an open answer.

Evaluating the questionnaire is straightforward. We calculate the score for usability itself, or a given indicator, as the arithmetic mean of the answers in a given question and in the group of questions pertaining to a given indicator. We convert the resulting value to a percentage score. The result will be the compliance rate of the given usability indicator. In this way, we can also measure the usability of the entire system, as, at the end, we calculate the arithmetic mean of all indicators. In order to be able to express usability verbally, not just numerically, we can also take the evaluation from the SUS questionnaire, as we mentioned in the evaluation with medical professionals. The SUS method considers the type of question and whether it is asked positively or negatively. The questions in our questionnaire are asked only in a positive way, so, in this case, we can apply the calculation of the score by the arithmetic mean of the answer weights. The result of the evaluation of the questionnaire is considered average or good if its score is about 68%. If the result of the questionnaire is above the threshold of 80.3%, the evaluated application is considered above average. Conversely, if the score is less than 51%, the application is considered insufficient and is deemed not usable.

7.1. *Usability of the Application When Working with Association Rules*

In this section, we created five questions to find out how respondents worked with the features that the application offers when working with association rules. The option to choose the type of dataset was useful for the vast majority of respondents, which can be judged by the high score value of 93%. According to the high score of respondents, 86%, the function of filtering records was also useful, as they were able to simplify their work and thus work with fewer records. When asked to limit the right side of the rule, respondents' answers were more uncertain, as the answers "I am not sure" and "I rather agree" were predominant. This may be because the tasks were simple and only partially covered this feature. This feature is more designed for professionals who use it frequently. The visualizations, as well as the actual understanding of the association rules, were well understood by respondents. The overall usability score for association rules featuresd in the app is 83.8%.

7.2. Usability of the Application When Applying the Clustering Methods

For tasks with hierarchical clustering, we created four questions to which respondents gave us their feedback. The most useful feature was the dataset selection feature, which reached a score of 88%. The visualizations of hierarchical clustering were not so well understood by non-professional users, as this feature only reached a value of around 77%. For a better understanding, it would be necessary to describe in more detail what the graph shows and which values are relevant to the user. The graphs can become very crowded and hard to read, even for experts in data analytics, so this part is absolutely understandable. Despite the fact that the evaluation of the success of tasks when working with hierarchical clustering came out better than tasks with association rules, the usability of this part of the app reached a slightly lower score of 80%.

7.3. Usability of the DESSFOCA App

For clearer usability measurements, we divided the questions asked into groups depending on which indicator they belonged to. We focused mainly on the new indicators from the proposal of the Van Welie layered model of usability extension (see Section 6).

7.3.1. Ease of Use and Overview in the System

The ease of use for respondents depends mainly on whether the controls and functions are easy to understand. Since respondents were guided by the description of the steps to use the application in order to reach the result, they considered the DESSFOCA app to be relatively simple to use. As soon as respondents in the second part of the tasks (hierarchical clustering application) had to work without assistance, it became more difficult for them to complete the tasks. The tasks to be answered by respondents were designed mainly for non-experts, a group which includes the testers. The response to the apps ease of use would probably be lower if it worked with more demanding features. In contrast, the design of the application and clarity were understandable to respondents, and they considered the application to be consistent. The overall rating of this indicator is 68.46%, which means that it needs improvement in this area. We would recommend adding explanatory notes to help better understand the attributes, or a more detailed description of how to work with the record filter and attribute tagging.

7.3.2. Explainability

When it comes to explainability, we mainly focused on whether the respondent needs a user manual and possibly the help of an expert, as well as whether it is necessary to already have some knowledge to work with the application or if it is self-explanatory. Up to 82.3% of respondents perceived that it is necessary to have instructions for using the application, with the help of a technician/expert being found not to be necessary. The respondents stated that even the initial quick training was sufficient for understanding the basic functions with which they worked. When asked whether it would be necessary to have previously acquired knowledge before using the application, the answers were very diverse; therefore, we assume that it depends on what functions the user will work with. Explainability based on answers is 48.25%, which is a very low value. This indicator was not fulfilled correctly by the application. In practice, we would recommend adding explanations to the application's functions, as well as illustrative examples of how to work with the application. We also recommend adding explainability features on descriptive models used within the app. We assume that the value of this indicator would be higher if the respondents were experts from practice and not non-expert users.

7.3.3. Accuracy of Results and Time of the Response

After examining the results of the respondents' responses to this indicator, we found that it had the best rating so far, reaching 82.93%. The response speed is acceptable, and respondents did not experience any problems with generating tables or graphs. According to them, the functions with which the respondents worked are well integrated. The visualization types that were applied to the application were rendered quickly and looked clear. The positive response to this question can also be attributed to the fact that the respondents had what to look for in the graphs explained to them. The application itself does not provide a description of the graphs, which is important for users and should have been implemented. The question asking whether, according to the respondents, the results are accurate had the most neutral answers, as the results were not compared to anything.

7.3.4. SYSTEM Features and Sustainability

The respondents could not assess whether the preset dataset in the DESSFOCA application contains all the necessary attributes, as this question should be asked of an expert in the given domain. As the respondents used the basic functions, they needed to work with the preloaded dataset, and the answers to the question about the use of functions were mostly positive. To improve this function, it would be advisable to insert the possibility of adding a custom attribute according to the user's needs. The ability to export results in several formats, such as .xlsx or .csv, is very useful, as the user can work with these results and format them outside the application. The application has a well-described procedure of steps in which user should proceed to achieve the result, so the response to the logical sequence was high, reaching 82.35%. The overall rating of this indicator was 82.63%.

7.3.5. Trust in the System

The questions on this indicator were directed mainly at whether the respondents find this application easier and more efficient, whether they would use it in the future, and whether they would recommend it to colleagues/classmates/researchers. Trust in the application had an average value of 68.90%. We assume that this average value is mainly influenced by the fact that the respondents' frequent answers regarding future use of the application were neutral, as all the respondents, while testing the application, were non-experts and will probably not work in the medical field in the future. We can see from testing with medical experts that, when we conducted the testing with domain experts, the response to the usefulness of the application was much higher. We conclude, based on the answers from our respondents from testing with experts, that the application will significantly simplify the work in comparison to if the evaluation of the results was performed manually. This statement had the highest value in the group focusing on the system trust indicator, reaching 82.35%.

7.3.6. Trust in Result

The trust in result indicator consisted of two questions that focused on whether respondents felt confident using the app. Since they worked only with a predetermined dataset, the trust in the results was high at 88.23%. The answers to the feeling of self-confidence when using the application were different, which we can see in the final value of this question (69.41%). This value was preceded by whether respondents had a good understanding of the procedure and explanation of how to navigate the application. The total value of this indicator was 78.82%.

7.3.7. Influence and Influence on Decision

With this indicator, most of the respondents perceived that the system simplifies and facilitates their work with data, as the application provides several functions for working with data; therefore, this question had a score of 78.82%. Since the respondents will not continue to work with this data and the application, the influence of a user's decision has an average value of 69.41%. The error rate was higher than we expected for some of the entered tasks, which may also be due to the fact that the application only provided a limited amount of help messages, such as how to proceed and correct the error in the event of it. Of course, we only noticed a reading comprehension problem in a small amount of answers. The value of this indicator reached 72.54% during testing.

7.3.8. Preference and Quality of the Solution

The last indicator we focused on during testing was whether the application provided everything the user expected from it. We have seen comments indicating that, when downloading a chart as an image, the chart was not downloaded correctly or the downloaded document was empty. However, the application met most of the users' expectations and therefore had a response value of 77.64%. Since the application provided only minimal notifications in cases of bad filtering of records or other functions, we recommend adding error messages to indicate why the step cannot be performed by the application and how the user can fix it. This feature would make the application faster and easier to use. This indicator acquired a value of 72.93%.

7.3.9. Discussion and Results Summary

The "Explainability" indicator had the lowest value. As we already mentioned in regard to the characteristics of this indicator, it is very important to focus on the quality of the explanation, as this indicator also affects the overall impression of the application/system and trust in the system. This statement is also supported by the study [46], where the authors noted that a good explanation should be understandable for people and accurately describe the model behavior in the entire feature space, which increases the value of the fidelity indicator among users. Explainability is very subjective, and the perceived quality of the explanation is contextual and depends on the expertise of users, the explanation itself, as well as the type of information the user is interested in. In practice, we would recommend adding a user manual on how to navigate the application, illustrative examples of how to work with the application, explanations of the application's functions, as well error messages and help on how to fix them.

The indicators "System functions and sustainability" and "Accuracy of results and part of response" had the highest values, and the percentage difference was minimal. The high value of the first mentioned indicator can be attributed to the fact that the tasks that the respondents had to process mainly covered the basic functions of the system. Therefore, respondents encountering a problem with a lack of functions in working with the application could not happen. Since they were working with a small amount of data, the response of the system and the generation of tables and graphs was acceptable.

The overall usability of the application reached a value of 73.61%. Based on the aforementioned articles and studies, we consider the application to be average, and it received an evaluation as being "good", with a score close to 68%. Table 1 shows us the values of individual usability indicators. We also evaluated the usability of the system from the point of view of the most basic aspects, such as "Effectiveness", "Efficiency" and "Satisfaction", based on the newly created usability model. We found that the indicators "Explainability" and "Fidelity in the result", which were poorly fulfilled according to the respondents, cause low user satisfaction with the application/system. Conversely, the indicators "Control", "Influence" and "Preferences", which belong to the aspect of effectiveness, had the highest values. The indicators "Accuracy" and "Response time" also had well-evaluated values, and therefore we consider this system to be effective. The overview of all usability indicators can be seen in Table 1.

Table 1. Usability indicators score overview.

Usability Indicators	Value of Usability Indicator
Accuracy of result and response time	82.93%
System functions and sustainability	82.63%
Fidelity (trust) in result	78.82%
Preferences and quality of solution	72.93%
Influence and Influence on decision	72.54%
Trust in system	68.90%
Easy control and easy system overview	68.46%
Explainability	48.25%

8. Conclusions

The use of data analytics in the field of medicine can be considered to be a beneficial and effective approach. The information and knowledge gained has a wide range of applications, whether in the early diagnosis of diseases or their treatment and prevention. We have implemented a new app named DESSFOCA (Decision Support System for Cardiologists). The implemented software consists of three main parts that focus on finding association rules, applying clustering methods, and finding association rules amongst defined clusters. By appropriately interpreting the obtained results, either in textual form or in the form of visualizations, it was possible to verify their relevance from a medical point of view, a task that also involved the participation of a doctor who specialized in cardiology. The software solutions usability was tested with a target user group consisting mainly of doctors or people working in a medical environment. The results obtained can be evaluated as positive, with an average score of 74.64 points, which corresponds to a grade B, which can be interpreted as good usability of the application.

We continued our research with tasks that consisted mainly of examining aspects, indicators, methods, and metrics according to which we could more easily determine which methods are suitable for measuring usability based on our requirements. We discovered that they were based on the well-known Van Welie model, a layered model of usability. We proposed an extension of this model with specific indicators of aspects of usability, which have their own measurement methods. We tested the DESSFOCA app again, now with non-professional users, in order to determine the general usability of the app and to approve the proposed extension of the Van Welie model. We prepared tasks with which the respondents had to work and describe their evaluation based on testing out the system. The evaluation of the system reached 73.61% in its average score of usability; even so, the system is usable, and it has appropriately selected and sufficient functions. During the evaluation, we focused on each newly proposed indicators in our extension of the Van Welie model separately, so that we could find out which indicators need to be improved. At the end of the work, after evaluating the usability of the DESSFOCA, we proposed recommendations for improving both the value of the indicators and also the overall usability, such as adding a manual for users and improving error messages for improvement and acquisition of higher usability value. Additionally, we also conclude that our extension of the Van Welie layered model of usability is sufficient and correct.

Author Contributions: Conceptualization, O.L. and J.P.; methodology, J.P.; software, A.P., O.L. and Z.P.; validation, O.L., Z.P., D.P. and J.P.; formal analysis, O.L.; investigation, Z.P. and D.P.; resources, D.P.; data curation, O.L., A.P. and Z.P.; writing—original draft preparation O.L. and J.P.; writing—review and editing, O.L. and J.P.; visualization, O.L. and A.P.; supervision, J.P.; project administration, J.P.; funding acquisition, J.P. All authors have read and agreed to the published version of the manuscript.

Funding: This research was funded by the Slovak Research and Development Agency under the contracts No. APVV-22-0414, APVV-20-0232, APVV-16-0213 and by the Scientific Grant Agency of the Ministry of Education, Research, Development and Youth of the Slovak Republic and the Slovak Academy of Sciences under grants No. 1/0685/21 and 1/0259/24.

Institutional Review Board Statement: The trial was approved on 24 January 2019 by the Ethical Committee of the Faculty of Medicine, Pavol Jozef Safarik University in Kosice, and the Ethical Committee of the East Slovak Institute for Cardiovascular Diseases (VÚSCH) in Kosice.

Informed Consent Statement: Informed consent was obtained from all subjects involved in the collection of data.

Data Availability Statement: Due to patient confidentiality, it is impossible to provide access to the data used in the development of the app, even though they are anonymized.

Conflicts of Interest: The authors declare no conflicts of interest.

References

1. WHO. Cardiovascular Diseases (Cvds). 2021. Available online: https://www.who.int/news-room/fact-sheets/detail/cardiovascular-diseases-(cvds) (accessed on 6 April 2024).
2. Cardiovascular Diseases—World Health Organization. Available online: https://www.who.int/health-topics/cardiovascular-diseases#tab=tab_1 (accessed on 9 April 2024).
3. European Heart Network. Available online: https://ehnheart.org (accessed on 9 April 2024).
4. State of Health in the EU—Slovakia. Health Profile of the Country. 2023. Available online: https://health.ec.europa.eu/system/files/2024-01/2023_chp_sk_slovak.pdf (accessed on 9 April 2024). (In Slovak).
5. World Heart Federation. FACTSHEET: Cardiovascular Diseases in China. 2017. Available online: https://world-heart-federation.org/wp-content/uploads/2017/05/Cardiovascular_diseases_in_China.pdf (accessed on 6 April 2024).
6. Vidinský, B. Slovakia—Country with One of the Highest Mortality Rates from Cardiovascular Diseases within the OECD. Zalepsiezdravotnictvo.sk. 2019. Available online: https://zalepsiezdravotnictvo.sk/blogy/36-slovensko-krajina-s-jednou-z-najvaecsich-umrtnosti-na-srdcove-ochorenia-v-ramci-oecd (accessed on 6 April 2024). (In Slovak).
7. Kochanek, K.D.; Murphy, S.L.; Xu, J.Q.; Arias, E. *Deaths: Final Data for 2017*; National Vital Statistics Reports; National Center for Health Statistics: Hyattsville, MD, USA, 2019; Volume 68.
8. Fryar, C.D.; Chen, T.; Li, X. *Prevalence of Uncontrolled Risk Factors for Cardiovascular Disease: United States, 1999–2010*; NCHS Data Brief, no. 103; National Center for Health Statistics: Hyattsville, MD, USA, 2012.
9. Islam, M.S.; Hasan, M.M.; Wang, X.; Germack, H.D.; Noor-E-Alam, M. A Systematic Review on Healthcare Analytics: Application and Theoretical Perspective of Data Mining. *Healthcare* **2018**, *6*, 54. Available online: https://www.ncbi.nlm.nih.gov/pmc/articles/PMC6023432/ (accessed on 6 April 2024). [CrossRef] [PubMed]
10. Simonov, M.; Ugwuowo, U.; Moreira, E.; Yamamoto, Y.; Biswas, A.; Martin, M.; Testani, J.; Wilson, F.P. A Simple Real-Time Model for Predicting Acute Kidney Injury in Hospitalized Patients in the US: A Descriptive Modeling Study. *Plos Med.* **2019**, *16*, e1002861. [CrossRef] [PubMed]
11. Towfek, S.K. A Semantic Approach for Extracting the Medical Association Rules. *J. Artif. Intell. Metaheuristics* **2023**, *5*, 46–52.
12. Patil, M.; Joshi, R.C.; Toshniwal, D. Association Rule for Classification of Type-2 Diabetic Patients, Second International Conference on Machine Learning and Computing. 2010. Available online: https://ieeexplore.ieee.org/document/5460715 (accessed on 6 April 2024).
13. Tandan, M.; Acharya, Y.; Pokharel, S.; Timilsina, M. Discovering Symptom Patterns of COVID-19 Patients Using Association Rule Mining. *Comput. Biol. Med.* **2021**, *131*, 104249. Available online: https://www.sciencedirect.com/science/article/pii/S001048252 1000433 (accessed on 6 April 2024). [CrossRef] [PubMed]
14. Awad, F.H.; Hamad, M.M.; Alzubaidi, L. Robust Classification and Detection of Big Medical Data Using Advanced Parallel k-Means Clustering, yolov4, and Logistic Regression. *Life* **2023**, *13*, 691. [CrossRef] [PubMed]
15. Guo, Q.; Lu, X.; Gao, Y.; Zhang, J.; Yan, B.; Su, D.; Song, A.; Zhao, X.; Wang, G. Cluster Analysis: A New Approach for Identification of Underlying Risk Factors for Coronary Artery Disease in Essential Hypertensive Patients. *Sci. Rep.* **2017**, *7*, 43965. Available online: https://doi.org/10.1038/srep43965 (accessed on 6 April 2024). [CrossRef] [PubMed]
16. Bevan, N.; Kirakowski, J.; Maissel, J. What is usability. In Proceedings of the 4th International Conference on HCI, Stuttgart, Germany, 1–6 September 1991; Available online: https://www.researchgate.net/publication/239609271 (accessed on 6 April 2024).
17. McNamara, N.; Kirakowski, J. Defining Usability: Quality of Use or Quality of Experience? In Proceedings of the IPCC 2005, Proceedings. International Professional Communication Conference, Limerick, Ireland, 10–13 July 2005; pp. 200–204.
18. Bevan, N.; MacLeod, M. Usability Measurement in Context. *Behav. Inf. Technol.* **1994**, *13*, 132–145. [CrossRef]
19. *ISO 9241-11:2018*; Ergonomic Requirements for Office Work with Visual Display Terminals: Part 11: Guidance on Usability. ISO: Geneva, Switzerland, 2018. Available online: https://www.iso.org/standard/63500.html (accessed on 6 April 2024).
20. Lohaj, O. Aspects of Usability in Clinical Decision Support Systems. In Proceedings of the 22nd Scientific Conference of Young Researchers (SCYR): Proceedings from Conference, Virtual Conference, Košice, Slovakia, 8 April 2022; Technical University of Košice: Košice, Slovakia, 2022; pp. 48–51.
21. Güiza, D.C. *The Real Importance of Usability and User Experience*; Medium Design Community, 2017. Available online: https://medium.com/@bluehair.co/the-real-importance-of-usability-and-user-experience-c7bfd4cef11 (accessed on 6 April 2024).

22. Doshi-Velez, F.; Kim, B. Towards a Rigorous Science of Interpretable Machine Learning. *arXiv* **2017**, arXiv:1702.08608.
23. Zytek, A.; Liu, D.; Vaithianathan, R.; Veeramachaneni, K. Sibyl: Understanding and Addressing the Usability Challenges of Machine Learning in High-Stakes Decision Making. *IEEE Trans. Vis. Comput. Graph.* **2022**, *28*, 1161–1171. [CrossRef] [PubMed]
24. Lewis, J.R. Usability Testing. In *Handbook of Human Factors and Ergonomics*; 2012; pp. 1267–1312. Available online: https://books.google.sk/books?hl=sk&lr=&id=WxJVNLzvRVUC&oi=fnd&pg=PA3&dq=In+Handbook+of+Human+Factors+and+Ergonomics;+&ots=p_lnEYTwn7&sig=i7fv_BMbkZpldPf1X7hcpShMTfs&redir_esc=y#v=onepage&q=In%20Handbook%20of%20Human%20Factors%20and%20Ergonomics;&f=false (accessed on 6 April 2024).
25. Barnum, C.M. *Usability Testing Essentials: Ready, Set... Test!* Morgan Kaufmann: Burlington, MA, USA, 2020.
26. Maramba, I.; Chatterjee, A.; Newman, C. Methods of Usability Testing in the Development of eHealth Applications: A Scoping Review. *Int. J. Med. Inform.* **2019**, *126*, 95–104. [CrossRef] [PubMed]
27. Geisen, E.; Bergstrom, J.R. *Usability Testing for Survey Research*; Morgan Kaufmann: Burlington, MA, USA, 2017.
28. Brooke, J. SUS—A Quick and Dirty Usability Scale. In *Usability Evaluation in Industry*; Jordan, B., Thomas, B., Weerdmeester, A., McClelland, I.L., Eds.; Taylor & Francis: London, UK, 1996; pp. 189–194.
29. Beyan, N. The MUSiC Methodology for Usability Measurement. In Proceedings of the Posters and Short Talks of the 1992 SIGCHI Conference on Human Factors in Computing Systems, Monterey, CA, USA, 3–7 May 1992; pp. 123–124.
30. Hertzum, M.; Frøkjær, E. Browsing and Querying in Online Documentation: A Study of User Interfaces and the Interaction Process. *ACM Trans. Comput.-Hum. Interact. (TOCHI)* **1996**, *3*, 136–161. [CrossRef]
31. Van Welie, M. Breaking Down Usability. 2000. Available online: https://www.academia.edu/27688099/Breaking_down_usability (accessed on 6 April 2024).
32. Shneiderman, B.; Plaisant, C. Designing the User Interface: Strategies for Effective Human-Computer Interaction: Sixth Edition, Pearson. 2016. Available online: http://www.cs.umd.edu/hcil/DTUI6 (accessed on 6 April 2024).
33. Hardenbol, A.; Knols, B.; Louws, M. Usability Aspects of Medication Related Decision Support Systems in the Outpatient Setting: A Systematic Literature Review. 2018. Available online: https://www.semanticscholar.org/paper/Usability-aspects-of-medication-related-decision-in-Hardenbol-Knols/46c9c3f51a5dcb6a8d316b8bfb0c7cec7e52c877 (accessed on 6 April 2024).
34. Agyapong, K.B.; Hayfron-Acquah, J.B.; Asante, M. An Overview of Data Mining Models (Descriptive and Predictive). *Int. J. Softw. Hardw. Res. Eng.* **2016**, *4*, 53–60. Available online: https://ijournals.in/wp-content/uploads/2017/07/11.4517-Agyapong.compressed.pdf (accessed on 6 April 2024).
35. Paralič, J. *Knowledge Discovery in Databases*; Elfa: Košice, Slovakia, 2003; ISBN 80-89066-60-7. (In Slovak)
36. Maimon, O.; Rokach, L. *Data Mining and Knowledge Discovery Handbook*; Springer: New York City, NY, USA, 2005; ISBN 978-0-387-25465-4.
37. Bouguettaya, A.; Yu, Q.; Liu, X.; Zhou, X.; Song, A. Efficient Agglomerative Hierarchical Clustering. *Expert Syst. Appl.* **2015**, *4*, 2785–2797. [CrossRef]
38. Roux, M. A Comparative Study of Divisive and Agglomerative Hierarchical Clustering Algorithms. *J. Classif.* **2018**, *35*, 345–366. [CrossRef]
39. Charu, C. *Aggarwal: Data Mining: The Textbook*; Springer: Berlin/Heidelberg, Germany, 2015.
40. Reynolds, A.P.; Richards, G.; Rayward-Smith, V.J. The Application of k-medoids and PAM to the Clustering of Rules. In Proceedings of the Intelligent Data Engineering and Automated Learning–IDEAL 2004: 5th International Conference, Exeter, UK, 25–27 August 2004; Springer: Berlin/Heidelberg, Germany, 2004; pp. 173–178.
41. Divyanshu, A. Gower's Distance. Available online: https://medium.com/analytics-vidhya/gowers-distance-899f9c4bd553 (accessed on 6 April 2024).
42. Nielsen, J.; Landauer, T.K. A Mathematical Model of the Finding of Usability Problems. In Proceedings of the ACM INTERCHI'93 Conference, Amsterdam, The Netherlands, 24–29 April 1993; pp. 206–213.
43. Explainability—C3 AI. Available online: https://c3.ai/glossary/machine-learning/explainability (accessed on 6 April 2024).
44. Veenendaal, E. Questionnaire Based Usability Testing. 1998. Available online: https://www.semanticscholar.org/paper/Questionnaire-based-usability-testing-Veenendaal/03de9324b0130cfb5e5cfd50c15ab98b89dc9b31 (accessed on 6 April 2024).
45. Granollers, T. Usability Effort: A New Concept to Measure the Usability of an Interactive System Based on UCD. 2006. Available online: https://www.researchgate.net/publication/225864466_Usability_Effort_a_new_concept_to_measure_the_usability_of_an_interactive_system_based_on_UCD (accessed on 6 April 2024).
46. Zhou, J. Evaluating the Quality of Machine Learning Explanations: A Survey on Methods and Metrics. 2021. Available online: https://www.researchgate.net/publication/349912446_Evaluating_the_Quality_of_Machine_Learning_Explanations_A_Survey_on_Methods_and_Metrics (accessed on 6 April 2024).

diagnostics

MDPI

Article

A Novel Respiratory Rate Estimation Algorithm from Photoplethysmogram Using Deep Learning Model

Wee Jian Chin [1,2], Ban-Hoe Kwan [1,2], Wei Yin Lim [3], Yee Kai Tee [1,2], Shalini Darmaraju [1], Haipeng Liu [4] and Choon-Hian Goh [1,2,*]

1 Department of Mechatronics and Biomedical Engineering, Lee Kong Chian Faculty of Engineering and Science, Universiti Tunku Abdul Rahman, Kajang 43000, Selangor, Malaysia; chinwj97@1utar.my (W.J.C.); kwanbh@utar.edu.my (B.-H.K.); teeyk@utar.edu.my (Y.K.T.); shalinid@utar.edu.my (S.D.)

2 Centre for Healthcare Science and Technology, Universiti Tunku Abdul Rahman, Kajang 43000, Selangor, Malaysia

3 Electrical and Computer Systems Engineering, School of Engineering and Advanced Engineering Platform, Monash University Malaysia, Bandar Sunway 47500, Selangor, Malaysia; weiyin.lim@monash.edu

4 Centre for Intelligent Healthcare, Coventry University, Coventry CV1 5RW, UK; haipeng.liu@coventry.ac.uk

* Correspondence: gohch@utar.edu.my

Abstract: Respiratory rate (RR) is a critical vital sign that can provide valuable insights into various medical conditions, including pneumonia. Unfortunately, manual RR counting is often unreliable and discontinuous. Current RR estimation algorithms either lack the necessary accuracy or demand extensive window sizes. In response to these challenges, this study introduces a novel method for continuously estimating RR from photoplethysmogram (PPG) with a reduced window size and lower processing requirements. To evaluate and compare classical and deep learning algorithms, this study leverages the BIDMC and CapnoBase datasets, employing the Respiratory Rate Estimation (RRest) toolbox. The optimal classical techniques combination on the BIDMC datasets achieves a mean absolute error (MAE) of 1.9 breaths/min. Additionally, the developed neural network model utilises convolutional and long short-term memory layers to estimate RR effectively. The best-performing model, with a 50% train–test split and a window size of 7 s, achieves an MAE of 2 breaths/min. Furthermore, compared to other deep learning algorithms with window sizes of 16, 32, and 64 s, this study's model demonstrates superior performance with a smaller window size. The study suggests that further research into more precise signal processing techniques may enhance RR estimation from PPG signals.

Keywords: photoplethysmogram; respiratory rate; deep learning; neural network

1. Introduction

Respiratory rate (RR) is a vital sign that provides basic information about a patient's respiratory condition and health status. Despite the development of new technical approaches, in clinical practice, RR is still manually counted or estimated using electrocardiogram (ECG) signals [1,2]. However, counting by hand is subjective and prone to mistakes, particularly in patients who are seriously ill or using mechanical breathing [2,3]. ECG signals need a lot of patient probing, which is inappropriate in many situations, including surgery and ambulance transfers [4].

Photoplethysmography (PPG) has gained popularity as a promising non-invasive method to calculate RR in recent years [3]. Photoplethysmography is easily acquired using a straightforward finger probe and measures variations in blood volume in peripheral tissues [5]. Photoplethysmography enables the simultaneous estimation of multiple physiological parameters using a single peripheral sensor, e.g., heart rate variability, oxygen saturation (SpO_2), blood pressure, and RR [5,6]. Photoplethysmography signals are modulated by respiration in amplitude, baseline, and frequency, enabling RR extraction in different approaches. PPG-based RR estimation has gained user acceptance by reducing the burden of wearing devices with several probes that record ECG to monitor RR for out-of-hospital patients continuously [2]. However, PPG signals are sensitive to noises and are influenced by multiple physiological factors, including age, measurement site, blood pressure, and neural activities [7–9]. The challenge towards clinical application lies in the accurate, reliable, and robust estimation of RR from PPG data.

Several algorithms have been developed for RR estimation, each employing distinct techniques to achieve accurate results. Filter-based techniques form one approach, incorporating digital filtering [10,11], low-pass finite impulse response filter [12,13], Butterworth band-pass filter [10,14], and adaptive infinite impulse response notch filter [15,16]. These methods have demonstrated performance with a mean absolute error (MAE) of 2.2 breaths/min, utilising a 5 s window size on the BIDMC datasets. Another approach involves feature-based techniques, encompassing Variable-Frequency Complex Demodulation [10,17], continuous wavelet transforms [10,17,18], Autoregressive Modeling [18–20], Beat Detection [21,22], and pulse segmentation [11,21]. State-of-the-art classical algorithms utilising feature-based techniques and time domain RR estimation techniques have achieved an MAE of 2.08 breaths/min with the same 5 s window size. Additionally, feature fusion and selection techniques can be implemented to consolidate the results from different feature-based methods or choose the most suitable technique for each dataset, potentially reducing the MAE to 1.95 breaths/min with a 5 s window size [23].

Numerous features in the signal domain have been extensively employed for RR estimation. However, the investigation into classical methods encounters a bottleneck, as enhancing the performance of RR estimation through signal features proves challenging [23]. Despite several studies employing deep learning algorithms, their performance has yet to surpass that of classical methods. Notably, current deep learning algorithms necessitate large window sizes for optimal performance, indicating a substantial gap in achieving superior results compared to classical methods [24]. Considering these observations, there is a compelling need to delve deeper into exploring deep learning algorithms for RR estimation, aiming to bridge the existing performance gap and potentially surpass the efficacy of classical methodologies.

In addition to that, ongoing research focuses on deep learning neural networks, with notable models like RRWaveNet ResNet, RespWatch, CycleGan and RespNet [24–28]. The state-of-the-art neural network is RRWaveNet, which has pushed the performance even further, achieving an MAE of 1.62 breaths/min and 1.59 breaths/min for BIDMC and CapnoBase Datasets, respectively. However, it utilises a longer 32 and 64 s window size [24]. The RRWaveNet algorithm employs a sophisticated architecture featuring multiple convolutional neural networks (CNNs). Signal data for its training and evaluation are sourced from the BIDMC and CapnoBase datasets through oximeter recordings [24]. This algorithm demonstrates commendable performance in terms of MAE, showcasing its efficacy in accurately estimating respiratory rates. Using diverse datasets enhances its robustness and generalisability, making RRWaveNet a promising and effective model for respiratory rate estimation.

The present study focused on RR estimation from PPG signals. A vital parameter is the window size, which represents the signal duration required for RR estimation. For instance, a 10 s window size provides RR readings after 10 s of applying the PPG sensor, offering real-time estimations with the average RR over the specified window. Shorter window sizes contribute to more immediate results, particularly crucial when a patient's breathing rate undergoes sudden changes. In contrast, larger window sizes, such as 32 or 64 s, delay the identification of abrupt respiratory rate drops. Given the importance of timely RR monitoring, especially for pulmonary patients at risk of brain damage after 4 min of oxygen deprivation, achieving a low window size becomes paramount in terms of developing an effective RR estimation algorithm [2].

The predominant optimisation of convolutional neural networks (CNNs) characterised algorithms aiming to comprehend signal patterns for RR estimation, with various architectures explored for signal learning and RR estimation. In the current study, a distinctive approach has been taken by introducing a Long Short-Term Memory (LSTM) layer to the architectural configuration [29]. Notably, the preceding studies had yet to investigate the performance of the LSTM layer in this context. As a type of recurrent neural network, the LSTM layer possesses the ability to analyse sequential input, optimising and preserving long-term relationships within the sequence. This is particularly crucial to our study's objective to reduce the window size. Consequently, our study focuses on exploring a combined neural network, integrating both CNN and LSTM layers, to uncover its potential impact on improving RR estimation accuracy through a nuanced analysis of signal patterns and long-term relationship recognition within the data sequence.

The performance of PPG-based RR estimation can still be improved by reducing the MAE. Additionally, there is potential to decrease the window size for more precise and real-time estimation. The potential for neural networks to improve RR estimation is vast and largely untapped in current research. Numerous neural network architectures and configurations are yet to be explored for this purpose. Moreover, there is room for innovation in reducing the window size for more real-time and responsive estimation, which could significantly enhance performance. The field of RR estimation remains open to a wide array of neural network possibilities, underscoring the need for further exploration and innovation in this domain to achieve more accurate and efficient results.

This study aims to develop an algorithm that can estimate RR accurately and reliably from PPG signals, where the best algorithmic parameters are identified. The performance in different window lengths was assessed to provide a reference for PPG-based real-time measurement of RR.

2. Materials and Methods

In this study, both classical and deep learning algorithms were optimised to estimate RR from PPG signals. A classical algorithm was employed to assess the performance of different classic techniques, determining the optimal combination for RR estimation through rigorous evaluation. Following the performance evaluation of the classical method, a deep learning algorithm was developed to enhance performance compared to previous iterations. The entire process, encompassing the classical and deep learning algorithms, is visually represented in Figure 1, illustrating the roadmap followed in this study for RR estimation.

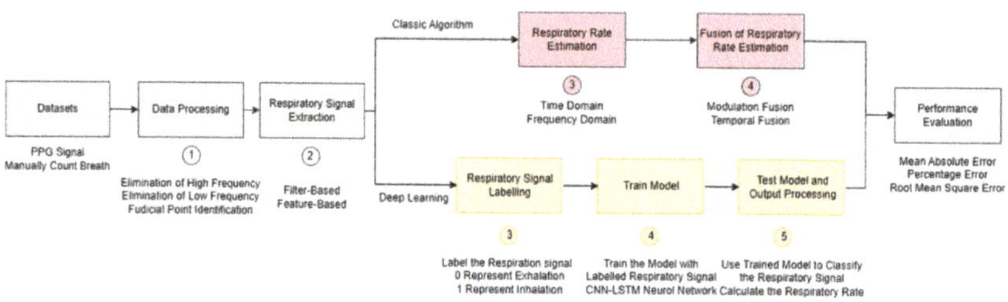

Figure 1. The process of classical algorithm and deep learning algorithm.

2.1. *Source of Data*

2.1.1. BIDMC Dataset

The BIDMC datasets are a collection of respiratory signals from intensive care patients' ECG, PPG, and impedance plethysmography [30,31]. The datasets are meant to be used to analyse RR algorithm performance, representing how well they might work in a real-world critical care setting [30,31].

The primary data for this study were obtained from severely ill patients at the Beth Israel Deaconess Medical Centre in Boston, MA, USA. For each recording in the dataset, individual breaths were meticulously annotated by two individuals using the impedance respiratory signal. The dataset comprises 53 recordings from different patients, each lasting

8 min [32]. These datasets encompass various physiological signals sampled at 125 Hz, including the ECG, impedance respiratory signal, and PPG. Physiological parameters such as heart rate, RR, and SpO_2 are sampled at 1 Hz. Additionally, the datasets contain manually annotated breaths and fixed parameters like age and gender [21]. The manually annotated breaths were used as reference RR in this study.

2.1.2. CapnoBase Dataset

The CapnoBase dataset comprises 8 min recordings from 42 subjects, including 13 adults and 29 children and neonates, which were obtained during elective surgeries and routine anaesthesia [33]. These recordings include ECG and PPG signals and capnometry signals measuring inhaled and exhaled carbon dioxide (CO_2), serving as a reference for breathing rate. The CapnoBase TBME RR benchmark dataset provides CO_2 waveforms (capnogram) and PPG data, along with expert-provided labels for pulse peaks in PPG and breaths in CO_2 [33]. Researchers can use this benchmark dataset to test and compare algorithms.

2.2. *Classical Respiratory Rate Estimation Algorithm*

2.2.1. Respiratory Rate Estimation Toolbox

The Respiratory Rate Estimation (RRest) toolbox is a MATLAB-based toolbox offering a comprehensive array of algorithms designed for estimating RR from physiological signals [32]. This toolbox is an integral part of the larger RR Estimation project and encompasses a wide range of algorithms that have been documented in prior literature. Rrest is adept at estimating RR using windows of ECG and pulse oximetry (PPG) signals while providing a reference RR derived from concurrent respiratory signals, such as Impedance Pneumography data. It serves as a valuable resource, particularly when researchers wish to compare their novel RR algorithms with established ones [21]. Rrest offers a library of validated algorithms and streamlines the statistical analysis of algorithm performance, facilitating the computation of various statistics commonly used for performance evaluation [32].

2.2.2. Respiratory Signal Extraction

There are two main respiratory signal extraction method categories: feature-based extraction and filter-based extraction. Techniques used in filter-based extraction include wavelet decomposition and band-pass filtering to remove non-respiratory frequencies. A measurement, such as pulse wave amplitude, is taken out of each cardiac cycle as part of feature-based extraction [34] (Figure 2).

The original signal was extracted to produce a respiration signal, which is a time series dominated by breathing modulation. Eliminating extremely low frequencies was the first stage, which was universal to all strategies. A high-pass filter with a 3 dB cutoff frequency of 4 breaths/min was used to achieve this [35]. Both feature-based and filter-based respiratory signal extraction techniques were employed, which dictated the intermediate steps. The next stage, which was the same for all approaches, was employing a band-pass filter with 3 dB thresholds of 4 and 60 breaths/min to exclude frequency unrelated to respiration from the reconstructed signal [36]. The following phases comprised the feature-based extraction process, which involved extracting a temporal sequence of beat-by-beat feature readings. Using low-pass filters with 3 dB thresholds from 100 and 35 Hz, extremely high frequencies were removed for the PPG signal [35].

This algorithm applied adaptive pulse segmentation and artifact detection for PPG. To accomplish automated heart rate and artifact detection, the periodic component of the PPG signal must be removed. The maximal volume peak optimisation of the regular heartbeat pulsations is sometimes followed by a second peak known as the dicrotic notch [37]. A line segmentation method processes the waveform and divides the PPG into pulses. This approach effectively enables the desired trend computation since PPG pulse components are based on a morphological shape that successive lines can identify.

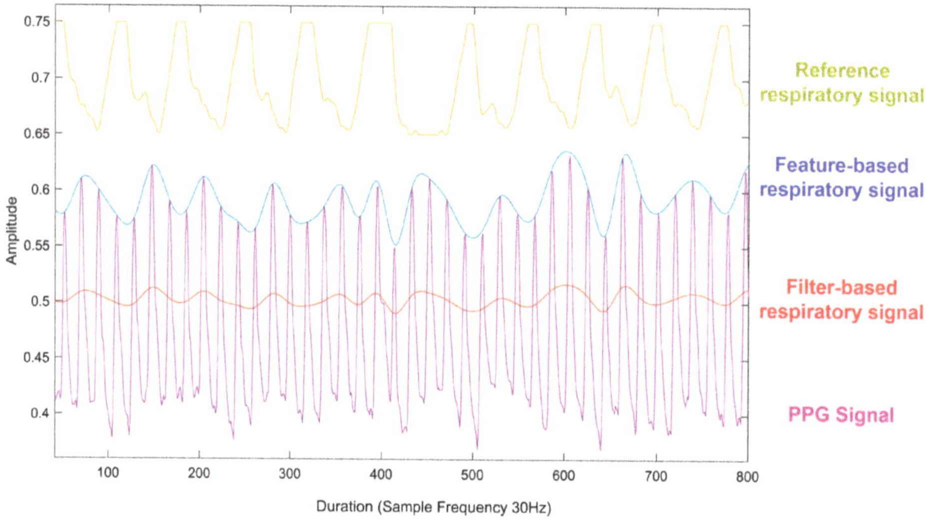

Figure 2. Respiratory signal extraction.

The Iterative-End-Point-Fit and Incremental algorithms that make up the Incremental-Merge Segmentation (IMS) technique were initially created for computer vision and mobile robotics applications [36]. The segmentation-assisted ECG signal compression method has also been applied. The IMS algorithm may be implemented in real time and is simple, quick, and based on a sliding-window structure [37]. The irregularly sampled feature-based respiratory signals are then resampled using linear, interpolation, cubic spline interpolation, or Berger's technique at a constant sampling rate. Band-pass filtering is an option that can be used after these methods if desired [21].

Linear interpolation is a method for fitting the points to a curve using a linear polynomial, such as the line equation. This is equivalent to drawing a line between two points in a dataset to connect them. The process of fitting the curve by connecting those points with a higher degree polynomial is known as polynomial interpolation. Low-degree polynomials are used in each of the intervals in spline interpolation, which is like polynomial interpolation x′ in that it selects the polynomial parts to fit together smoothly [38]. The outcome is a function known as a spline. Using cubic spline interpolation, finding a curve that connects data points with a degree of three or fewer is possible. Splines are polynomials that have continuous first and second derivatives and are smooth and continuous across a specified plot [38].

2.2.3. Respiratory Rate Estimation

Respiratory Rate Estimation was performed after RR signal extraction using frequency-based techniques and time domain detection. Frequency-based pinpoint the frequency component associated with respiration. Time domain detection includes peak detection on respiratory signal and identifying zero-crossings with a positive gradient along the RR signal. The RR was found to correlate to the frequency of the spectral peak with the largest magnitude from 4 to 60 breaths/min using spectral analysis.

The fusion stage has drawn much attention lately because of the gains in algorithm performance seen when it is used. To combine simultaneous RR estimates for each modulation that were generated using the feature-based methodology, two modulation fusion algorithms were optionally optimised. Smart fusion is the anticipated RR signal's quality, which is assessed using the BW, AM, and FM respiratory signals. If the standard deviation is less than 4 breaths/min, the RR is calculated as the mean; otherwise, no RR is produced [35].

Spectral peak-conditioned averaging optimises the Welch periodogram, and frequency spectra generated from BW, AM, and FM respiratory signals are fused to provide a mean spectrum. Only spectra that include a specific percentage of their spectral power in a frequency range centred on the frequency corresponding to their maximum spectral power are included. The frequency that corresponds to the mean spectrum's highest power is thought to be RR [35]. It was possible to employ temporal fusion to smooth out subsequent RR estimations from the same person. It can be used with or without modulation fusion beforehand [35].

2.3. Neural Network for Respiratory Rate Estimation

A deep learning algorithm has been developed to estimate RR. BIDMC datasets have been used to develop the deep learning algorithm. The datasets were cut from 53 eight-minute data to 424 one-minute data for training and testing. The algorithm combines a CNN and a LSTM neural network. The convolutional layer was used as the first layer to classify the RR signal. A convolutional neural network can extract features from both spatial and temporal dimensions. A Long Short-Term Memory network is a type of recurrent layer that can improve the prediction ability of the deep learning model. Long Short-Term Memory network operates on sequential data by iterating through time steps and capturing intricate relationships between them, facilitating the learning of long-range dependencies within the sequence. The input of the deep learning model was a 30 Hz RR signal, and the output was the labelling of inhalation and exhalation.

2.3.1. Deep Learning Algorithm

Similar data processing and RR signal extraction were applied before the deep learning algorithm to obtain a clean RR signal. To reduce the training time, the RR signal was resampled from 125 Hz to 30 Hz, which retained the signal's pattern and reduced the input data's size. Then, a severe baseline difference was found due to the different blood pressures of the patients. Therefore, the zero-score method was applied to remove the baseline difference and rescale all the RR signals between negative two and positive two.

The labelling process is carried out by cutting the zero as the midline. Any signal higher than zero is considered an inhalation signal and is labelled as one. Meanwhile, signals lower than zero are considered exhalation signals and are labelled as zero. Many RR signals have artifacts and fluctuations. Therefore, the labelling should be modified for the training model to learn all the artifacts and fluctuations. All the train and validation RR signals should be labelled for training.

After that, the labelled RR signal will be used as the input to train the deep learning model. The RR signal data were split into three parts: training, validation, and testing. The train and validation splits are 80% train data and 20% validation data. The output of the training model shall classify inhalation and exhalation signals. Next, RR was computed based on the number of inhalation signals detected. The performance of this deep learning algorithm can be evaluated by comparing the computed RR to the reference RR. The performance evaluation method includes MAE, RMSE, and percentage error. The whole process of deep learning is shown in Figure 3. This process was applied similarly to the testing data.

The CapnoBase datasets serve as a crucial testing ground for neural network models initially trained on the BIDMC datasets. This evaluation allows researchers to assess and ascertain the model's performance when applied to diverse datasets beyond its training environment. The objective is to ensure that the model can demonstrate its efficacy across a range of conditions and with various patient demographics, ultimately verifying its suitability for broader clinical applications.

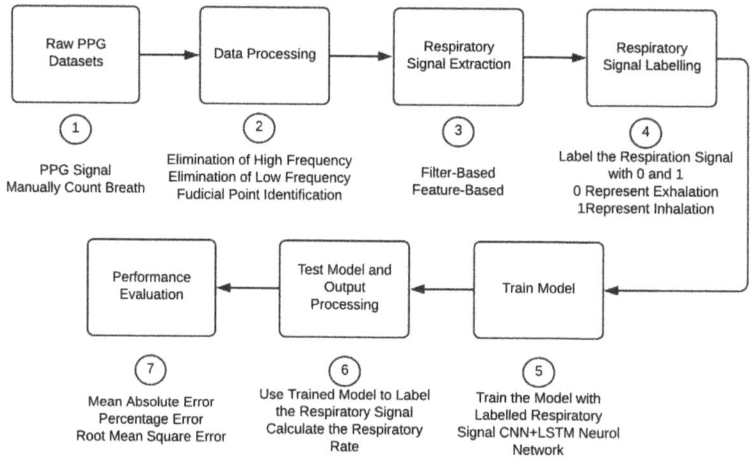

Figure 3. Deep learning algorithm flow chart.

2.3.2. Deep Learning Neural Network Architecture

Clean and scaled 30 Hz RR signals and the labels were fed as input to the CNN-LSTM model. The architecture of the CNN-LSTM neural network is shown in Figure 4.

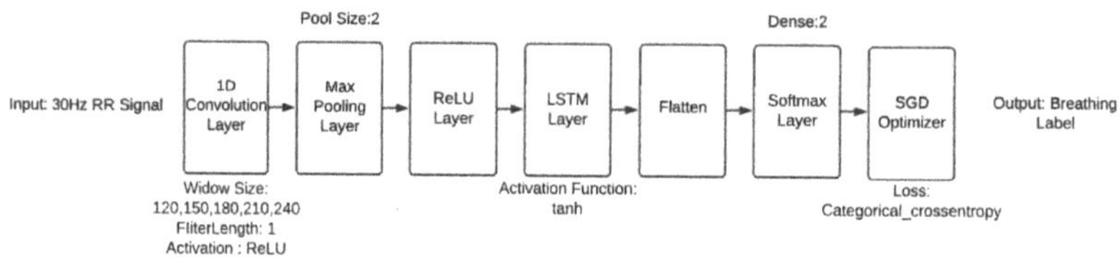

Figure 4. CNN-LSTM neural network architecture.

The neural network takes a 30 Hz labelled RR signal as input. The convolution layer learns from the input signal. Pooling layers offer a technique for reducing the dimensionality of feature maps by condensing information about the presence of features within patches of the feature map. Max pooling will capture the most activated presence of a feature. The Rectified Linear Activation Function, commonly referred to as ReLU, is a mathematical function that behaves in a piecewise linear manner. When the input is positive, it directly outputs the same value; otherwise, it outputs zero. ReLU has gained widespread popularity as the default activation function in various neural network architectures due to its ease of training and its tendency to yield improved performance in many cases.

The model uses LSTM as the recurrent neural network. It is a fundamental building block to analyse sequential input, such as time series or text written in natural language. It can optimise and retain long-term relationships between sequences. The flattened layer is a vital bridge between CNNs and artificial neural networks. It facilitates the seamless integration of CNNs with ANNs, enabling the neural network models to acquire a deep understanding of intricate patterns in data and subsequently make accurate predictions.

The final layer of the mode is the SoftMax layer. It is a significant layer of the neural network. It transforms the prior layer's raw scores or logits into probability distributions across many classes. This deep learning model gives the label probability on two classes: inhalation and exhalation. Then, the output layer gives the label as 1 (inhalation) or 0 (exhalation). After the classification, the output processing step estimates RR by calculating the number of inhalations and exhalations.

The last part of the model is stochastic gradient descent (SGD) optimisation, a widely used optimisation algorithm used in deep learning. It deals with the classic Gradient Descent methods' computational inefficiencies while working with huge datasets in deep learning. Utilising SGD, the enormous expense of backpropagation over the whole training set is the driving force behind neural networks in this context. SGD can offset this expense and result in rapid convergence to the optimal.

2.3.3. Deep Learning Neural Network Optimisation Parameters

Several parameters can be modified to optimise the deep learning model performance. The first parameter to optimise is the train–validation and test split. In this study, train–validation data and test data are split by three ratios, which are 20:80, 50:50, and 80:20. Within the train–validation, the train data would be 80%, and the validation data would be 20%. In addition to that, five window sizes were used in the optimisation, which are 120, 150, 180, 210 and 240, corresponding to 4 to 8 s signals, respectively.

In addition, various parameters were optimised to achieve the best model. The CNN layers were tested from one to three, revealing that while additional layers extended training and testing times, they did not significantly enhance performance. Filter lengths ranging from 1 to 5 were examined, with shorter lengths demonstrating better results without substantially increasing processing time because of the downsampling from 125 Hz to 30 Hz in data preprocessing. Pool sizes of 2 to 5 were tested, with larger sizes reducing processing time but adversely impacting performance. Lastly, batch sizes within the range of 5, 10, 15, 20, 25, and 30 were examined, with the determination that the most favourable balance between computational efficiency and optimal performance in respiratory rate estimation was provided by a batch size of 20. This exhaustive exploration across various batch sizes aimed to identify the specific configuration that maximises computational resource utilisation while ensuring excellence in the estimation process.

Training and evaluation were carried out on Intel(R) Core (TM) i5-10300H CPU with 8 GB of RAM hosted by Nvidia GeForce GTX 1650 GPU. In each experimental run, 100 training epochs with a batch size of 20, a pool size of 2, and a filter length of 1 were applied. To mitigate over-fitting, an early stopping technique was implemented during the training phase. The model that achieved the lowest validation loss was preserved and subsequently employed for testing.

2.4. *Performance Evaluation*

Three parameters were used to evaluate the algorithm's performance. It includes MAE, percentage error, and root mean square error (RMSE). For the performance evaluation, the lower the value, the better the performance in RR estimation. The formula of the performance evaluation parameter is shown in the equation below.

$$\mathrm{MAE} = \frac{1}{\mathrm{N}}\sum_{i=1}^{\mathrm{N}}|x_i - \hat{x}_i|(\mathrm{breaths/min}) \quad (1)$$

$$\text{Percentage error} = \frac{\text{Estimate RR} - \text{Reference RR}}{\text{Reference RR}} \times 100\% \quad (2)$$

$$\mathrm{RMSE} = \sqrt{\frac{1}{\mathrm{N}}\sum_{i=1}^{\mathrm{N}}|x_i - \hat{x}_i|^2}(\mathrm{breaths/min}) \quad (3)$$

where x_i is a reference value and $\hat{x}_i$ is an estimated value of the signal, and N is the total number of samples in the signal.

3. Results

3.1. *RRest Toolbox Algorithm That Utilised Fusion Method*

In the 150 combinations that included fusion techniques, the MAE values ranged from 1.95 breaths/min in the best case to 35.06 breaths/min in the worst. The measurements show a percentage error of 10.89%, an RMSE of 2.82 breaths/min, and a CP2 of 66%.

These combinations include mean Peak Amplitude as a feature selection technique, breath detection via combined trough and peak detection as the RR estimation techniques and temporal smoothing as fusion techniques (Table 1).

Table 1. Top 10 technique combination performance with fusion.

Rank	Combination	MAE	Percentage Error	RMSE
1	Peak Amplitude, Trough and Peak Detection, Temporal Fusion	1.95	10.89	2.82
2	Peak Amplitude, Positive Gradient Zero-Crossing Detection, Temporal Fusion	2.03	11.35	2.88
3	Trough and Peak Detection, Smart Fusion, Temporal Fusion	2.04	11.38	2.62
4	Consecutive Trough Mean Value, Trough and Peak Detection, Temporal Fusion	2.04	11.40	2.88
5	Consecutive Trough Mean Value, Positive Gradient Zero-Crossing Detection, Temporal Fusion	2.04	11.43	2.89
6	Peak Amplitude, Positive Gradient Zero-Crossing Detection, Temporal Fusion	2.10	11.77	3.35
7	Peak Amplitude, Trough and Peak Detection, Temporal Fusion	2.13	11.90	3.39
8	Consecutive Trough Mean Value, Detrend and Detect Trough and Peak, Temporal Fusion	2.15	12.00	3.11
9	Consecutive Trough Mean Value, Spikes and Drop Detection, Temporal Fusion	2.16	12.08	2.87
10	Positive Gradient Zero-Crossing Detection, Smart Fusion, Temporal Fusion	2.21	12.37	2.87

MAE—mean absolute error. RMSE—root mean square error.

3.2. Deep Learning Neural Network Optimisation

The hyperparameter for the CNN-LSTM neural network has been optimised. The final neural network contains a 1-D convolution layer, a max pooling layer, a ReLU layer, an LSTM layer activated by tanh, a flatten layer, a SoftMax layer and an SGD optimiser layer. The optimal model uses a pool size of 2, a filter length of 1, a window size of 210 and a 50% train–test split. The performance of the best model has achieved an MAE of 2.02 breaths/min, a percentage error of 11.82% and an RMSE of 2.6 breaths/min.

In Figure 5, the chart shows the variation in loss of validation data over the training epochs with 210 data points or 7 s as the window size. It is worth noting that the loss starts to exhibit fluctuations after the 30th epoch, indicating the stabilisation of the model's performance. To select the most suitable model for testing, consider the model with the lowest loss over the course of 100 epochs. This rigorous methodology led to the identification of an optimal model, achieving an impressively low loss of 0.098; this exceptional performance milestone was reached precisely during the 74th epoch.

3.2.1. Performance of Different Window Sizes

Table 2 illustrates the relationship between window size, performance, and processing time in the context of RR estimation. Notably, it reveals that as the window size increases, performance also improves, reaching an optimum at a 210 window size. However, it is essential to note that this improvement comes at the cost of increased training and test times. A saturation point is observed in the training process when the window size is extended to 210 and further to 240. Consequently, the decision was made to halt the window size expansion at 240, recognising that pushing it beyond this point yielded diminishing returns, likely due to computational constraints or other limiting factors.

Table 2. Performance of different window sizes.

Window Size	MAE	Percentage Error (%)	Training Time (50%)	Testing Time (50%)
120 (4 s)	3.94	23.9	14 h	3 min
150 (5 s)	3.45	20.39	20 h	5 min
180 (6 s)	2.76	16.16	25 h	7 min
210 (7 s)	2.42	14.42	36 h	9 min
240 (8 s)	2.49	15.09	55.5 h	13 min

MAE—mean absolute error.

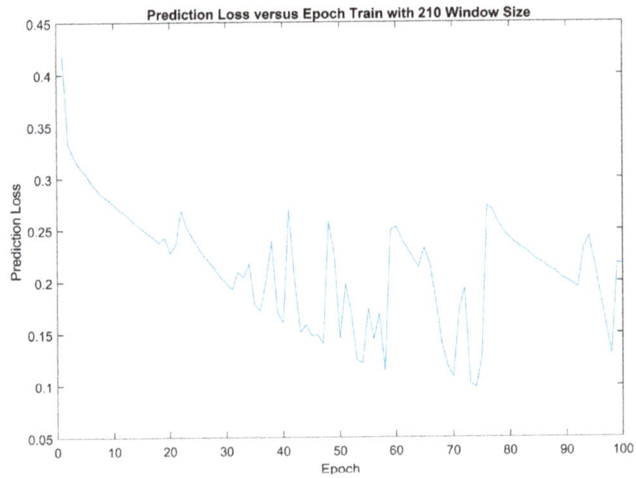

Figure 5. Prediction loss versus epoch train with a 210 window size.

The model with the lowest loss over the course of 100 epochs was selected as the model for testing. Therefore, the optimal model was selected during the 74th epoch of a 210 window size, which achieved a loss of 0.098. This underscores the model's stability and its readiness for thorough testing and evaluation.

3.2.2. Performance of Different Train–Test Split

Interestingly, it was observed that splitting the data into a 50% training and 50% testing configuration produced outcomes comparable to those achieved with an 80% training split. Consequently, opting for a 50% training split streamlines the training process and reduces computational demands, making it a practical choice for model optimisation. This fine calibration of the train–test split parameter is critical for ensuring effective learning and the feasibility of the model development effort. Table 3 shows different train–test split performance on 210 window sizes.

Table 3. Performance of different train–test split.

	Training		Testing	
Train:Test	**MAE**	**Time**	**MAE**	**Time**
20:80	1.34	16 h	4.27	3 min
50:50	1.3	36 h	2.42	9 min
80:20	1.47	52 h	2.57	12 min

MAE—mean absolute error.

4. Discussion

4.1. Performance of Classical Algorithm

The initial phase of respiratory rate (RR) signal extraction involves employing either feature-based or filter-based techniques. Feature-based methods encompass key aspects like Amplitude Modulation, Frequency Modulation, Baseline Wander, Peak Amplitude, and Trough Amplitude [23,39]. On the other hand, filter-based techniques encompass band-pass filtering, filtering using the centred-correntropy function, and wavelet extraction [23]. Subsequently, RR estimation is performed on the extracted RR signal utilising both frequency domain techniques (e.g., Fourier transform and auto-regressive spectral analysis) and time domain techniques (e.g., peak detection and Zero-crossing breath detection). Fusion methods, including smart fusion and temporal fusion, play a crucial role in refining the RR signal. Smart fusion adaptively selects from Amplitude Modulation, Frequency Modulation, and Baseline Wander to combine their features for enhanced RR signal estimation [23]. Temporal fusion is employed to smooth out the RR signal from the same individual, ensuring a more stable and reliable respiratory rate estimation.

This study's comprehensive exploration of respiratory rate estimation using classical techniques involved testing 274 combinations. Out of these, 150 combinations incorporated fusion methods, revealing that using fusion techniques significantly enhanced performance. The evaluation of various classical method combinations allowed for identifying the most effective technique, shedding light on the optimal approaches for accurate RR estimation. This systematic assessment underscores the importance of fusion methods in refining classical techniques and highlights the specific combinations that yield superior performance in RR estimation. Table 1 shows the top 10 best-performing algorithms with the fusion method ranked with MAE value. All top 10 algorithms used the time domain techniques for RR estimation, indicating that the time domain techniques have better performance than frequency domain techniques in RR estimation with the fusion method in this dataset.

4.2. Performance Analysis in Deep Learning Model

Compared with window size, it is worth noting that this 7 s window size aligns closely with conventional algorithms, which typically employ a 5 s window. However, it is vital to recognise that the choice of window size impacts both the model's training and testing times significantly. For instance, a 120-sized window necessitates approximately 14 h for training and a mere 3 min for predictions on 200 min of signal data. In contrast, a 240-sized window extends the training time considerably, requiring roughly 55.5 h, with predictions for the same 200 min of signal data taking approximately 13 min. This underscores the intricate trade-off between window size, training duration, and the responsiveness of the algorithm in real-time scenarios.

In comparison with the train–test split, the allocation of data into training and testing sets is a crucial parameter in model development. Striking the right balance is essential, as having too little data for training can hinder the model's ability to grasp intricate patterns and may result in missing out on essential scenarios not adequately represented in the training set [25]. Conversely, an overly large training dataset can significantly extend the training time, necessitating substantial computational power and potentially leading to kernel disconnections in platforms like Visual Studio Code and runtime errors in platforms like Google Colab.

4.3. Performance Comparison with the State of the Art

Three contemporary algorithms were found to estimate RR. The first algorithm was the RRWaveNet Model, a novel deep learning technique recently published in the field and achieved an MAE of 1.62 with a 32 s window size. The second algorithm was the RRest Toolbox, a powerful tool designed for the comprehensive assessment of classic techniques in RR estimation. Through 274 different combinations of these techniques, the lowest MAE of 1.95 was attained. Lastly, within the scope of this project, an algorithm was developed that combines CNN with LSTM networks. This unique model delivered an MAE of 2.42 under optimal conditions.

However, the testing phase encountered several challenges. As compared to a full signal set of 424 min applied to the RRest Toolbox, only 217 min of signals were utilised during the deep learning testing phase due to the fact that a 50:50 train–test split was applied, which is considerably less than the 424 min used for the RRest Toolbox. Additionally, 12 non-consecutive 1 min signal segments exhibited severe fluctuations, significantly affecting the model's performance. Strikingly, upon removing these troublesome signal segments, a substantial enhancement in the model's accuracy was observed, where MAE was reduced from 2.42 to 2.02 with a low window size of 7 s.

To ensure an equitable comparison, identical data preprocessing steps were applied to all datasets and the classic methods were evaluated using the RRest Toolbox. These classical techniques achieved an MAE of 1.90, which held its own in contrast to the outcomes after eliminating fluctuating signal segments. This observation suggests that deep learning algorithms harbour untapped potential for further exploration. In summary, the findings in this study imply that, with meticulous data preprocessing and rigorous signal quality control, deep learning methods can perform on par with well-established classical techniques in this specific domain. Expanding the variety of datasets during the training of deep learning methods has been shown to enhance model performance [25]. The model exhibits improved capabilities when exposed to a more diverse dataset. Additionally, having better computational power enables the incorporation of more extensive datasets, fostering a deeper understanding and refinement of the model. Moreover, using synthetic data becomes feasible when there is sufficient computational power, contributing to the augmentation of the training dataset [25] (Table 4).

Table 4. Performance of different algorithms on the BIDMC dataset.

Study	Method	Window Size (Seconds)	MAE	Processing Time
[26]	RespNet	16, 32, 64	2.45, 2.07, 2.06	-
[24]	RRWaveNet	16, 32, 64	1.87, 1.62, 1.66	-
[25]	ResNet	16, 32, 64	2.25, 2.46, 2.16	-
[27]	RespWatch	16, 32, 64	1.88, 1.96, 1.66	-
[28]	CycleGan	32	1.9	-
This Study	RRest Toolbox	5	1.95	2.5 h
This Study	RRest Toolbox (Removed)	5	1.90	1.5 h
This Study	CNN-LSTM Model	7	2.42	36 h, 9 min (Train, Test)
This Study	CNN-LSTM Model (Removed)	7	2.02	36 h, 9 min (Train, Test)

MAE—mean absolute error. RRest Toolbox (Removed)—Fluctuation signals are removed before applying to the RRest Toolbox. CNN-LSTM Model (Removed)—Fluctuation signals are removed before applying to the CNN-LSTM Model.

To assess the model's performance across diverse datasets, CapnoBase datasets were employed for testing, maintaining consistent data preprocessing and scaling techniques. The BIDMC dataset focuses on adults, predominantly adults over 50 years of age, in intensive care units. In contrast, the CapnoBase dataset encompasses a variety of ages, including both adults and children undergoing elective surgery and anesthesia. Both datasets are inclusive of both male and female patients. The CapnoBase dataset, with its broader representation of patient demographics and medical contexts, provides a more comprehensive testing ground for model robustness compared to the BIDMC dataset, which focuses more on a specific age group.

Remarkably, the model demonstrated effective RR estimation with a commendable MAE of 1.99 breaths/min. Further refinement was achieved by removing segments with severe signal fluctuations to mitigate artifacts, resulting in an even more impressive MAE of 1.24 breaths/min. The MAE achieved by this model exceeded the performance of the state-of-the-art deep learning algorithm, which is 1.59 breaths/min. This model exhibits several strengths, including its ability to operate with low window size, minimal processing time, and adaptability to perform well when applied to other datasets, underscoring its robustness and suitability for various applications (Table 5).

Table 5. Performance of different algorithms on CapnoBase datasets.

Study	Method	Window Size (Seconds)	MAE	Processing Time
[26]	RespNet	16, 32, 64	5.4, 4.86, 4.48	-
[24]	RRWaveNet	16, 32, 64	1.79, 1.86, 1.59	-
[25]	ResNet	16, 32, 64	2.62, 2.36, 2.29	-
[27]	RespWatch	16, 32, 64	1.87, 2.09, 1.82	-
This Study	RRest Toolbox	5	0.93	2.5 h
This Study	RRest Toolbox (Removed)	5	0.90	1.5 h
This Study	CNN-LSTM Model	7	1.99	7 min (Test)
This Study	CNN-LSTM Model (Removed)	7	1.24	7 min (Test)

MAE—mean absolute error. RRest Toolbox (Removed)—fluctuation signals are removed before applying to the RRest Toolbox. CNN-LSTM Model (Removed)—fluctuation signals are removed before applying to the CNN-LSTM Model.

When comparing various deep learning algorithms, it becomes evident that window size and MAE are pivotal parameters signifying algorithm performance, with improvements noted as more studies emerge. While increasing the window size tends to result in a better MAE, the significance of a smaller window size for rapid responsiveness to changes in RR cannot be understated. Despite RRWaveNet's higher MAE, optimal performance requires a larger window size, only achieving lower MAE values when using 32 and 64 s windows [24]. In this study, although the MAE is slightly higher than other deep learning algorithms, the window size is remarkably reduced to only 7 s, providing efficient RR estimation within a shorter timeframe for faster response. Balancing window size and MAE poses a significant challenge in algorithm development, often requiring trade-offs. The current study incrementally improves window size and MAE, recognising the ongoing opportunity for exploration in uninvestigated neural network layers and architectures with the potential for enhanced algorithm performance.

4.4. Fluctuation Signal Elimination

Eliminating fluctuation signals is crucial for ensuring the accuracy of respiratory rate (RR) estimation, as motion artifacts, light disruptions, and noise can adversely affect signal quality, leading to the emergence of fluctuation signals. In the BIDMC datasets, 12 one-minute signal segments with severe fluctuation signals were identified and removed due to their potential to introduce extremely large errors. Such errors could result in a reported RR of 30 breaths/min, while the reference RR was only 14 breaths/min. To attain a more precise estimation, it is imperative to eliminate fluctuation signals that may arise during instances of motion artifacts, light disruptions, or noise, thereby enhancing the overall accuracy of RR measurements. Figure 6 shows the comparison between the normal signal and the fluctuation signal.

4.5. Clinical Application, Limitations, and Future Directions

4.5.1. Clinical Application

The clinical application of RR estimation from PPG signals, as demonstrated by the findings in this study, holds promising implications for patient monitoring in various medical conditions [40]. A single-probe oximeter recording PPG signals proves to be an efficient tool for long-term patient condition monitoring, offering a distinct advantage over ECG, which typically requires multiple probes. The PPG-based device's simplicity, portability, and user-friendly nature make it a practical choice in clinical settings.

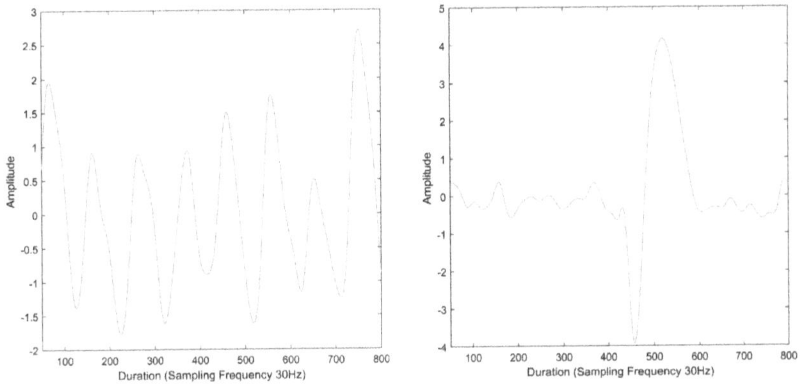

Figure 6. Normal signal (**Left**); fluctuation signal (**Right**).

In a clinical setting, RR estimation burdens patients and healthcare providers, often requiring multiple probes from ECG or manual recording and monitoring, particularly in ambulatory services. However, implementing an accurate, fast-response, and low-processing requirement RR estimation algorithm into a bedside monitor and a simple oximeter offers a transformative solution [1]. With this approach, all RR data can be easily recorded and monitored, streamlining the process and significantly reducing the time and effort involved [1]. Such an algorithm enables the swift identification of abnormal RR, facilitating prompt responses and interventions when needed, ultimately improving the efficiency and effectiveness of RR monitoring in clinical scenarios.

For chronic pulmonary diseases such as Chronic Obstructive Pulmonary Disease and obstructive sleep apnea, continuous and reliable RR monitoring is crucial for disease management and early detection of exacerbation. In these cases, the low window size and real-time responsiveness of the deep learning model in the study can significantly contribute to effective patient care. In acute pulmonary diseases like acute respiratory distress syndrome or pneumonia, rapid and accurate monitoring of RR is vital for timely intervention. The fast and reliable response the PPG-based device provides with the developed algorithm becomes particularly critical in such acute scenarios, ensuring the prompt detection of RR changes.

4.5.2. Limitations

In this study, limitations exist in the domain of RR estimation from PPG signals, with computational constraints being a notable challenge. Model training and development are hindered by limitations in computational power, often resulting in interruptions such as kernel disconnections that halt the training process. With improved computational capabilities, model training could be more robust and efficient, allowing for the inclusion of more extensive datasets and achieving better performance with reduced errors.

Another limitation arises from the availability of open-source datasets, which may not be sufficiently comprehensive to fully develop and refine the algorithm. The need for a wider variety of datasets is evident, as a more diverse range of data is essential for enhancing the algorithm's generalisation and overall accuracy. In essence, overcoming these limitations hinges on advancements in computational resources and expanding diverse datasets to foster the continual improvement and applicability of respiratory rate estimation algorithms based on PPG signals.

4.5.3. Future Directions

Guiding the future progression of the project for respiratory rate estimation based on PPG signals involves strategic considerations. Addressing computational constraints by exploring enhanced resources or refined algorithms is a pivotal focus for advancing model training. This may encompass utilising advanced hardware, cloud-based computing, or implementing streamlined parallel processing methods to surmount the existing limitations.

Future advancements in artifact identification and reconstruction are essential for refining signal quality and optimising RR estimation performance. To address the impact of artifacts on signal accuracy, innovative approaches may involve advanced signal processing and dedicated deep learning methods for effective artifact detection and reconstruction. Developing robust algorithms capable of distinguishing PPG signals from artifacts and efficient reconstruction techniques is critical for enhancing RR estimation.

Equally essential is the expansion of the dataset repository. Establishing partnerships with healthcare institutions or gaining access to more diverse datasets is imperative for comprehensive algorithmic development. This approach fortifies the model's resilience and adaptability, ensuring its efficacy across diverse patient profiles and health conditions.

Furthermore, continual enhancement of the algorithm, particularly in terms of real-time responsiveness and reduced window size, is critical for practical applicability. Integrating more sophisticated signal processing techniques or exploring emerging technologies can provide avenues to further optimise the algorithm's efficacy.

5. Conclusions

In conclusion, the project's goal of creating a reliable algorithm for accurately estimating RRs from PPG was achieved through a series of well-defined objectives. These objectives encompassed fine-tuning algorithmic parameters, validating accuracy using diverse PPG datasets collected under varying physiological and environmental conditions, enhancing robustness against disturbances such as noise and motion artifacts, and optimising real-time processing to ensure swift and precise RR predictions. This project's results are of utmost significance for healthcare and monitoring applications, where the accurate estimation of RRs is pivotal in patient care and assessment. In evaluating classical methods, the performance was accurately assessed with 274 different technique combinations. The optimal algorithm emerged, featuring mean Peak Amplitude as the chosen feature selection method, combined trough and peak detection for RR estimation, and temporal smoothing as the fusion technique. Remarkably, this configuration achieved an impressive MAE of 1.9.

A CNN + LSTM neural network was also successfully developed, yielding a slightly higher MAE of 2.02. The model can also be applied to other datasets. The model has been applied to CapnoBase datasets and achieved an MAE of 1.24. This verified the model performance on new PPG raw data. These results underline the comprehensive nature of our project and its ability to excel in both classical and advanced techniques, reinforcing the significance of our findings for healthcare and monitoring applications where precise RR estimation is paramount.

Author Contributions: Conceptualization, B.-H.K. and C.-H.G.; Methodology, W.J.C. and C.-H.G.; Software, Y.K.T.; Validation, W.Y.L. and C.-H.G.; Formal analysis, W.J.C. and Y.K.T.; Investigation, W.J.C., S.D. and H.L.; Resources, B.-H.K., W.Y.L., H.L. and C.-H.G.; Data curation, W.J.C., W.Y.L. and S.D.; Writing—original draft, W.J.C.; Writing—review & editing, B.-H.K., W.Y.L., Y.K.T., S.D., H.L. and C.-H.G.; Visualization, W.J.C. and W.Y.L.; Supervision, B.-H.K. and C.-H.G.; Project administration, C.-H.G. All authors have read and agreed to the published version of the manuscript.

Funding: This work was supported by the Ministry of Higher Education Malaysia under the Fundamental Research Grant Scheme (No. FRGS/1/2023/SKK05/UTAR/02/1 to CHG).

Institutional Review Board Statement: No Institutional Review Board approval is needed for this study. All data are obtained from online public databases.

Informed Consent Statement: No specific informed consent statement are obtained, but all data including informed consent should be conducted from the data source.

Data Availability Statement: Publicly available datasets were analyzed in this study.

Acknowledgments: The authors have full appreciation for the availability of public datasets from BIDMC PPG and Respiration Dataset and CapnoBase Dataset.

Conflicts of Interest: The authors declare that they have no known competing financial interests or personal relationships that could have appeared to influence the work reported in this paper.

References

1. Liu, H.; Allen, J.; Zheng, D.; Chen, F. Recent development of respiratory rate measurement technologies. *Physiol. Meas.* **2019**, *40*, 07TR01. [CrossRef]
2. Meredith, D.J.; Clifton, D.; Charlton, P.; Brooks, J.; Pugh, C.W.; Tarassenko, L. Photoplethysmographic derivation of respiratory rate: A review of relevant physiology. *J. Med. Eng. Technol.* **2011**, *36*, 1–7. [CrossRef]
3. Baker, S.; Xiang, W.; Atkinson, I. Determining respiratory rate from photoplethysmogram and electrocardiogram signals using respiratory quality indices and neural networks. *PLoS ONE* **2021**, *16*, e0249843. [CrossRef]
4. Stankoski, S.; Kiprijanovska, I.; Mavridou, I.; Nduka, C.; Gjoreski, H.; Gjoreski, M. Breathing Rate Estimation from Head-Worn Photoplethysmography Sensor Data Using Machine Learning. *Sensors* **2022**, *22*, 2079. [CrossRef]
5. Zhao, D.; Sun, Y.; Wan, S.; Wang, F. SFST: A robust framework for heart rate monitoring from photoplethysmography signals during physical activities. *Biomed. Signal Process. Control* **2017**, *33*, 316–324. [CrossRef]
6. Shuzan, M.N.I.; Chowdhury, M.H.; Chowdhury, M.E.H.; Murugappan, M.; Hoque Bhuiyan, E.; Arslane Ayari, M.; Khandakar, A. Machine Learning-Based Respiration Rate and Blood Oxygen Saturation Estimation Using Photoplethysmogram Signals. *Bioengineering* **2023**, *10*, 167. [CrossRef]
7. Hartmann, V.; Liu, H.; Chen, F.; Hong, W.; Hughes, S.; Zheng, D. Toward Accurate Extraction of Respiratory Frequency From the Photoplethysmogram: Effect of Measurement Site. *Front. Physiol.* **2019**, *10*, 732. [CrossRef]
8. Khalid, S.G.; Ali, S.M.; Liu, H.; Qurashi, A.G.; Ali, U. Photoplethysmography temporal marker-based machine learning classifier for anesthesia drug detection. *Med. Biol. Eng. Comput.* **2022**, *60*, 3057–3068. [CrossRef] [PubMed]
9. Gajbhiye, P.; Tripathy, R.K.; Bhattacharyya, A.; Pachori, R.B. Novel Approaches for the Removal of Motion Artifact From EEG Recordings. *IEEE Sens. J.* **2019**, *19*, 10600–10608. [CrossRef]
10. Iqbal, T.; Elahi, A.; Ganly, S.; Wijns, W.; Shahzad, A. Photoplethysmography-Based Respiratory Rate Estimation Algorithm for Health Monitoring Applications. *J. Med. Biol. Eng.* **2022**, *42*, 242–252. [CrossRef] [PubMed]
11. Karlen, W.; Raman, S.; Ansermino, J.M.; Dumont, G.A. Multiparameter Respiratory Rate Estimation From the Photoplethysmogram. *IEEE Trans. Biomed. Eng.* **2013**, *60*, 1946–1953. [CrossRef]
12. Icazatti, F.; Dell'Aquila, C.; Leber, E.L. Design and validation of a respiratory rate estimation algorithm based on photoplethysmography (PPG) signal. In Proceedings of the 2021 XIX Workshop on Information Processing and Control (RPIC), San Juan, Argentina, 3–5 November 2021. [CrossRef]
13. Tun, H.M. Photoplethysmography (PPG) Scheming System Based on Finite Impulse Response (FIR) Filter Design in Biomedical Applications. *Int. J. Electr. Electron. Eng. Telecommun.* **2021**, *10*, 272–282. [CrossRef]
14. Khreis, S.; Ge, D.; Rahman, H.A.; Carrault, G. Breathing Rate Estimation Using Kalman Smoother With Electrocardiogram and Photoplethysmogram. *IEEE Trans. Biomed. Eng.* **2020**, *67*, 893–904. [CrossRef] [PubMed]
15. Kim, H.; Kim, J.-Y.; Im, C.-H. Fast and Robust Real-Time Estimation of Respiratory Rate from Photoplethysmography. *Sensors* **2016**, *16*, 1494. [CrossRef]
16. Park, C.; Lee, B. Real-time estimation of respiratory rate from a photoplethysmogram using an adaptive lattice notch filter. *BioMedical Eng. OnLine* **2014**, *13*, 170. [CrossRef] [PubMed]
17. Chon, K.H.; Dash, S.; Kihwan, J. Estimation of Respiratory Rate From Photoplethysmogram Data Using Time–Frequency Spectral Estimation. *IEEE Trans. Biomed. Eng.* **2009**, *56*, 2054–2063. [CrossRef] [PubMed]
18. Dash, S.; Shelley, K.H.; Silverman, D.G.; Chon, K.H. Estimation of Respiratory Rate From ECG, Photoplethysmogram, and Piezoelectric Pulse Transducer Signals: A Comparative Study of Time–Frequency Methods. *IEEE Trans. Biomed. Eng.* **2010**, *57*, 1099–1107. [CrossRef] [PubMed]
19. Lin, Y.-D.; Chien, Y.-H.; Chen, Y.-S. Wavelet-based embedded algorithm for respiratory rate estimation from PPG signal. *Biomed. Signal Process. Control* **2017**, *36*, 138–145. [CrossRef]
20. Pimentel, M.A.F.; Johnson, A.E.W.; Charlton, P.H.; Birrenkott, D.; Watkinson, P.J.; Tarassenko, L.; Clifton, D.A. Toward a Robust Estimation of Respiratory Rate From Pulse Oximeters. *IEEE Trans. Biomed. Eng.* **2017**, *64*, 1914–1923. [CrossRef]
21. Charlton, P.H.; Celka, P.; Farukh, B.; Chowienczyk, P.; Alastruey, J. Assessing mental stress from the photoplethysmogram: A numerical study. *Physiol. Meas.* **2018**, *39*, 054001. [CrossRef]
22. Yang, H.; Minteer, S.D.; He, D.; Che, X.; Qin, X. Respiratory Rate Estimation from the Photoplethysmogram Combining Multiple Respiratory-induced Variations Based on SQI. In Proceedings of the 2019 IEEE International Conference on Mechatronics and Automation (ICMA), Tianjin, China, 4–7 August 2019. [CrossRef]
23. Charlton, P.H.; Birrenkott, D.A.; Bonnici, T.; Pimentel, M.A.F.; Johnson, A.E.W.; Alastruey, J.; Tarassenko, L.; Watkinson, P.J.; Beale, R.; Clifton, D.A. Breathing Rate Estimation From the Electrocardiogram and Photoplethysmogram: A Review. *IEEE Rev. Biomed. Eng.* **2021**, *11*, 2–20. [CrossRef]
24. Osathitporn, P.; Sawadwuthikul, G.; Thuwajit, P.; Ueafuea, K.; Mateepithaktham, T.; Kunaseth, N.; Choksatchawathi, T.; Punyabukkana, P.; Mignot, E.; Wilaiprasitporn, T. RRWaveNet: A Compact End-to-End Multiscale Residual CNN for Robust PPG Respiratory Rate Estimation. *IEEE Internet Things J.* **2023**, *10*, 15943–15952. [CrossRef]

25. Bian, D.; Mehta, P.; Selvaraj, N. Respiratory Rate Estimation using PPG: A Deep Learning Approach. In Proceedings of the 2020 42nd Annual International Conference of the IEEE Engineering in Medicine & Biology Society (EMBC), Montreal, QC, Canada, 20–24 July 2020. [CrossRef]
26. Ravichandran, V.; Murugesan, B.; Balakarthikeyan, V.; Ram, K.; Preejith, S.; Joseph, J.; Sivaprakasam, M. RespNet: A deep learning model for extraction of respiration from photoplethysmogram. In Proceedings of the 41st Annual International Conference of the IEEE Engineering in Medicine and Biology Society (EMBC), Berlin, Germany, 23–27 July 2019; pp. 5556–5559. [CrossRef]
27. Dai, R.; Lu, C.; Avidan, M.; Kannampallil, T. RespWatch. In Proceedings of the International Conference on Internet-Of-Things Design and Implementation, Charlottesvle, VA, USA, 18–21 May 2021. [CrossRef]
28. Aqajari, S.A.H.; Cao, R.; Zargari, A.H.A.; Rahmani, A.M. An End-to-End and Accurate PPG-based Respiratory Rate Estimation Approach Using Cycle Generative Adversarial Networks. In Proceedings of the 2021 43rd Annual International Conference of the IEEE Engineering in Medicine & Biology Society (EMBC), Guadalajara, Mexico, 1–5 November 2021. [CrossRef]
29. Lampier, L.C.; Coelho, Y.L.; Caldeira, E.M.O.; Bastos-Filho, T.F. A Deep Learning Approach to Estimate the Respiratory Rate from Photoplethysmogram. *Ingenius* **2022**, *27*, 96–104. [CrossRef]
30. Goldberger, A.L.; Amaral, L.A.N.; Glass, L.; Hausdorff, J.M.; Ivanov, P.C.; Mark, R.G.; Mietus, J.E.; Moody, G.B.; Peng, C.-K.; Stanley, H.E. PhysioBank, PhysioToolkit, and PhysioNet. *Circulation* **2000**, *101*, e215–e220. [CrossRef] [PubMed]
31. physionet.org., n.d. BIDMC PPG and Respiration Dataset v1.0.0. Available online: https://physionet.org/content/bidmc/1.0.0/ (accessed on 30 April 2023).
32. Peterhcharlton. A Toolbox of Respiratory Rate Algorithms, GitHub. 2017. Available online: https://github.com/peterhcharlton/RRest/wiki (accessed on 25 April 2023).
33. Karlen, W. CapnoBase IEEE TBME Respiratory Rate Benchmark. *Sch. Portal Dataverse* **2021**. [CrossRef]
34. Charlton, P.H.; Bonnici, T.; Tarassenko, L.; Alastruey, J.; Clifton, D.A.; Beale, R.; Watkinson, P.J. Extraction of respiratory signals from the electrocardiogram and photoplethysmogram: Technical and physiological determinants. *Physiol. Meas.* **2017**, *38*, 669–690. [CrossRef]
35. Charlton, P.H.; Bonnici, T.; Tarassenko, L.; Clifton, D.A.; Beale, R.; Watkinson, P.J. An assessment of algorithms to estimate respiratory rate from the electrocardiogram and photoplethysmogram. *Physiol. Meas.* **2016**, *37*, 610–626. [CrossRef]
36. Charlton, P.H.; Bonnici, T.; Tarassenko, L.; Clifton, D.A.; Beale, R.; Watkinson, P.J.; Alastruey, J. An impedance pneumography signal quality index: Design, assessment and application to respiratory rate monitoring. *Biomed. Signal Process. Control* **2021**, *65*, 102339. [CrossRef]
37. Karlen, W.; Ansermino, J.M.; Dumont, G.A. Adaptive pulse segmentation and artifact detection in photoplethysmography for mobile applications. In Proceedings of the 2012 Annual International Conference of the IEEE Engineering in Medicine and Biology Society, San Diego, CA, USA, 28 August–1 September 2012. [CrossRef]
38. GeeksforGeeks. Cubic Spline Interpolation. 2021. Available online: https://www.geeksforgeeks.org/cubic-spline-interpolation/ (accessed on 28 April 2023).
39. Liu, H.; Chen, F.; Hartmann, V.; Khalid, S.G.; Hughes, S.; Zheng, D. Comparison of different modulations of photoplethysmography in extracting respiratory rate: From a physiological perspective. *Physiol. Meas.* **2020**, *41*, 094001. [CrossRef]
40. Nicolò, A.; Massaroni, C.; Schena, E.; Sacchetti, M. The Importance of Respiratory Rate Monitoring: From Healthcare to Sport and Exercise. *Sensors* **2020**, *20*, 6396. [CrossRef]

diagnostics

MDPI

Opinion

Machine Learning for Predicting Biologic Agent Efficacy in Ulcerative Colitis: An Analysis for Generalizability and Combination with Computational Models

Philippe Pinton

Clinical and Translational Sciences, International PharmaScience Center Ferring Pharmaceuticals, Amager Strandvej 405, 2770 Kastrup, Denmark; philippe.pinton@ferring.com; Tel.: +45-28-78-74-50

Abstract: Machine learning (ML) has been applied to predict the efficacy of biologic agents in ulcerative colitis (UC). ML can offer precision, personalization, efficiency, and automation. Moreover, it can improve decision support in predicting clinical outcomes. However, it faces challenges related to data quality and quantity, overfitting, generalization, and interpretability. This paper comments on two recent ML models that predict the efficacy of vedolizumab and ustekinumab in UC. Models that consider multiple pathways, multiple ethnicities, and combinations of real-world and clinical trial data are required for optimal shared decision-making and precision medicine. This paper also highlights the potential of combining ML with computational models to enhance clinical outcomes and personalized healthcare. Key Insights: (1) ML offers precision, personalization, efficiency, and decision support for predicting the efficacy of biologic agents in UC. (2) Challenging aspects in ML prediction include data quality, overfitting, and interpretability. (3) Multiple pathways, multiple ethnicities, and combinations of real-world and clinical trial data should be considered in predictive models for optimal decision-making. (4) Combining ML with computational models may improve clinical outcomes and personalized healthcare.

Keywords: biologic agents; machine learning; artificial intelligence; computational models; prediction; ulcerative colitis; precision medicine

---

Key Questions

Question

What are the benefits of using machine learning in predicting clinical outcomes?

Answer

Machine learning can provide precision, personalization, efficiency, and automation, and it can improve decision support in predicting clinical outcomes. It can analyze large and diverse datasets, identify intricate patterns, automate data analysis, and assist clinicians in making evidence-based predictions.

Question

What are the challenges in using machine learning for predicting clinical outcomes?

Answer

Challenges in machine learning predictions include low data quality and data scarcity, overfitting, and poor generalizability and interpretability. Inadequate or biased data may lead to inaccurate predictions. Overfitting occurs when models learn noise from training data, and some models lack interpretability, which is crucial for clinicians to trust and understand the predictions.

Question

What factors should predictive models notably consider for optimal decision-making in ulcerative colitis?

Answer

Predictive models of ulcerative colitis should consider multiple pathways, multiple ethnicities, and combinations of real-world and clinical trial data. By incorporating these factors, the models can ensure optimal shared decision-making and precision medicine in the treatment of ulcerative colitis.

Question

Why is it important to validate machine learning models across different populations?

Answer

Validating machine learning models across different populations, especially for treating ulcerative colitis, is important to ensure the translatability and generalizability of the models. Differences in efficacy and the response to biologic agents may occur between ethnicities, and validating models across diverse populations can help to determine their applicability and effectiveness in different patient groups.

1. Introduction

Ulcerative colitis (UC), a chronic inflammatory bowel disease, poses significant therapeutic hurdles. While biologic agents have revolutionized UC treatment, optimizing their use remains crucial to surpass efficacy limitations. Machine learning (ML) and computational modeling present promising avenues for addressing these challenges. By analyzing clinical trials and real-world data, incorporating multiple omics and markers, improving diagnostic accuracy, predicting patient responses, and personalizing therapy, ML can enhance treatment outcomes and reduce reliance on steroids. Additionally, computational modeling enables us to simulate disease progression and predict both short-term and long-term treatment responses, leading to transformative approaches in UC management.

2. Machine Learning

ML is an interdisciplinary field that combines principles from computer science, mathematics, and statistics. It involves the development of algorithms and models that allow computers to learn from data and make predictions or decisions without requiring explicit programming to that end [1].

Conventional risk stratification methods have traditionally relied on statistical techniques. However, in the last decade, advanced computational approaches—such as machine learning—have gained prominence for risk prediction and prognosis across various conditions. Specifically, artificial intelligence methods for analyzing medical images (e.g., endoscopy, radiology, histology) have garnered significant attention in managing inflammatory bowel disease (IBD). Nonetheless, whether machine-learning-based prediction models using routine clinical data or clinical trial data provide any clear advantages over conventional risk prediction methods remains debated, especially when it comes to personalized medicine in daily practice.

The main benefits of ML in predicting clinical outcomes include (1) precision and personalization, (2) efficiency and automation, and (3) improved decision support. The main pitfalls and challenges of ML in this task include (4) low data quality and data scarcity, (5) overfitting and low generalizability, and (6) lack of interpretability [2–9]. These characteristics are described as follows:

(1) ML models can analyze large and diverse datasets and identify intricate patterns that may not be apparent to human clinicians. Personalized predictions allow for tailoring treatments to individual patients and improve clinical outcomes.

(2) ML automates the data analysis and reduces manual effort and time. Predictive models can quickly process vast amounts of information, thereby aiding clinical decision-making.

(3) ML assists clinicians in providing evidence-based predictions. It can complement medical expertise and enhance diagnostic accuracy and treatment planning.

(4) ML models require high-quality data. Inadequate or biased data can result in inaccurate predictions. Data scarcity may limit model performance, especially for rare conditions.

(5) Overfitting occurs when a model learns noise from the training data, leading to poor generalization to new data. Balancing model complexity and generalizability is crucial.
(6) Some ML models (e.g., deep neural networks) lack interpretability. Clinicians must trust and understand ML predictions; improving the interpretability of ML models is thus necessary.

ML has recently emerged as a powerful tool for predicting outcomes and personalizing treatments for inflammatory bowel disease (IBD), including ulcerative colitis (UC). Its use in predicting the efficacy and safety of therapeutic agents is expected to help practitioners with their prescriptions. In fact, ML techniques allow practitioners to aggregate and analyze vast quantities of data as well as identify and quantify features that may lead to novel insights into disease management. Various studies have incorporated clinical, laboratory, and omics data to predict significant outcomes in IBD [10]. These data include hospitalization records, corticosteroid use, biologic responses, and refractory disease after colectomy. In addition, encouraging results from various medical fields (e.g., histopathology, endoscopy, and radiology) support the use of ML image analysis to advance outcome prediction for IBD [11].

ML models using baseline patient data alone have achieved an area under the receiver operating characteristic curve of 0.62 in predicting the remission for vedolizumab in UC. When the data from week 6 were included, the area under the curve improved to 0.73. Remarkably, patients predicted to be remitters had a higher probability of achieving corticosteroid-free endoscopic remission [11]. The Paddington international virtual chromoendoscopy score histologic remission index and a related artificial intelligence system have been found to be a simple scoring system and predictor of histological remission in UC, respectively [12]. Moreover, ML models based on serological markers have demonstrated high accuracy in predicting UC relapse. These models may help determine personalized treatment plans [13].

For this opinion paper, we conducted a search in PubMed for eligible studies published between 1 January 2014 and 31 December 2023. Our focus was on articles describing prediction modeling using machine learning techniques, specifically related to UC and biologic agents. To be eligible, publications needed to describe the development or validation of at least one multivariable prediction model aimed at individualized efficacy outcome prediction. We did not impose restrictions based on study design, data source, or patient-related health outcomes. We considered a study to involve machine learning when a non-regression statistical technique was used to create or validate a prediction model. Excluded were publications reporting single predictors, tests, or biomarkers, as well as those using machine learning solely for image or signal enhancement or relying exclusively on genetic traits or molecular markers as predictors. We also excluded systematic reviews, methodological articles, conference abstracts, and publications unavailable through our institution. Our search was limited to studies involving human participants and articles written in English.

Additionally, as we wanted to reflect on the opportunities of developing models with one or multiple pathways, and of using real-world data or clinical trials data, we limited ourselves to biologic agents that were included in modeling exercises with such features. Vedolizumab and ustekinumab were the only biologic agents matching the mentioned criteria [14–16].

This opinion paper focuses then on two recent models and suggests different or complementary approaches for predicting the efficacy and safety of biologic agents for treating UC [14,15]. It includes a generative artificial intelligence and computational modeling combination, serving to exploit the potential of artificial intelligence but still benefit from the traceability back to causality that the computational modeling approach provides.

3. Example of Efficacy Prediction Using ML: Vedolizumab and Ustekinumab for Treating UC

Morikubo et al. [14] and Miyoshi et al. [15] recently proposed ML models for predicting the efficacy of ustekinumab and vedolizumab in patients with UC. They developed and tested prediction models using real-world clinical data as the baseline from Japanese patients who received ustekinumab or vedolizumab for treating UC. Their models achieved high predictive values of 88.9% and 92.3%, respectively. However, for the same two biologic agents in Crohn's disease (CD), the model for vedolizumab could not be applied to ustekinumab and required the selection of specific features using the random forest algorithm [16]. The difference in the mechanism of action between ustekinumab and vedolizumab could be the main reason why the model initially developed for vedolizumab does not work for ustekinumab. This is well-supported by the inclusion in the models of the fraction of monocytes (vedolizumab) and of lymphocytes (ustekinumab).

The model developed by Morikubo et al. [14] could have also benefited from a longer follow-up period to evaluate the long-term efficacy of biologic agents and their impact on disease progression and remission. This is a critical aspect to envision, knowing that a loss of response may occur after several months of treatment. CM can help, as shown by Venkatapurapu et al. [16].

The vedolizumab clinical decision support tool (VDZ-CDST) was developed to predict the treatment efficacy in patients with refractory CD who were intolerant to anti-TNF (tumor necrosis factor) drugs. The predictive capabilities of this tool were assessed using vedolizumab and ustekinumab. Although successful prediction of clinical and steroid-free clinical remissions was obtained for vedolizumab, the tool did not demonstrate the same efficacy for ustekinumab in this patient population. The specific ML approach used to create VDZ-CDST was not mentioned by Alric et al. [17]. In addition, the reason for ustekinumab not demonstrating the same efficacy as vedolizumab was not stated. However, we can make the following speculations based on the available information:

(1) Mechanisms of action: Vedolizumab and ustekinumab have different mechanisms of action, which may contribute to the varying responses in patients with refractory CD. This is broadly in line with previous works [18].
(2) Patient heterogeneity: CD is a heterogeneous disease, and the varied patient profiles in the study population may have led to substantial differences in the responses to biologic agents.
(3) Sample size and study design: The study sample size (insufficient statistical power) and design (study duration) possibly undermined the results.

The latter are well-identified issues with ML studies. Numerous investigations into ML-based prediction models reveal suboptimal methodological rigor and a substantial risk of bias. Factors contributing to this bias include participant exclusions, small sample sizes, inadequate handling of missing data, and a failure to address overfitting. To enhance the utility of supervised ML models in clinical practice and minimize research inefficiencies, efforts should focus on improving the study design, execution, reporting, and validation [19].

4. Analysis and Discussion

We recommend taking up the developments by Morikubo et al. [14] and Miyoshi et al. [15] given their innovative and useful predictive tools, which may help clinicians select the optimal treatment for patients with UC in the near future [20]. We provide perspectives regarding the generalizability and applicability of their models to other populations or settings. The predictive tools developed by Morikubo et al. [14] and Miyoshi et al. [15] may support decision-making and precision medicine. These tools may assist clinicians and patients by providing personalized insights, optimizing treatment choices, and improving patient outcomes based on individual characteristics and preferences. Tools for the optimal prediction of IBD should balance complexity and interpretability.

Integrating diverse omics strategies within a network framework can enhance our understanding of the pathogenesis involved in IBD. By analyzing various molecules—such as genomic, transcriptomic, proteomic, microbiome, epigenetic, and metabolomic data—simultaneously, we can develop multiomics models that provide valuable insights. These models may help identify disease subgroups, optimize therapy regimens, and discover promising predictive biomarkers, ultimately facilitating early diagnosis [21].

Although incorporating multiple immune pathways can enhance accuracy, the models must remain clinically applicable and transparent for shared decision-making. To enhance choice and decision, the models underlying predictive tools can consider a ubiquitous path instead of a specific target, such as vedolizumab and the $\alpha4\beta7$ receptor. For example, STAT 3 modulation intersects with several therapeutic agents that have distinct signatures correlated with clinical benefits [22,23].

Rogers et al. [24] reported the potential of a pharmacological model with dynamic quantitative systems for considering the immune system in IBD. Although modeling IBD with multiple pathways and layers is highly complicated and requires large datasets, a simulation model of various immune networks allows for the consideration of different therapeutic strategies based on the IBD pathoetiology and on the patient's phenotype at the treatment onset.

The model proposed by Rogers et al. [24] can simulate both CD and UC using shared underlying biological mechanisms but distinct initial conditions, which more accurately represent the specific subtype of IBD. The model has been designed in a modular manner, incorporating each biological interaction as a Michaelis–Menten-type reaction. This modular structure allows for easy addition, removal, or modification of interactions based on new biological insights.

Numerous approved or under-development drugs target the mechanisms, e.g., recruitment of leukocytes to sites of inflammation or Th cell differentiation, considered in the model of Rogers et al. [24]. Ustekinumab is one of those, and the combination of this approach with the one proposed by Morikubo et al. is an opportunity [25].

Tools like the abovementioned ones represent a step forward in precision medicine. Currently, computational models (CMs), platforms, digital twins, and quantum computers can enable the implementation of such solutions [16,26–30].

4.1. ML and CMs

ML and CMs present complementary characteristics, and their combination may lead to powerful hybrid approaches that leverage their strengths. Combining these two approaches requires a thoughtful design, a full understanding of the problem domain, and careful integration. These considerations are essential to balance data-driven learning and domain-specific knowledge. Precision medicine can benefit from such a combination for improving clinical outcomes and personalized healthcare.

ML is a specialized branch within the broader field of artificial intelligence. It is focused on creating algorithms that can learn from data without explicit programming. Popular ML models include decision trees, naïve Bayes, random forests, and support vector machines. These models analyze features in the available data to create decision boundaries for prediction or classification. ML models learn from data by adjusting their internal parameters based on observed patterns. Based on training data, they generalize to make predictions on previously unseen data. ML is well-suited for processing unstructured data.

CMs encompass a wide range of approaches, including traditional mathematical models, simulations, and rule-based systems. These models are designed to represent and simulate real-world phenomena and are often based on mathematical equations or logical rules. In fact, they are typically based on mathematical formulations or predefined rules. CMs do not learn from data but rely on explicit instructions. Hence, CMs are suitable for handling structured data.

Weather forecasting provides a fair illustration of the potential benefit of such a combination. Similar behaviors can be applied to UC and biologic agents.

4.1.1. Traditional Mathematical Model (CM)

In this case, for the weather forecast, we create a mathematical model based on physical principles. We consider factors like temperature, humidity, wind speed, and pressure. The model incorporates differential equations that describe how these variables interact. By solving these equations, we simulate how weather conditions evolve over time. This approach relies on predefined rules derived from scientific knowledge. For instance, the Navier–Stokes equations describe fluid flow, which can be applied to air movement in the atmosphere. By solving these equations, we predict wind patterns and temperature changes. CMs like this are suitable for structured data because they rely on explicit mathematical formulations.

For UC, similar to weather models, CM models rely on predefined rules based on scientific knowledge. In UC treatment, e.g., we can consider immune pathways (such as cytokine signaling, e.g., TNF-α, IL-23), disease duration (chronicity affecting treatment response), and genomic factors (including HLA variants that influence susceptibility and response to biologics). Differential equations describe how these variables interact, reflecting disease progression and treatment outcomes. By solving these equations, we simulate treatment responses over time. Clinicians can use CMs to predict the efficacy of specific biologic agents based on individual patient characteristics.

4.1.2. Machine Learning Approach

ML models, on the other hand, learn from historical data. We collect past weather data (temperature, humidity, etc.) and corresponding outcomes (rain, sunshine, etc.). Using algorithms like decision trees, neural networks, or regression, the ML model identifies patterns and relationships in the data. Once trained, the ML model can predict tomorrow's weather based on the current conditions (input features). However, ML models may struggle with extreme events or sudden changes if not enough relevant data are available.

For UC, we collect, e.g., data on past biologic responses (remission rates, mucosal healing) and patient-specific features (genetic markers, disease duration). Algorithms like decision trees or neural networks identify patterns. The ML model predicts the efficacy of a biologic agent based on current patient data. However, ML models may struggle with extreme cases (e.g., refractory UC) due to limited relevant data.

To improve accuracy, we can combine both approaches, and keeping the weather forecast illustration: Use the CM to simulate large-scale weather patterns (e.g., global air circulation). Then, fine-tune the predictions using ML based on local data (e.g., historical weather records for a specific city). This hybrid approach leverages the strengths of both CMs (domain-specific knowledge) and ML (data-driven learning).

In summary, CMs encompass various techniques, including traditional mathematical models and simulations. While CMs rely on explicit rules, ML learns from data. Combining these approaches can enhance precision and personalized outcomes in fields like weather forecasting or healthcare.

ML and CMs can be combined in multiple ways [31,32].

(1) Ensemble methods:

- Boosting: Boosting algorithms, such as AdaBoost or XGBoost, combine multiple weak ML models, often decision trees, to create a strong ensemble. These models mitigate the errors of their constituent models to improve the overall performance.
- Stacking: Stacking involves training multiple ML models (base learners) and using their predictions as inputs for another model (meta-learner). The meta-learner learns to appropriately combine the base model outputs.

(2) Physics-informed ML:

- By combining ML techniques with domain-specific knowledge, such as fluid dynamics principles, encoded in a CM, ML models can predict turbulence based on data, whereas computational fluid dynamics models provide physical insights. Integrating both approaches can enhance prediction accuracy.

(3) Data preprocessing:
- ML techniques preprocess data before feeding them into CMs.
- ML can manage missing data and perform feature engineering and noise reduction, thereby improving the quality of CM inputs.

(4) Hybrid models:
- Models that blend ML and computational components can be constructed. For example, an artificial neural network (ML) combined with a differential equation solver (CM) can be used to model chemical reactions.

(5) Uncertainty quantification:
- ML models often lack uncertainty estimates.
- ML predictions can be combined with CMs for including uncertainty bounds (e.g., Bayesian models).

(6) Transfer learning:
- An ML model can be trained on a task to then be fine-tuned for a specific computational problem.
- Transfer learning can proceed from an ML model to CM.

4.2. Randomized Clinical Trials and Real-Word Datasets

Combining individual patient data collected from randomized controlled clinical trials with real-world datasets can enhance the development of predictive tools and guide the integration of multiple treatments in shared decision-making and precision medicine for IBD. Clinical trial datasets have become more accessible via open-access platforms, and real-world-evidence trial datasets have gained accuracy, as demonstrated by Morikubo et al. [14] and Miyoshi et al. [15].

Platforms such as Vivli can help design models for multiple treatments using predictive tools. Vivli provides access to anonymized individual-level data collected during clinical trials. An example is a predictive model for vedolizumab using its pivotal phase 3 clinical program in UC. The VARSITY and tofacitinib trial datasets were integrated, allowing the model to identify patients' responses to adalimumab, tofacitinib, and vedolizumab [33].

4.3. Ethnicity Considerations

Using a dataset collected from a population of a single country to train and validate a model causes uncertainty in the translatability to other countries and populations. However, existing clinical trial data on vedolizumab in UC seem to confirm translatability to the general public. The pharmacokinetic parameters in Asian and non-Asian patients with UC are similar, and the results of a clinical trial performed in Japanese patients with UC confirmed the suitability of a specific dose of vedolizumab for application in patients worldwide [34,35]. In contrast, translatability is not obvious for ustekinumab, and the performance of the model developed by Morikubo et al. [14] for non-Asian patients with UC is uncertain owing to some differences in the efficacy of ustekinumab between Asian and non-Asian patients. According to the UNIFI program, a global phase 3 trial of ustekinumab for treating UC, the clinical remission rates at weeks 8 and 44 were lower in East Asian patients than in the overall population; however, the differences were not statistically significant [36–38]. In detail, a post hoc subpopulation analysis of the UNIFI program showed that clinical remission rates during both the induction and maintenance phases were lower in patients from East Asia (i.e., China, Japan, Korea, and Taiwan) than in the overall population. The clinical remission rates at week 8 were 11.4% (6 mg/kg) and 11.1% (130 mg) in East Asian patients and 15.6% (6 mg/kg) and 15.5% (130 mg) in the overall population, and those at week 44 were 47.6% (90 mg q12 w) and 19.2% (90 mg q8w) in East Asian patients and 38.4% (90 mg q12 w) and 43.8% (90 mg q8w) in the overall population [38]. As the differences in the efficacy of ustekinumab between Asian and non-Asian patients with UC were not statistically significant, they may have occurred by chance, whereas the safety profiles of ustekinumab are likely similar.

While current evidence suggests that ustekinumab is generally effective and safe for treating UC in patients from different regions and ethnicities, the observed variations in the response rates between Asian and non-Asian patients may bias prediction, if not anticipated, when developing a model. The differences may be attributable to various factors, such as genetics, including the genetic expression of IL-12 and IL-23 receptors, pharmacokinetics, disease characteristics, or concomitant medications, which may influence the response to ustekinumab. Population pharmacokinetics and exposure–response analyses conducted in patients with CD showed a higher clearance of ustekinumab in Asian patients than in Caucasian patients. This effect was observed after controlling for weight [39].

In recent years, various studies have demonstrated that numerous IL-23 receptor polymorphisms are correlated with the susceptibility of UC and CD. For example, the dominant model of the rs10889677 polymorphism is associated with the risk of developing UC in Asians but not in Caucasians [40]. Likewise, differences potentially exist when considering single-nucleotide polymorphisms of the IL-12B gene [41,42].

Currently, no clinical predictive tools dedicated to ustekinumab that consider ethnic differences are available. Most existing studies are based on clinical and biochemical variables, such as disease location, prior anti-TNF exposure, serum albumin, and fecal calprotectin. The available tools have been developed and evaluated in different populations and settings. However, no tool has specifically addressed the differences between Asian and non-Asian patients with UC. Therefore, assessing existing models in diverse regions or across ethnicities and comparing the results with those of Japanese cohorts is a valuable future research direction.

4.4. Dosage Alternatives

The model's applicability to patients who receive different doses or regimens of ustekinumab should be evaluated considering the variety of dosing and monitoring strategies for patients with UC. In the UNIFI program, the patients received an intravenous induction dose of ustekinumab based on their body weight (either 130 mg or 6 mg/kg), followed by a subcutaneous maintenance dose of 90 mg every 8/12 weeks. In clinical practice, some patients require dose escalation, interval shortening, or therapeutic drug monitoring to achieve or maintain the clinical response. Therefore, a model for patients receiving different doses or regimens of ustekinumab should be evaluated, and the impact of drug exposure or pharmacokinetic parameters on the model performance should be explored.

5. Conclusions

In conclusion, a predictive tool should involve different pathways to help physicians achieve the most suited and beneficial decision for the patient and should consider different omics and their impact on accuracy and translatability. A combination of ML and CMs can be helpful. The parameters identified by Morikubo et al. [14] have the potential to align with the model developed by Rogers et al. [24].

The contributions of Morikubo et al. [14] and Miyoshi et al. [15] to personalized medicine for UC are highly valuable and deserve continued development. We expect that their models will be further and prospectively validated and refined in different populations and settings toward optimizing the use of ustekinumab and other biologic agents for treating patients with UC.

Predictive models, combining machine learning and computational modeling, offer personalized treatment optimization. Clinicians can tailor therapies based on patient-specific data, detect disease progression early, optimize dosages, assess safety profiles, and predict treatment responses. These models enhance shared decision-making, inform clinical trial design, and integrate real-world evidence. However, challenges include model validation and clinician training. Overall, this approach holds promise for precision medicine in UC management.

Precision medicine requires balancing innovation (treatment, diagnostic, generative artificial intelligence, quantum computers, cloud), equity, individual benefit and societal impact, ethics, and scalability. It needs evidence from clinical trials and real-world practice. It can also incorporate lifestyle, environmental, and social determinants of health. Ensuring affordability and cost-effectiveness will determine its sustainability. As we move forward, collaboration among researchers, clinicians, policymakers, payers, and patients will shape the future of precision medicine.

Funding: This research received no external funding.

Conflicts of Interest: I am an employee of Ferring Pharmaceuticals, belong to the board of directors of PharmaBiome, and own stocks of Takeda Pharmaceutical Company, Limited.

References

1. Weng, W.H. Machine Learning for Clinical Predictive Analytics. In *Leveraging Data Science for Global Health*; Celi, L., Majumder, M., Ordóñez, P., Osorio, J., Paik, K., Somai, M., Eds.; Springer: Cham, Switzerland, 2020. [CrossRef]
2. Weissler, E.H.; Naumann, T.; Andersson, T.; Ranganath, R.; Elemento, O.; Luo, Y.; Freitag, D.F.; Benoit, J.; Hughes, M.C.; Khan, F.; et al. The role of machine learning in clinical research: Transforming the future of evidence generation. *Trials* **2021**, *22*, 537. [CrossRef]
3. Rajkomar, A.; Oren, E.; Chen, K.; Dai, A.M.; Hajaj, N.; Hardt, M.; Liu, P.J.; Liu, X.; Marcus, J.; Sun, M.; et al. Scalable and accurate deep learning with electronic health records. *NPJ Digit. Med.* **2018**, *1*, 18. [CrossRef]
4. Mijderwijk, H.J.; Steiger, H.J. Predictive Analytics in Clinical Practice: Advantages and Disadvantages. In *Machine Learning in Clinical Neuroscience*; Acta Neurochirurgica Supplement; Staartjes, V.E., Regli, L., Serra, C., Eds.; Springer: Cham, Switzerland, 2022; Volume 134. [CrossRef]
5. Kernbach, J.M.; Staartjes, V.E. Foundations of Machine Learning-Based Clinical Prediction Modeling: Part I—Introduction and General Principles. In *Machine Learning in Clinical Neuroscience*; Acta Neurochirurgica Supplement; Staartjes, V.E., Regli, L., Serra, C., Eds.; Springer: Cham, Switzerland, 2022; Volume 134. [CrossRef]
6. Adlung, L.; Cohen, Y.; Mor, U.; Elinav, E. Machine learning in clinical decision making. *Med* **2021**, *2*, 642–665. [CrossRef]
7. Zhang, X.; Wang, X.; Xu, L.; Liu, J.; Ren, P.; Wu, H. The predictive value of machine learning for mortality risk in patients with acute coronary syndromes: A systematic review and meta-analysis. *Eur. J. Med. Res.* **2023**, *28*, 451. [CrossRef]
8. Wang, W.; Kiik, M.; Peek, N.; Curcin, V.; Marshall, I.J.; Rudd, A.G.; Wang, Y.; Douiri, A.; Wolfe, C.D.; Bray, B. A systematic review of machine learning models for predicting outcomes of stroke with structured data. *PLoS ONE* **2020**, *15*, e0234722. [CrossRef]
9. Shamout, F.; Zhu, T.; Clifton, D.A. Machine Learning for Clinical Outcome Prediction. *IEEE Rev. Biomed. Eng.* **2021**, *14*, 116–126. [CrossRef]
10. Peng, J.; Jury, E.C.; Dönnes, P.; Ciurtin, C. Machine Learning Techniques for Personalised Medicine Approaches in Immune-Mediated Chronic Inflammatory Diseases: Applications and Challenges. *Front. Pharmacol.* **2021**, *12*, 720694. [CrossRef] [PubMed]
11. Javaid, A.; Shahab, O.; Adorno, W.; Fernandes, P.; May, E.; Syed, S. Machine Learning Predictive Outcomes Modeling in Inflammatory Bowel Diseases. *Inflamm. Bowel Dis.* **2022**, *28*, 819–829. [CrossRef]
12. Gui, X.; Bazarova, A.; del Amor, R.; Vieth, M.; de Hertogh, G.; Villanacci, V.; Zardo, D.; Parigi, T.L.; Røyset, E.S.; Shivaji, U.N.; et al. PICaSSO Histologic Remission Index (PHRI) in ulcerative colitis: Development of a novel simplified histological score for monitoring mucosal healing and predicting clinical outcomes and its applicability in an artificial intelligence system. *Gut* **2022**, *71*, 889–898. [CrossRef]
13. Pang, W.; Zhang, B.; Jin, L.; Yao, Y.; Han, Q.; Zheng, X. Serological Biomarker-Based Machine Learning Models for Predicting the Relapse of Ulcerative Colitis. *J. Inflamm. Res.* **2023**, *16*, 3531–3545. [CrossRef]
14. Morikubo, H.; Tojima, R.; Maeda, T.; Matsuoka, K.; Matsuura, M.; Miyoshi, J.; Tamura, S.; Hisamatsu, T. Machine learning using clinical data at baseline predicts the medium-term efficacy of ustekinumab in patients with ulcerative colitis. *Sci. Rep.* **2024**, *14*, 4386. [CrossRef] [PubMed]
15. Miyoshi, J.; Maeda, T.; Matsuoka, K.; Saito, D.; Miyoshi, S.; Matsuura, M.; Okamoto, S.; Tamura, S.; Hisamatsu, T. Machine learning using clinical da-ta at baseline predicts the efficacy of vedolizumab at week 22 in patients with ulcerative colitis. *Sci. Rep.* **2021**, *11*, 16440. [CrossRef] [PubMed]
16. Venkatapurapu, S.P.; Iwakiri, R.; Udagawa, E.; Patidar, N.; Qi, Z.; Takayama, R.; Kumar, K.; Sato, Y.; Behar, M.; Offner, P.; et al. A Computational Platform Integrating a Mechanistic Model of Crohn's Disease for Predicting Temporal Progression of Mucosal Damage and Healing. *Adv. Ther.* **2022**, *39*, 3225–3247, Erratum in *Adv. Ther.* **2023**, *40*, 721. [CrossRef] [PubMed]
17. Alric, H.; Amiot, A.; Kirchgesner, J.; Tréton, X.; Allez, M.; Bouhnik, Y.; Beaugerie, L.; Carbonnel, F.; Meyer, A. Vedolizumab clinical decision support tool predicts efficacy of vedolizumab but not ustekinumab in refractory Crohn's disease. *Inflamm. Bowel Dis.* **2022**, *28*, 218–225. [CrossRef] [PubMed]

18. Sharip, M.T.; Nishad, N.; Pillay, L.; Goordyal, N.; Goerge, S.; Subramanian, S. Ustekinumab or Vedolizumab after Failure of Anti-TNF Agents in Crohn's Disease: A Review of Comparative Effectiveness Studies. *J. Clin. Med.* **2024**, *13*, 2187. [CrossRef] [PubMed]
19. Navarro, C.L.A.; Damen, J.A.A.; Takada, T.; Nijman, S.W.J.; Dhiman, P.; Ma, J.; Collins, G.S.; Bajpai, R.; Riley, R.D.; Moons, K.G.M.; et al. Risk of bias in studies on prediction models developed using supervised machine learning techniques: Systematic review. *BMJ* **2021**, *375*, n2281. [CrossRef] [PubMed]
20. Pinton, P. Impact of artificial intelligence on prognosis, shared decision-making, and precision medicine for patients with inflammatory bowel disease: A perspective and expert opinion. *Ann. Med.* **2023**, *55*, 2300670. [CrossRef]
21. Agamah, F.E.; Bayjanov, J.R.; Niehues, A.; Njoku, K.F.; Skelton, M.; Mazandu, G.K.; Ederveen, T.H.A.; Mulder, N.; Chimusa, E.R.; Hoen, P.A.C.'t. Computational approaches for network-based integrative multi-omics analysis. *Front. Mol. Biosci.* **2022**, *9*, 967205. [CrossRef] [PubMed]
22. Pickert, G.; Neufert, C.; Leppkes, M.; Zheng, Y.; Wittkopf, N.; Warntjen, M.; Lehr, H.-A.; Hirth, S.; Weigmann, B.; Wirtz, S.; et al. STAT3 links IL-22 signaling in intestinal epithelial cells to mucosal wound healing. *J. Exp. Med.* **2009**, *206*, 1465–1472. [CrossRef]
23. Schreiber, S.; Aden, K.; Bernardes, J.P.; Conrad, C.; Tran, F.; Höper, H.; Volk, V.; Mishra, N.; Blase, J.I.; Nikolaus, S.; et al. Therapeutic Interleukin-6 Trans-signaling Inhibition by Olamkicept (sgp130Fc) in Patients with Active Inflammatory Bowel Disease. *Gastroenterology* **2021**, *160*, 2354–2366. [CrossRef]
24. Rogers, K.V.; Martin, S.W.; Bhattacharya, I.; Singh, R.S.P.; Nayak, S. A Dynamic Quantitative Systems Pharmacology Model of Inflammatory Bowel Disease: Part 1-Model Framework. *Clin. Transl. Sci.* **2021**, *14*, 239–248. [CrossRef] [PubMed]
25. Gottlieb, Z.S.; Sands, B.E. Personalised Medicine with IL-23 Blockers: Myth or Reality? *J. Crohn's Colitis* **2022**, *16* (Suppl. S2), ii73–ii94. [CrossRef] [PubMed]
26. Bretto, E.; Ribaldone, D.G.; Caviglia, G.P.; Saracco, G.M.; Bugianesi, E.; Frara, S. Inflammatory Bowel Disease: Emerging Therapies and Future Treatment Strategies. *Biomedicines* **2023**, *11*, 2249. [CrossRef] [PubMed]
27. Breitwieser, L.; Hesam, A.; de Montigny, J.; Vavourakis, V.; Iosif, A.; Jennings, J.; Kaiser, M.; Manca, M.; Di Meglio, A.; Al-Ars, Z.; et al. BioDynaMo: A modular platform for high-performance agent-based simulation. *Bioinformatics* **2022**, *38*, 453–460. [CrossRef] [PubMed]
28. Boulos, M.N.K.; Zhang, P. Digital Twins: From Personalised Medicine to Precision Public Health. *J. Pers. Med.* **2021**, *11*, 745. [CrossRef] [PubMed]
29. Flöther, F.F. The state of quantum computing applications in health and medicine. *Res. Dir. Quantum Technol.* **2023**, *1*, e10. [CrossRef]
30. Shams, M.; Choudhari, J.; Reyes, K.; Prentzas, S.; Gapizov, A.; Shehryar, A.; Affaf, M.; Grezenko, H.; Gasim, R.W.; Mohsin, S.N.; et al. The Quantum-Medical Nexus: Understanding the Impact of Quantum Technologies on Healthcare. *Cureus* **2023**, *15*, e48077. [CrossRef] [PubMed]
31. Fan, K.; Fan, K.; Cheng, L.; Li, L. Artificial intelligence and machine learning methods in predicting anti-cancer drug combination effects. *Briefings Bioinform.* **2021**, *22*, bbab271. [CrossRef]
32. Wu, L.; Wen, Y.; Leng, D.; Zhang, Q.; Dai, C.; Wang, Z.; Liu, Z.; Yan, B.; Zhang, Y.; Wang, J.; et al. Machine learning methods, databases and tools for drug combination prediction. *Briefings Bioinform.* **2022**, *23*, bbab355. [CrossRef]
33. Dulai, P.S.; Singh, S.; Jairath, V.; Wong, E.; Narula, N. Integrating Evidence to Guide Use of Biologics and Small Molecules for Inflammatory Bowel Diseases. *Gastroenterology* **2024**, *166*, 396–408.e2. [CrossRef]
34. Okamoto, H.; Dirks, N.L.; Rosario, M.; Hori, T.; Hibi, T. Population pharmacokinetics of vedolizumab in Asian and non-Asian patients with ulcerative colitis and Crohn's disease. *Intest. Res.* **2021**, *19*, 95–105. [CrossRef] [PubMed]
35. Motoya, S.; Watanabe, K.; Ogata, H.; Kanai, T.; Matsui, T.; Suzuki, Y.; Shikamura, M.; Sugiura, K.; Oda, K.; Hori, T.; et al. Vedolizumab in Japanese patients with ulcerative colitis: A Phase 3, randomized, double-blind, placebo-controlled study. *PLoS ONE* **2019**, *14*, e0212989, Erratum in *PLoS ONE* **2019**, *14*, e0215491. [CrossRef]
36. Sands, B.E.; Sandborn, W.J.; Panaccione, R.; O'Brien, C.D.; Zhang, H.; Johanns, J.; Marano, C. Ustekinumab as induction and maintenance therapy for ul-cerative colitis. *N. Engl. J. Med.* **2019**, *381*, 1201–1214. [CrossRef] [PubMed]
37. Afif, W.; Arasaradnam, R.P.; Abreu, M.T.; Danese, S.; Sandborn, W.J.; Miao, Y.; Zhang, H.; Panaccione, R.; Hisamatsu, T.; Scherl, E.J.; et al. Efficacy and Safety of Ustekinumab for Ulcerative Colitis Through 4 Years: Final Results of the UNIFI Long-Term Maintenance Study. *Am. J. Gastroenterol.* **2024**, *119*, 910–921. [CrossRef] [PubMed]
38. Hisamatsu, T.; Kim, H.J.; Motoya, S.; Suzuki, Y.; Ohnishi, Y.; Fujii, N.; Matsushima, N.; Zheng, R.; Marano, C.W. Efficacy and safety of ustekinumab in East Asian patients with moderately to severely active ulcerative colitis: A subpopulation analysis of global phase 3 induction and maintenance studies (UNIFI). *Intest. Res.* **2021**, *19*, 386–397. [CrossRef] [PubMed]
39. Adedokun, O.J.; Xu, Z.; Gasink, C.; Jacobstein, D.; Szapary, P.; Johanns, J.; Gao, L.-L.; Davis, H.M.; Hanauer, S.B.; Feagan, B.G.; et al. Pharmacokinetics and Exposure Response Relationships of Ustekinumab in Patients With Crohn's Disease. *Gastroenterology* **2018**, *154*, 1660–1671. [CrossRef] [PubMed]
40. Zhu, Y.; Jiang, H.; Chen, Z.; Lu, B.; Li, J.; Shen, X. Genetic association between IL23R rs11209026 and rs10889677 polymorphisms and risk of Crohn's disease and ulcerative colitis: Evidence from 41 studies. *Inflamm. Res.* **2020**, *69*, 87–103. [CrossRef]

41. Tang, L.; Xu, M. Candidate polymorphisms and susceptibility to inflammatory bowel disease: A systematic review and meta-analysis. *Gene* **2020**, *753*, 144814. [CrossRef] [PubMed]
42. Wang, J.; Liu, H.; Wang, Y.; Wu, J.; Wang, C.; Liu, K.; Qin, Q. The Polymorphisms of Interleukin-12B Gene and Susceptibility to Inflammatory Bowel Diseases: A Meta-analysis and Trial Sequential Analysis. *Immunol. Investig.* **2021**, *50*, 987–1006. [CrossRef]

MDPI AG
Grosspeteranlage 5
4052 Basel
Switzerland
Tel.: +41 61 683 77 34

Diagnostics Editorial Office
E-mail: diagnostics@mdpi.com
www.mdpi.com/journal/diagnostics

www.ingramcontent.com/pod-product-compliance
Lightning Source LLC
Chambersburg PA
CBHW061445210926
51850CB00008B/1617